## Aims and Scope

The goal of the JCMSE is to publish new research results on computational methods employed in all sciences and engineering. Common experience proves that computational methods originally developed in sciences can be of crucial importance to other research areas, including engineering. This potentially beneficial practice of cross-fertilization will be continuously and systematically endorsed by the JCMSE, which shall also simultaneously stimulate similar initiatives within computational methods for knowledge transfer from engineering to sciences.

Currently, there is a communication gap between sciences and engineering. The JCMSE is aimed at bridging this gap. The main rationale for creating the JCMSE was a common observation that there is a genuinely universal link which naturally binds together sciences and engineering. This link is a joint and large set of computational methods that are often invented and reinvented in the two artificially separated fields with an insufficient overlap. The JCMSE was born out from the common experience acquired over the years indicating that progress in the sciences could have been significantly enhanced by a judicious intertwining between the mathematical and the engineering literatures. Inspecting e.g. the abundant and versatile engineering literature one may readily realize that the computational methods developed and successfully used there could be of a direct and great importance to the sciences. The JCMSE shall be the home of such interactions with expected benefit to both the sciences and engineering.

The scope of the JCMSE is primarily within the realm of computational methods, but the aim is also to encompass thorough reporting on the development of theory and models that lead to numerical implementations. The JCMSE will encourage efforts on this indispensable liaison between mathematical modelling and the corresponding computational methods. Moreover, the JCMSE shall provide a continuous support for publishing numerical codes, algorithms and software of either a general purpose or a particular tailoring to certain special tasks with the potential of broader ramifications and applications. Computational methods published by the JCMSE are expected to undergo detailed testing with other existing competitive strategies, as well as with experimental data, whenever possible.

The JCMSE is a journal on theoretical methods, but computational results from real measurements will also be published. A given modern measurement in the sciences may yield a large data base of observables that are customarily digitised and stored for post-experimental studies. The use of various computational methods to analyse these data may well be so powerful that the ensuing results could turn out to be almost equivalent to re-running the original experiment many times, with exploration of new possibilities that might not be accessible to a real measurement. The JCMSE will publish original results along these lines in experimental research, as well.

Ironically, when computers were small, programming was not a problem. However, programming eventually became a serious problem with the availability of large computing facilities. Bridging this latter gap with the advent of new programming languages as well as novel computing modalities, e.g. quantum computers and the like, will also be welcomed by the JCMSE. Overall, the JCMSE is open to all sciences and engineering to facilitate fruitful dialogue among interdisciplinary research fields reporting on new results obtained by means of computational methods.

---

The journal of Computational Methods in Sciences and Engineering is recognized by the European Society of Computational Methods in Sciences and Engineering (ESCMSE). More information about the ESCMSE can be found on the website *www.uop.gr/escmse/.*

## CONTENTS

**Special Issue: Computational aspects of electric polarizability calculations: Atoms, Molecules and clusters, Part II**

Journal of Computational Methods in Sciences and Engineering 4 (2004) 517–549
IOS Press


# Dynamical and static dipole polarizability of metal clusters studied within different density functional approximations

M.B. Torres[a] and L.C. Balbás[b]
[a]*Departamento de Matemáticas y Computación. Universidad de Burgos, E-09006 Burgos, Spain*
[b]*Departamento de Física Teórica. Universidad de Valladolid, E-47011 Valladolid, Spain*

**Abstract.** We study the linear response of pure and doped metal clusters to spin-dependent and spin-independent external fields within the framework of the time-dependent density functional theory. As test cases we have considered the response of clusters with spin-saturated configurations ($Na_9^+$, $Na_8$, $Na_{21}^+$, $Na_{20}$, $Na_{40}$, $Na_4Pb$ and $Na_6Pb$) and with spin-polarized configurations ($Na_6^+$, $Na_5Pb$, $Al_{11}Fe$, $Al_{18}$ and $Al_{18}Fe$). For spin-dependent excitations we have obtained in all the pure and doped sodium clusters a strongly collective spin mode of surface type lying at lower energies than the unperturbed particle-hole excitations. This mode uncouples cleanly from the electric dipole mode in the case of spin-saturated clusters, but for spin-polarized clusters these two modes turn out to be intertwined in the responses to spin-dependent and to spin-independent fields. We have performed the calculations within two different exchange-correlation potentials, the first constructed from the local density approximation by Perdew and Wang and the second constructed by Parr and Ghosh starting from the strongly non-local Amaldi Approximation. The latter incorporates the correct $-1/r$ long-range behaviour leading to calculated static polarizabilities and photoabsorption spectra closer to experiments that the results from local density calculations. For the doped clusters $Na_nPn$ and $Al_nFe$ we have compared the results obtained within two descriptions of the atoms surrounding the central impurity atom, namely the jellium model and the spherical average of the pseudopotentials method. The polarizabilities of $Al_nFe$ clusters with $n \leqslant 18$, calculated within the two ionic models are similar and for $n > 9$ are also similar to those of pure $Al_n$ clusters. The main difference between jellium and pseudopotential models when describing doped $Al_nFe$ clusters is obtained for the dipole response at several tens of eV for certain sizes where the last occupied d orbital has different occupation number in the pseudopotential model than in the jellium model.

Keywords: Metal clusters, photoabsorption spectrum, exchange-correlation functional

Mathematics Subject Classification: 68-04, 78-04, 65C20

## 1. Introduction

Shortly after the discovery of electronic shell structure in alkali-metal clusters [1], measurements of their static dipole polarizabilities [2] and photoabsorption spectra [3] were performed. These methods still today constitute some of the most powerful probes of the structure and electromagnetic response properties of metal clusters (for reviews see references [4–7]). A large number of theoretical calculations based on various methodologies have been directed to these properties [8–24]. In view of the fact that the static polarizability of clusters is generally obtained as a byproduct of the dynamical polarizability

*Corresponding author. E-mail: balbas@lcb.fam.cie.uva.es.

(needed to compute the photoabsorption spectra), we will concentrate in the following on the theoretical description and computation of the linear response (complex polarizability) of small metal clusters. For other less common methods see for example refs. [10,22].

Despite the fact that the absorption spectra have been known for a long time, their theoretical interpretation is still being discussed. The typical experiments of photoabsorption shows a few strong absorption lines exhausting most of the oscillator strength. The interpretation of these lines leads to two different points of view: i) in the language of quantum chemistry, understanding the excitations as transitions between distinct molecular electronic states, and ii) in terms of collective oscillations of the valence electron density against the inert ionic background, similar to the surface plasmon modes in metal spheres or the giant resonances in atomic nuclei. Both pictures have proved to be compatible with experimental data [25]. The dominance of collective states in the photoabsorption of metal clusters was predicted by W. Ekardt [8] before its experimental confirmation [3] using the time-dependent-local density approximation (TDLDA) within the jellium model [26]. The TDLDA method, first introduced intuitively by A. Zangwill and P. Soven [27], was generalized and rigorously proved by E.K.U. Gross and colaborators [28]. TDLDA tends to underestimate the experimentally measured static polarizability of metal clusters, and different corrections have been proposed to deal with this discrepancy, which remain unresolved. For example, recent theoretical work has addressed the possibility of a strong influence of metal cluster temperature on the electrical polarizability [23].

In the primitive jellium model [26], appropriate for nearly free electron metal clusters, the ions are replaced by a continuous positive background which is constant inside a sphere of radius $R_{jel} = r_s N^{1/3}$ and zero outside, being $N$ the number of delocalized valence electrons and $r_s$ the Wigner-Seitz radius of the electron gas. This provides a size-dependent ionic potential, $V_{ion}$, that binds the valence electron cloud, with density $n(r)$such that $\int n(\mathbf{r})\, d\mathbf{r} = N$, in the ground state. This valence density is determined, according to density functional theory (DFT) [29,30], by minimizing the total electronic energy functional

$$E[n] = \int d\mathbf{r}\, V_{ion}(\mathbf{r})\, n(\mathbf{r}) + G[n] \tag{1}$$

with

$$G[n] = E_{kin}[n] + E_H[n] + E_{xc}[n] \tag{2}$$

$E_{kin}$describes the kinetic energy of the system of correlated electrons, $E_H$ is the classical coulomb repulsion (Hartree term) and $E_{xc}$ is the unknown exchange-correlation (xc) energy. The traditional approximation to the unknown xc part of the functional is the local density approximation (LDA) [30]. The exact density, which determines the total electronic energy of the ground state, is found by solving the so-called Kohn-Sham equations [30]. The total density $n(\mathbf{r})$ can be considered as the sum of two electronic densities $n^{\uparrow}(\mathbf{r})$ and $n^{\downarrow}(\mathbf{r})$, each one describing electrons with different spin polarization. We will use the DFT which includes the spin polarization as introduced in ref. [31]. The existence of spin modes deserves a detailed study in metal clusters, with $N$ electrons, where the search for low-energy excitations has been a central subject in the exploration of their collective properties. Multipole resonances of spin character are caused by the external field

$$V_{ext}^{\sigma} = \sum_{j=1}^{N} r_j^L Y_{L0} \sigma_j^z \tag{3}$$

($r$ and $\sigma^z$ are the radial position and the third component of the spin vector, respectively). For the dipole case, $L = 1$, the operator $V^{\sigma}_{ext} \approx \sum_{j=1}^{N} z_j \sigma^z_j$ provides the opposite shifts of spin-up and spin-down electrons in the $z$-direction, where $z$ is the third component of the position vector.

The theoretical framework for the description of the ground state of metal clusters is outlined briefly in Section 2.1, and for the description of the dynamical and static cluster polarizability we will use the time dependent density functional theory (TDDFT) [28], which is outlined briefly in Section 2.2. In Section 3 we present results for the dynamical and static polarizability of sodium clusters. In Subsection 3.1 we give results for spin and electrical dipolar excitations within the TDLSDA approach and the spherical jellium model. In Subsection 3.2 we show how to improve the TDLDA results for the ground state, polarizabilities and photoabsorption spectra of jellium-like clusters by using an exchange-correlation potential with the correct asymptotic behaviour, $\sim -1/r$, instead of the exponential decay displayed by the LDA xc potential.

In Section 4 we present results for the dipole response of doped metallic clusters $Na_nPb$ (spin and electrical responses) and $Al_nFe$ (electrical response) taking into account the ionic structure beyond the jellium model, by the so-called SAPS method [32]. In the SAPS model the ground state ionic and electronic structures of a cluster are obtained by minimizing the total energy, $E_T$, with respect to the positions of all ions. $E_T$ can be written as $E_T = E[n^{\uparrow}, n^{\downarrow}] + E_{ion-ion}$, where $E[n^{\uparrow}, n^{\downarrow}]$ is the energy of the valence electrons in the field created by ions and $E_{ion-ion}$ gives the ion-ion interaction energy. For a given configuration $(\mathbf{R}_1, \mathbf{R}_2, \ldots, \mathbf{R}_N)$ of the ions, $E[n^{\uparrow}, n^{\downarrow}]$ is obtained from the electronic density calculated by the Eqs (1) and (2), where the ionic potential, $V_{ion}^{SAPS}(\mathbf{r})$ is constructed by first forming the total ionic potential $V_{ion}(\mathbf{r}) = \sum_j^N v_{ps}(\mathbf{r} - \mathbf{R}_j)$ and then calculating the spherical average $V_{ion}^{SAPS}(\mathbf{r})$ of $V_{ion}(\mathbf{r})$ around the center of mass of the cluster. $E_{ion-ion}$ is evaluated as the repulsion between the ion cores and consequently takes into account the geometry. For the atomic pseudopotential we use the one of Fiolhais et al. [33]. In Section 5 we give a summary and conclusions.

## 2. Theory

The linear response theory to an external field with an uniform direction of magnetization, was derived by William and Von Barth [34] and first applied to unpolarized sodium clusters by Serra and co-workers [35]. In this work we extend somewhat the formulation of the TDLSDA given in [34] as to be able to study the response of spin-saturated clusters as well as that of strongly spin-polarized clusters to spin-dependent and spin-independent fields [36]. In practice we must construct some approximation for the functional dependence of the total energy on the density matrix $n^{\sigma\sigma'}(\mathbf{r})$ and we can obtain the ground state energy and the density matrix from a relatively simple method based on an equivalent one-electron formulation. Density-functional theory can also be used to obtain different time-dependent correlation functions such as the density-density response function needed for optical absorption [8,27,28] or the spin-spin response function giving information on the spin excitations [35]. This is due to the fact that the Hamiltonian, and therefore all expectation values with respect to its eigenvectors, are functionals of the density matrix.

### 2.1. Ground state of metal clusters

In order to demonstrate the basic ideas, we will here consider the case of a uniform direction of magnetization. The density matrix of a system with $N$ electrons is assumed to be diagonal, $n^{\sigma\sigma'}(\mathbf{r}) =$

$n^\sigma(\mathbf{r})\delta_{\sigma\sigma'}$, with the spin-up and spin-down densities $n^\sigma(\mathbf{r})$ (for $\sigma = \uparrow, \downarrow$ respectively) given by the equations

$$n^\sigma(\mathbf{r}) = \sum_i^{N_\sigma} |\phi_i^\sigma(\mathbf{r})|^2 \tag{4}$$

The exact density determines the total energy in the ground state, Eq. (1), and $\phi_i^\sigma(r)$ are the single-particle eigenfunctions which solve the Kohn-Sham equations [30]

$$\left[-\frac{1}{2}\nabla^2 + V_{eff}^\sigma(\mathbf{r})\right]\phi_i^\sigma(\mathbf{r}) = \varepsilon_i^\sigma\phi_i^\sigma(\mathbf{r}) \tag{5}$$

In these one-particle equations, the effective potential, $V_{eff}^\sigma(\mathbf{r})$

$$V_{eff}^\sigma(\mathbf{r}) = V_{ion}(\mathbf{r}) + V_H(\mathbf{r}) + V_{xc}^\sigma(\mathbf{r}) \tag{6}$$

is the sum of the ionic, $V_{ion}(\mathbf{r})$, electron-electron (Hartree), $V_H(\mathbf{r}) = \int d\mathbf{r}' n(\mathbf{r}')/|\mathbf{r} - \mathbf{r}'|$, and spin-dependent xc contributions, $V_{xc}^\sigma(\mathbf{r})$ (which contains all the many-electron effects missing in $V_H(\mathbf{r})$). Both, $V_H(\mathbf{r})$ and $V_{xc}^\sigma(\mathbf{r})$, are the functional derivatives with respect to density of functionals $E_H[n]$ and $E_{xc}[n]$, respectively, in Eq. (2).

The ionic potential, $V_{ion}(\mathbf{r})$, is constructed here within three different models, one for pure sodium clusters, other for doped $Na_nPb$ clusters and other for doped $Al_nFe$ clusters. For $Na_n$ clusters we use the spherical jellium approximation [26]. For the ionic potential of $Na_nPb$ clusters, we use the spherically averaged pseudopotential (SAPS) model [32], within the local atomic pseudopotential of Fiolhais et al. [33], because this model reproduces well [36] the trends for the observed stability [37] and the main geometrical and electronic features of the first-principles calculations [38] for these clusters. For the doped clusters, $Al_nFe$, we consider the iron atom with all its electrons, and the resulting $V_{ion}(\mathbf{r})$ is the sum of the ionic potential of pure $Al_n$ (using the jellium o SAPS models) and the atomic potential of iron.

On the other hand, the xc potential is determined as $V_{xc}^\sigma = \delta E_{xc}\left[n^\uparrow, n^\downarrow\right]/\delta n^\sigma$ after some approximation is assumed for the xc energy functional, $E_{xc}[n^\uparrow, n^\downarrow]$, a key ingredient-but unknown of the DFT. Within the local-spin-density approximation (LSDA), $E_{xc}$ reads

$$E_{xc}^{LDA}\left[n^\uparrow, n^\downarrow\right] = \int d\mathbf{r}(n^\uparrow + n^\downarrow)\varepsilon_{xc}(n^\uparrow, n^\downarrow) \tag{7}$$

where $\varepsilon_{xc}$ is the exchange-correlation energy density per particle of the homogeneous electron system with spin-densities $n^\uparrow$ and $n^\downarrow$. In our LSDA calculations we use for $\varepsilon_{xc}$ the parametrization of Perdew and Wang [39] except for $Al_nFe$, where we use the parametrization of Vosko-Wilk-Nusair [40].

Although there is ample evidence that LDA is accurate enough for many practical purposes in atomic, molecular, and solid-state calculations, there are problems in which the LDA is not good enough [41, 42]. The LDA can be improved mainly in two respects:

1) by constructing functionals that not only depend on the local density $n(\mathbf{r})$ at point $\mathbf{r}$, but also depend on the gradients of the density, which leads to the concept of the generalized gradient approximation (GGA) [42].
2) by explicitly removing the self-interaction error contained in the classical Coulomb energy (Hartree term in Eq. (2)), which is only partially cancelled by the LDA xc energy.

In order to illustrate these improvements over LSDA results we will apply option 2) in Section 3.2 below to the dipole electric response of electronic shell-closed clusters, where we need consider only the paramagnetic solution of Kohn-Sham equations. In this case the spin-up and spin-down densities are the same function and the total density, $n(\mathbf{r}) = n^{\uparrow}(r) + n^{\downarrow}(\mathbf{r})$, leads to the xc potential $V_{xc}(\mathbf{r}) = \delta E_{xc}[n]/\delta n$ entering in the $V_{eff}(\mathbf{r})$ of the Kohn-Sham Eq. (5).

Various efforts have been made to eliminate the self-interaction error in LDA and related density functional schemes. For example, two decades ago, Perdew and Zunger [41] developed methods for the self-interaction correction (SIC) of any energy density functional. Among other well known outcomes of using SIC, the more relevant for the purposes of this paper is that SIC leads to the correct long-range behaviour of the atomic potential, $\sim -1/r$, and the ground-state electronic density, $\sim e^{-2ar}$ (with $a = (2I)^{1/2}$, where $I$ is the first ionization potential). A different approach for the xc density functional, which is free of self-interaction and exhibits the asymptotic xc potential $\sim -1/2r$ for neutral atoms and clusters, is the weighted density approximation (WDA) [43]. In order to reproduce the $-1/r$ asymptotic behaviour of the exact xc potential, Przybylski and Borstel (PB) [44] introduced a further approximation in the WDA. Aiming to calculate the static polarizability of sodium clusters in the context of the so-called Sternheimer equation, Rubio et al. [22] introduced an *ad hoc* (WDA-PB) procedure, in which the virtues of both the original WDA and PB approaches are exploited simultaneously.

Specifically, in order to remove the self-interaction error we will use in this work the Parr and Ghosh approximation to the paramagnetic xc potential [45]. This approximation was first constructed for atoms and is based on a series of seminal papers by Zhao, Morrison and Parr (ZMP) [46]. These authors have shown how to calculate Kohn-Sham kinetic energies and orbitals, as well as DFT orbital energies, and xc potentials, $V_{xc}^{ZMP}[n]$, starting from accurate ground state electron densities. They have also shown that the accurate xc energy functional cannot be local. The scheme used to construct $V_{xc}^{ZMP}[n]$ is based on the Levy constrained-search method [47]. An important conclusion of the analysis of $V_{xc}^{ZMP}[n]$ by Parr and Ghosh is that it may be well to eschew the traditional resolution of exchange-correlation into exchange plus correlation, in favor of a resolution into two terms,

$$V_{xc}^{ZMP}[n] = V_{xc}^{FA}[n] + V_C[n], \tag{8}$$

both of them contributing many-body corrections to the Hartree electrostatic potential, $V_H$[n]. The first term in Eq. (8) is the self-interaction correction to $V_H$ introduced by Fermi and Amaldi (FA) in 1934 [48],

$$V_{xc}^{FA}[n] = -(1/N)V_H[n] \tag{9}$$

with $N$ (a parameter, not a functional) being the number of electrons. The FA term assures that the xc hole correctly normalizes to $-1$ electrons, a very important condition fulfilled by LDA, and guarantees the correct long-range behavior of the potential, in a natural way [48]. This is a significant improvement over the exponential decay of LDA. The second term in Eq. (8), $V_C[n]$, arises from the special constraint used in the ZMP method to minimize the kinetic energy and has been called the *constraint potential.* As said above one can compute accurate xc potentials from accurate electron densities following the ZMP procedure [46], and, by substracting $V_{xc}^{FA}[n]$, the constraint potential $V_C[n]$is obtained.

In this paper we will use the constraint potential $V_C[n]$ entering Eq. (8) as constructed by Parr and Ghosh (PG) [45],

$$V_{xc}^{PG} = V_C(0)F(R) \tag{10}$$

being determined, up to a constant $V_C(0)$, as a function of the density $n(r)$, with

$$F(R) = \frac{[kR - 1 + \exp(-kR)]\,R}{[k - 1 + \exp(-k)]} \tag{11}$$

being$R[n] = n(r)/n(0)$ and $k = 7.5$. The function $F(R)$ is determined from fits to accurate existing data for atoms $He$ to $Ar$.

When added to the Fermi-Amaldi component, it results a particular construction of the ZMP xc potential that we denoted as the Fermi-Amaldi-Parr-Ghosh (FAPG) xc potential

$$V_{xc}^{FAPG}[n] = V_{xc}^{FA} + V_{xc}^{PG} \tag{12}$$

The results for the static and dynamical properties of metal clusters using TDDFT with the FAPG potential will be compared to those arisen from LDA and other beyond LDA type of calculations. To this end we choose to fit $V_C(0)$ of Eq. (10) in such a way that $V_{xc}^{FAPG}(r=0) = V_{xc}^{LDA}(r=0)$ for each one of the clusters we are dealing with.

### 2.2. *Response properties of metal clusters*

We discuss the response of the system to a time dependent external field of the form $V_{ext}^{\sigma}(\mathbf{r},t) = V_{ext}^{\sigma}(\mathbf{r},\omega)e^{-i\omega t}$, with $V_{ext}^{\sigma}$ given in Eq. (3).

This external field will induce time-dependent variations, $\delta n^{\sigma}(\mathbf{r},t)$, in the electron spin densities. We may treat individually each Fourier component, $\delta n^{\sigma}(\mathbf{r},\omega)$, which is defined by

$$\delta n^{\sigma}(\mathbf{r},t) = \int \delta n^{\sigma}(\mathbf{r},\omega)e^{-i\omega t}d\omega \tag{13}$$

The change $\delta n^{\sigma}(\mathbf{r},\omega)$ induced in the density of electrons with spin $\sigma$ is easily obtained form first-order time-dependent perturbation theory as:

$$\delta n^{\sigma}(\mathbf{r},\omega) = \sum_{\sigma'} \int d\mathbf{r}' \chi_0^{\sigma\sigma'}(\mathbf{r},\mathbf{r}',\omega)\delta V_{eff}^{\sigma'}(\mathbf{r}',\omega), \tag{14}$$

where $\delta V_{eff}^{\sigma}$ is the change induced in the effective potential

$$\delta V_{eff}^{\sigma}(\mathbf{r},\omega) = V_{ext}^{\sigma}(\mathbf{r},\omega) + \sum_{\sigma'} \int d\mathbf{r}' \left[\frac{1}{|\mathbf{r}-\mathbf{r}'|} + \frac{\delta^2 E_{xc}}{\delta n_{\sigma}(\mathbf{r})\delta n_{\sigma'}(\mathbf{r}')}\right] \delta n^{\sigma'}(\mathbf{r}') \tag{15}$$

and $\chi_0^{\sigma\sigma'}(\mathbf{r},\mathbf{r}',\omega)$ is the response function for non-interacting electrons.

Thus, within the TDLSDA theory, the full response of the system is obtained by letting the electrons respond as free particles to an effective field which is the sum of the external one plus the induced variation of the ground state effective field.

The functions $\chi_0^{\sigma\sigma'}$ can be expressed in terms of the eigenfunctions, eigenvalues and single-particle Green's function, $G^{\sigma}(\mathbf{r},\mathbf{r}',E^{\sigma})$, of the LSDA ground state, as

$$\chi_0^{\sigma\sigma'}(\mathbf{r},\mathbf{r}',\omega) = \delta_{\sigma\sigma'} \sum_{i=1}^{occ} \left[\phi_i^{\sigma*}(\mathbf{r})\,\phi_i^{\sigma}(\mathbf{r}')\,G^{\sigma}(\mathbf{r},\mathbf{r}',\varepsilon_i^{\sigma}+h\omega) + \phi_i^{\sigma}(\mathbf{r})\,\phi_i^{\sigma*}(\mathbf{r}')\,G^{\sigma}(\mathbf{r},\mathbf{r}',\varepsilon_i^{\sigma}-h\omega)\right] \tag{16}$$

By definition, the linear spin-density response function, $\chi^{\sigma\sigma'}$, measures the proportionality between the perturbation $V^{\sigma}_{ext}$ and the resulting change $\delta n^{\sigma}(\mathbf{r},\omega)$ in the density

$$\delta n^{\sigma}(\mathbf{r},\omega) = \sum_{\sigma'} \int d\mathbf{r}' \chi^{\sigma\sigma'}(\mathbf{r},\mathbf{r}',\omega) V^{\sigma'}_{ext}(\mathbf{r}',\omega). \tag{17}$$

By comparing Eqs (14) and (17), we arrive at the following Dyson-type integral equations for the spin-density response function $\chi^{\sigma\sigma'}(\mathbf{r},\mathbf{r}',\omega)$ in terms of the non-interacting one $\chi_0^{\sigma\sigma'}(\mathbf{r},\mathbf{r}',\omega)$

$$\chi^{\sigma\sigma'}(\mathbf{r},\mathbf{r}',\omega) = \chi_0^{\sigma\sigma'}(\mathbf{r},\mathbf{r}',\omega) + \sum_{\sigma_1,\sigma_2} \int dr_1 d\mathbf{r}_2\, \chi_0^{\sigma\sigma_1}(\mathbf{r},\mathbf{r}_1,\omega) \left\{ \frac{1}{|\mathbf{r}_1 - \mathbf{r}_2|} + K_{xc}^{\sigma_1\sigma_2}(\mathbf{r}_1,\mathbf{r}_2) \right\} \chi^{\sigma_2\sigma'}(\mathbf{r}_2,\mathbf{r}',\omega) \tag{18}$$

In these integral equations, the kernel component,

$$K_{xc}^{\sigma\sigma'}(\mathbf{r}_1,\mathbf{r}_2) = \frac{\delta^2 E_{xc}}{\delta n^{\sigma}(\mathbf{r}_1)\,\delta n^{\sigma'}(\mathbf{r}_2)} \tag{19}$$

represents the residual two-body interaction due to exchange and correlation effects. The curly brackets in Eq. (18) contains the variation induced by the external field in the electronic terms of the Kohn-Sham effective potential, namely the spin-independent Coulomb potential and the spin-dependent exchange-correlation potential. The change induced in the ionic potential is assumed to be negligible.

In a previous TDLSDA study of spin resonances specialized to unpolarized clusters responding to spin dependent external fields, by Serra and coworkers [35], only one response which can be seen as the counterpart, in the spin channel, of the dipole surface plasmon [18] was found. In this paper, we have calculated, solving the integral Eq. (18) as matrix equations in coordinate space (developing the appropriate computational program) [19], the response of *polarized alkali clusters* to external fields, which may or may not include the dependence on the $z$ component of the electronic spin (see Eq. (3)). This is a general formalism for electronic system allowing us to investigate the dynamics of the electronic cloud in the electric channel as well as in the spin channel, and to search for the possible mixing between these two modes.

### 2.2.1. *Spin independent field*

Considering an external field $V_{ext}(\mathbf{r},\omega)$, which does not depend on the electron spin, the two spin electronic clouds are excited in the same way, resulting from Eq. (17)

$$\delta n^{\uparrow}(\mathbf{r},\omega) = \int \left(\chi^{\uparrow\uparrow} + \chi^{\uparrow\downarrow}\right) V_{ext}(\mathbf{r}',\omega) \tag{20}$$

$$\delta n^{\downarrow}(\mathbf{r},\omega) = \int \left(\chi^{\downarrow\uparrow} + \chi^{\downarrow\downarrow}\right) V_{ext}(\mathbf{r}',\omega) \tag{21}$$

Due to the coupling between the two spin channels, through the functions $\chi^{\sigma\sigma'}$ of Eq. (18) with $\sigma \neq \sigma'$, the response of the two electron clouds are not independent. This coupling is a consequence of the change in the effective correlation potential induced by the external field and it does not appears for non-interacting particles ($\chi_0^{\sigma\sigma'} = 0,\ \sigma \neq \sigma'$).

In terms of the total electron-density, $n(\mathbf{r}) = n^{\uparrow}(\mathbf{r}) + n^{\downarrow}(\mathbf{r})$, and the spin-density, $m(\mathbf{r}) = n^{\uparrow}(\mathbf{r}) - n^{\downarrow}(\mathbf{r})$, whose induced variations are written as

$$\delta n(\mathbf{r},\,\omega) = \delta n^{\uparrow}(\mathbf{r},\,\omega) + \delta n^{\downarrow}(\mathbf{r},\,\omega) = \int \chi_{nn}\, V_{ext}(\mathbf{r}',\,\omega)\, d\mathbf{r}' \tag{22}$$

$$\delta n(\mathbf{r},\,\omega) = \delta n^{\uparrow}(\mathbf{r},\,\omega) - \delta n^{\downarrow}(\mathbf{r},\,\omega) = \int \chi_{nn}\, V_{ext}(\mathbf{r}',\,\omega)\, d\mathbf{r}' \tag{23}$$

where the operators $\chi_{nn}$ and $\chi_{mn}$ are defined as

$$\chi_{nn} \;=\; \chi^{\uparrow\uparrow} \;+\; \chi^{\uparrow\downarrow} + \;\chi^{\downarrow\uparrow} +\; \chi^{\downarrow\downarrow} \tag{24}$$

$$\chi_{mn} \;=\; \chi^{\uparrow\uparrow} \;+\; \chi^{\uparrow\downarrow} - \;\chi^{\downarrow\uparrow} -\; \chi^{\downarrow\downarrow} \tag{25}$$

describing, respectively, the total density and the spin-density responses to an external field which does not depend on the spin.

### 2.2.2. *Spin dependent field*

Considering an external field which depends on the $z$ component of the electron spin, $V^{\sigma}_{ext}\,(\mathbf{r},\,\omega) = V_{ext}\,(\mathbf{r},\,\omega)\;\sigma^{z}$, the two spin-electronic clouds are excited in a different way. From Eq. (17), we have

$$\delta n^{\uparrow}(\mathbf{r},\,\omega) = \int \left(\chi^{\uparrow\uparrow} - \chi^{\uparrow\downarrow}\right)\, V_{ext}(\mathbf{r}',\,\omega) \tag{26}$$

and

$$\delta n^{\downarrow}(\mathbf{r},\,\omega) = \int \left(\chi^{\uparrow\uparrow} - \chi^{\uparrow\downarrow}\right)\, V_{ext}(\mathbf{r}',\,\omega) \tag{27}$$

where, as a consequence of the spin operator $\sigma^{z}$, the contributions of $\chi^{\uparrow\downarrow}$ and $\chi^{\downarrow\downarrow}$ to $\delta n^{\uparrow}$ and $\delta n^{\downarrow}$, respectively, have the opposite sign than in Eqs (20) and (21). The variations in the total electron-density and in the spin-density are, in this case,

$$\delta n(\mathbf{r},\,\omega) = \delta n^{\uparrow}(\mathbf{r},\,\omega) + \delta n^{\downarrow}(\mathbf{r},\,\omega) = \int \chi_{nm}\, V_{ext}(\mathbf{r}',\,\omega)\, d\mathbf{r}' \tag{28}$$

$$\delta m(\mathbf{r},\,\omega) = \delta n^{\uparrow}(\mathbf{r},\,\omega) - \delta n^{\downarrow}(\mathbf{r},\,\omega) = \int \chi_{mm}\, V_{ext}(\mathbf{r}',\,\omega)\, d\mathbf{r}' \tag{29}$$

where now the operators $\chi_{nn}$ and $\chi_{mn}$ are defined as

$$\chi_{nm} = \chi^{\uparrow\uparrow} - \chi^{\uparrow\downarrow} + \chi^{\downarrow\uparrow} - \chi^{\downarrow\downarrow} \tag{30}$$

$$\chi_{mm} = \chi^{\uparrow\uparrow} - \chi^{\uparrow\downarrow} - \chi^{\downarrow\uparrow} + \chi^{\downarrow\downarrow} \tag{31}$$

describing, respectively, the total density and the spin-density responses to a spin dependent external field.

In both cases, spin and no-spin dependent fields, and for a multipole field of the type $V_{ext} \approx r^L Y_L(\theta, \varphi)$ with a well defined angular momentum, the corresponding responses $R_n^L(r, \omega)$ and $R_m^L(r, \omega)$ are given by the integrals

$$R_n^L(\omega) = \int \delta n_L(r, \omega)\, r^L\, dr \tag{32}$$

$$R_m^L(\omega) = \int \delta m_L(r, \omega)\, r^L\, dr \tag{33}$$

which represent the multipole moments. In this work we deal with the dipole case $L = 1$, and the corresponding subscript will be dropped. The value of $R_n(\omega)$ for a static field, $\omega = 0$, gives the static polarizability, $\alpha = R_n(0)$. On the other hand, the imaginary part of $R_i(\omega)$ $(i = n, m)$ is related to the strength function $S_i(\omega)$ by

$$S_i(\omega) = \frac{1}{\pi} Im[R_i(\omega)] \tag{34}$$

*2.2.3. The paramagnetic case*

For spin saturated clusters, our formulation reduces to the one given in ref [35] and we can write, after a little bit of algebra [36]

$$K_{xc}^{\uparrow\uparrow} = K_{xc}^{\downarrow\downarrow} = K_{xc} + I_{xc} \tag{35}$$

$$K_{xc}^{\uparrow\downarrow} = K_{xc}^{\downarrow\uparrow} = K_{xc} - I_{xc} \tag{36}$$

with the operators $K_{xc}$ and $I_{xc}$ given by

$$K_{xc} = \frac{\delta^2 E_{xc}}{\delta n(\mathbf{r}_1)\, \delta n(\mathbf{r}_2)} \tag{37}$$

and

$$I_{xc} = \frac{\delta^2 E_{xc}}{\delta m(\mathbf{r}_1)\, \delta m(\mathbf{r}_2)} \tag{38}$$

respectively. In this case, the four Eqs 24–25, 30–31, reduce to

$$\chi_{nn} = \chi_{nn}^0 + \chi_{nn}^0 \left( \frac{1}{|\mathbf{r}_1 - \mathbf{r}_2|} + K_{xc} \right) \chi_{nn}, \tag{39}$$

$$\chi_{mm} = \chi_{mm}^0 + \chi_{mm}^0\, I_{xc}\, \chi_{mm}, \tag{40}$$

$$\chi_{nm} = \chi_{mn} = 0. \tag{41}$$

The correlation functions of free particles, in both spin-spin, $\chi_{mm}^0$, and density-density, $\chi_{nn}^0$, responses coincide, $\chi_{nn}^0 = \chi_{mm}^0 = \chi_0^{\uparrow\uparrow} + \chi_0^{\downarrow\downarrow}$. Moreover, for the unpolarized case, the response of the total density to a spin dependent field, $\chi_{nm}$, as well as the response of the spin-density to a no-spin dependent external potential, $\chi_{mn}$, are suppressed. The remaining responses, that is, the total density response to a no-spin dependent field, $\chi_{nn}$, and the spin-density response to a spin dependent field, $\chi_{mm}$, are not coupled one to the other.

In the paramagnetic case, the LDA results for $R_n(\omega)$ in the case of no-spin dependent field have been obtained by first calculating the LDA independent-particle susceptibility, $\chi_{nn}^{0\,LDA}$ and the local field correction, $K_{xc}^{LDA}$, then integrating Eq. (39) to obtain $\chi_{nn}^{LDA}$, and finally $R_n^{LDA}(\omega)$ from Eqs (22) and (32). A similar procedure leads to $R_n^{FAPG}(\omega)$. The FAPG kernel is calculated in this work from the FAPG potential as

$$K_{xc}^{FAPG} = \frac{\partial V_{xc}^{FAPG}[n(r)]}{\partial n(r')} \tag{42}$$

which is a non-diagonal *symmetric* matrix in the coordinates $r$, $r'$, as it should be.

We comment now on previous attempts to include the correct long-range behaviour of the xc potential and the corresponding kernel (local field correction) in TDDFT response calculations. The independent-particle susceptibility was obtained by Pacheco and Ekardt [24] using SIC corrections, and then the screening potential effects contained in the kernel $K$ were considered in two ways:

(1) neglecting SIC, which leads to a diagonal matrix like in the LDA case; and
(2) considering also the orbital-dependent SIC potentials, that is, the full SIC-TDDFT with the correct $K^{SIC}$kernel.

On the other hand, starting with the PB non-local xc potential, there results a kernel $K^{PB}$ which is not a symmetric matrix [44], unlike the exact $K(r, r')$, which must be symmetric. For this reason, in previous works [49] several *ad-hoc* approximations were introduced leading to an $\sim -1/r$ asymptotic $V_{xc}(r)$ and to a symmetrical $K(r, r')$. A mixed ad hoc WDA-PB procedure was used. In contrast, the use for TDLDA response calculations of $V_{xc}^{FAPG}$ will be consistent throughout this paper, in the sense that $K_{xc}^{FAPG}$ is constructed from $V_{xc}^{FAPG}$. Both facts, the correct asymptotic behaviour of $V_{xc}^{FAPG}$ and the corresponding symmetrical non-local kernel, modify substantially the structure of the other approximations to the response function $\chi_{nn}$, yielding improved static polarizabilities and photoabsorption.

## 3. Electrical and spin responses of neutral and charged sodium clusters

We can view the excitations of the initial cluster ground state as simple spatial displacements of the electronic cloud. When the external field does not acts on the electron spin a dipole mode is generated through a shift of the whole electronic cloud with respect to the center of the positive charge. In the case of a spin dependent field, as in Eq. (3), the spin-up and the spin-down electronic densities are displaced in opposite directions and a spin dipole mode is generated. These responses influences each to the other in a manner that we analyze for sodium clusters with closed and open electronic shells in Sections 3.1.1 and 3.1.2 respectively. In Section 3.2 we analyze the corrections on the static and dynamical polarizabilities of sodium clusters due to consider the correct long-range behaviour of the xc potential.

### *3.1. TDLDA results for spin-dependent excitations*

In order to illustrate the formalism developed in the previous section we apply it to jellium sodium clusters. For these systems we obtain, in this section, the response functions of the total density and of the spin density to dipole external fields with and without spin dependence. To look for a possible cross talk between the electric and the spin dipole excitation modes, we calculate the spin-density response function, $R_m$, in a no-spin dependent dipole excitation, and conversely, the total density response function, $R_n$, in a spin dependent excitation. These responses are related to the operators $\chi_{mn}$ (Eq. (25)) and $\chi_{nm}$ (Eq. (30)), respectively. In each case, we will calculate the imaginary part of the response as a more robust quantity related to the photoabsorption process (cross section). We distinguish two situations for the initial shell model ground state, namely, spin-saturated clusters with closed electronic shells, and open shell clusters with nonzero magnetization.

*3.1.1. Spin saturated clusters*

As test cases we have considered the "magic" clusters $Na_9^+$, $Na_8$, and $Na_{20}$ whose respective ground state spin up and spin down densities have identical electron distributions and their shapes come out to be spherical in the jellium model. We expect a clear separation of electric modes from spin modes due to the spin symmetry. There is no response of the spin-density, $m$, to a no-spin dependent field because their variation, $\delta m$, is zero, and, thus, the spin mode remains dormant. There does not also a response of the total density to a spin dependent field because the center of gravity of the whole electron cloud remains unaltered with respect to the one of the positive charge. The effect of a spin dependent field is to create a local magnetization. However, it does not induce a net magnetic moment in the whole cluster. The induced spin density, $\delta m$, is a spin wave, which, when spatially integrated, vanishes. Serra and coworkers [35] have observed that it is peaked at the surface of the cluster, allowing to identify these modes as surface spin excitations, which have been called "surface paramagnons" by analogy with the spin-wave excitations in the bulk system.

We consider first the case of $Na_9^+$ cluster, with electronic configuration $1s^2 1p^6$ in the spherical jellium model. The $\chi^0_{nn}$ and $\chi^0_{mm}$ non-interacting susceptibilities to no-spin and spin dependent fields, respectively, are identical. They have simple poles at the particle-hole excitation energies, whose strength, in general, decreases for increasing energy. In the case of $Na_9^+$ these excitations, lying at 1.56 eV and 2.18 eV, are well separated and they can be identified as the particle-hole (p-h) transitions $1p \to 1d$ and $1p \to 2s$, respectively. When we include the induced effective potential, the resulting full responses are substantially modified. In the upper panel of Fig. 1 the imaginary part of the full total density response function, $R_n$, to an external no-spin dependent field is shown (long-dashed curve). Note the logarithmic scale for the strength in the figure. The dominant peak at 3.06 eV is the well know Mie plasmon, which is blue-shifted with respect to the p-h excitations as a consequence of the electron correlations induced by the strong and repulsive electron-electron interaction. To analyze the p-h composition of the plasmon peak, we have calculated the response when a particular p-h transition is suppressed in the calculation of $\chi^0_{nn}$. By suppressing the transition $1p \to 1d$ (dotted line in Fig. 1(a)), results only a main peak at 2.34 eV. By eliminating the transition $1p \to 2s$, a resonance results at about 3.36 eV with a large strength (dotted-dashed line spectrum) which is very similar to the Mie plasmon. These calculations show that the plasmon peak takes nearly all its strength from the two p-h transitions, $1p \to 1d$ and $1p \to 2s$, being the contribution of the first larger, whereas $1p \to 2s$ transition provides the main contribution to the peak at about 2.2 eV observed in the full spectrum of Fig. 1 (upper panel). The full spin-density response function, $R_m$, of $Na_9^+$ to a spin dependent field is shown in the lower panel of Fig. 1. In this case, at variance with the p-h correlations in the total density response of the total density, the net residual interaction acting among the p-h excitations, is attractive and much smaller than in the dipole channel. That is because the induced Hartree potential leads to canceling oscillations of up and down electronic clouds, and the remaining induced potential is due to the exchange and correlation contribution, which is attractive. Thus, the spectral strength shifts to lower energies than the p-h transitions giving essentially

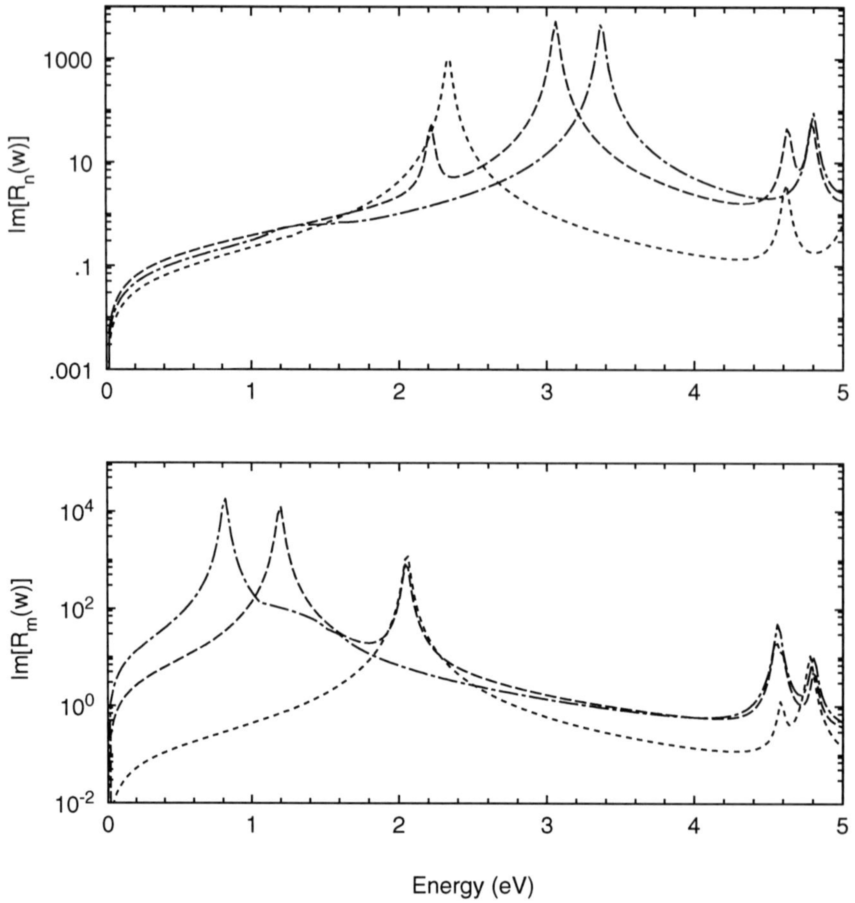


Fig. 1. The imaginary part of the response function of the total density, $R_n(w)$, to a dipole field (upper panel) and of the spin density, $R_m(w)$, to a spin-dipole field (lower panel), are plotted for $Na_9^+$ versus the excitation energy. The full responses are given as the long-dashed curves. Considering the dotted lines may be seen the responses with the $1p \rightarrow 1d$ transition eliminated and the responses with the $1p \rightarrow 2s$ transition eliminated are plotted as the dotted-dashed lines.

one peak at 1.2 eV, that we identify as a collective spin excitation, which exhausts a large fraction of the spin-dipole sum rule. This state washes out, almost completely, the low-energy p-h transitions. In Fig. 1(b) we also show the response functions when the transitions $1p \longrightarrow 1d$ (dotted line) and $1p \longrightarrow 2s$ (dotted-dashed line) are eliminated. In the first case, a peak appears at 2.16 eV and in the second case there is a main resonance at 0.82 eV close to the spin mode. Superimposing these two profiles, we obtain a spectrum similar to the full response, but shifted to lower energies. The contribution of the transition $1p \longrightarrow 1d$ to the strength of the spin mode is larger, as in the case of the Mie plasmon dipole mode of Fig. 1(a). The higher-lying p-h transitions are not so strongly affected by the interaction. They are slightly shifted in energy but still keep approximately their original strength. In the case of an external excitation, which includes spin and no-spin components, $V_{ext}^{\sigma} = \sum_{j=1}^{N} r_j^L Y_{L0}(1+\ \sigma_j^z)$, the electric and spin channels will be excited with comparable strength. The two different channels will develop independently and the resulting spectrum will be the linear superposition of the spectra shown in the upper and lower panels of Fig. 1.

For the sake of completeness we present in Fig. 2 the response functions of the total density and of the spin-density of $Na_8$ and $Na_{20}$ clusters. The non-interacting response of $Na_8$ has two peaks at 1.5 eV

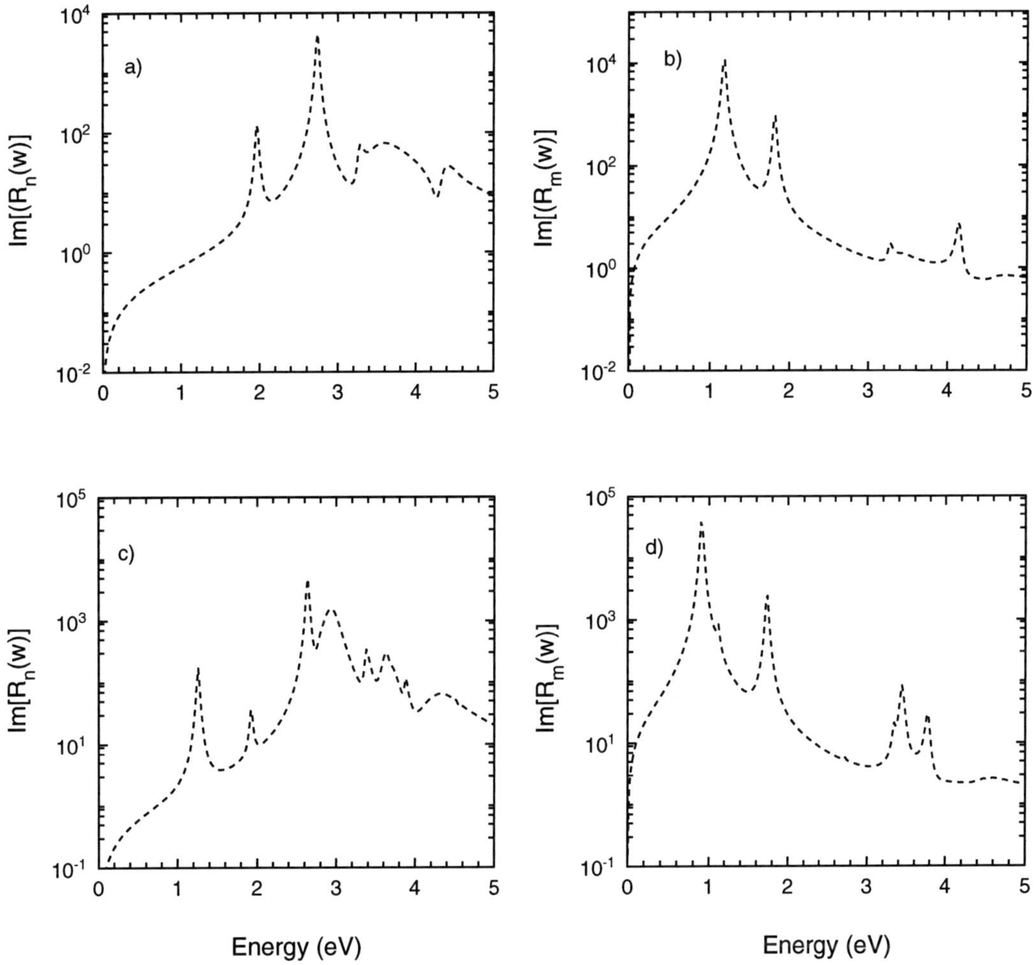


Fig. 2. The imaginary part of the response functions of $Na_8$ and $Na_{20}$ clusters is plotted versus the excitation energy. a) The full response, $R_n$, of $Na_8$ to an electric dipole field; b) the full response, $R_m$, of $Na_8$ to a spin dipole field; c) the full response, $R_n$, of $Na_{20}$ to an electric dipole field; d) the full response, $R_m$, of $Na_{20}$ to a spin dipole field.

and 1.92 eV, due to the transitions $1p \rightarrow 1d$ and $1p \rightarrow 2s$, respectively. The full response functions of $Na_8$ to electric and to spin dipole excitations are represented in Fig. 2(a) and 2(b), respectively. As in the case of $Na_9^+$, a perfect separation of the modes can be observed. The Mie plasmon resonance appears at 2.74 eV in the dipole channel, together with another peak at lower energy (1.98 eV) and with smaller strength. There is some structure at around 4 eV corresponding to a mixing of p-h transitions in the proximity of the ionization threshold. Photo-depletion experiments [50] show a main resonance at 2.53 eV with a width of 0.19 eV. The spin mode of $Na_8$ is seen, in Fig. 2(b), as a pronounced peak at 1.16 eV, very close to the first p-h excitation. This result agrees with the previous calculations by Serra and co-workers [35].

For $Na_{20}$ with the ground state electronic configuration $1s^2 1p^6 1d^{10} 2s^2$, the non-interacting response have several low-energy transitions that lie very close. The strength function for a dipole excitation, $R_n$, shows (Fig. 2(c)) a multipeaked structure with two main resonances at 2.64 eV and at 2.94 eV, somewhat below the threshold for electron detachment. Quantum size effects leading to the breaking of the strength (Landau fragmentation) are quite clear in this spectrum and they have been extensively discussed in the literature [22,24]. The measured photoabsorption cross section for $Na_{20}$ [50], shows two main transitions at 2.46 eV and 2.74 eV (widths of 0.2 eV and 0.4 eV, respectively). The calculated response function to the spin dependent field, $R_m$, shows a broad and strong resonance at 0.9 eV (spin mode) and another smaller resonance at 1.7 eV (Fig. 2(d)).

By comparing the $Na_8$ and $Na_{20}$ spectra we can asses the evolution of the spin modes with cluster size. The short range interaction causes the energy of the collective state to decrease with increasing size following the trend $\propto N^{-1/3}$, in a way that resembles the behavior of the analogous nuclear collective excitations, and that is opposite to that exhibited by the Mie plasmons (electric dipole response). For increasing sizes, the residual interaction is less and less effective in separating the collective spin state from the p-h excitations. Thus, for large clusters, we expect to approach the bulk picture in which the spin wave excitations are strongly damped by surrounding p-h excitations (Landau damping).

*3.1.2. Open shells clusters*

There are several examples of clusters where spontaneous spin polarization occurs in the ground state or in a low lying isomer [51]. In this section we study the response of selected spin polarized clusters, the spherical isomer of $Na_6^+$. We expect a natural mixing of the dipole and the spin dipole modes in the total density and the spin-density responses to external fields. The global motion of the whole electronic cloud will excite the spin mode and, conversely, the displacements of the up and down spin densities into opposite directions will excite also the electric dipole mode. The $Na_6^+$ cluster is triaxial in its ground state, but we consider instead the isomeric spherical state of $Na_6^+$, which is strongly spin-polarized [51]. This isomer is an example where spontaneous spin polarization (Hund's rule) occurs in the ground state, leading to the electronic configuration $1s^2 1p^3$, with the $p$ electrons having the same spin component. We can study, in this case, the cross talk between the dipole and the spin-dipole channels without bothering about additional features due to ionic degrees of freedom.

The main allowed p-h transitions are $1s^{\downarrow} \rightarrow 1p^{\downarrow}$ at 1.64 eV, $1p^{\uparrow} \rightarrow 1d^{\uparrow}$ at 1.94 eV and $1p^{\uparrow} \rightarrow 2s^{\uparrow}$ at 2.50 eV. The first one, which is not possible for closd-shell clusters, originates the main qualitative differences between the full responses of these clusters and the ones of $Na_6^+$ cluster. The calculated strength function for the dipole excitation, $R_n$, is shown in figure 3a. We can see the usual pattern of the Mie plasmon, caused by a residual interaction strongly repulsive, which shifts the peak to 3.18 eV, a much higher energy than the unperturbed p-h transitions. In Fig. 3(b) the response function for a spin-dipole excitation, $R_m$, is also shown. The strength is now red-shifted with respect to the relevant p-h transitions, due to the attractive residual interaction in the spin channel and the spin mode appears at 1.38 eV. Due to the spin polarization of the $Na_6^+$ spherical isomer, the corresponding dipole and spin dipole modes are mixed very efficiently. Both channels interact among themselves and a signal of the spin-dipole mode is visible in the spin-dipole spectrum at 3.08 eV (1.38 eV), as can be seen in the Fig. 3(a) and 3(b). Moreover, both electrical and spin perturbations excite channels with comparable strength. Thus, figure 3c, shows the full response function of the density response to a spin-dipole excitation, which coincides in the case of half-filled shell clusters as $Na_6^+$ with the strength function of the spin-density response for a dipole excitation. Now, these responses are not suppressed as in the case of unpolarized clusters and significant cross talk between dipole and spin dipole modes are detected this figure, where one can see a small copy of the spin mode and a response in the region of the dipole plasmon. On the other hand, it is interesting to note the coincidence of the positions of the peaks in the three panels of Fig. 3. Each excitation type is enhanced by its related mode.

The plasmon peak and the spin dipole mode derive their strength from the $1p^{\uparrow} \rightarrow 1d^{\uparrow}$ and $1p^{\uparrow} \rightarrow 2s^{\uparrow}$ p-h transitions, as in the case of the spin saturated cluster, $Na_9^+$ (see Fig. 4). By suppressing the $1p^{\uparrow} \rightarrow 1d^{\uparrow}$

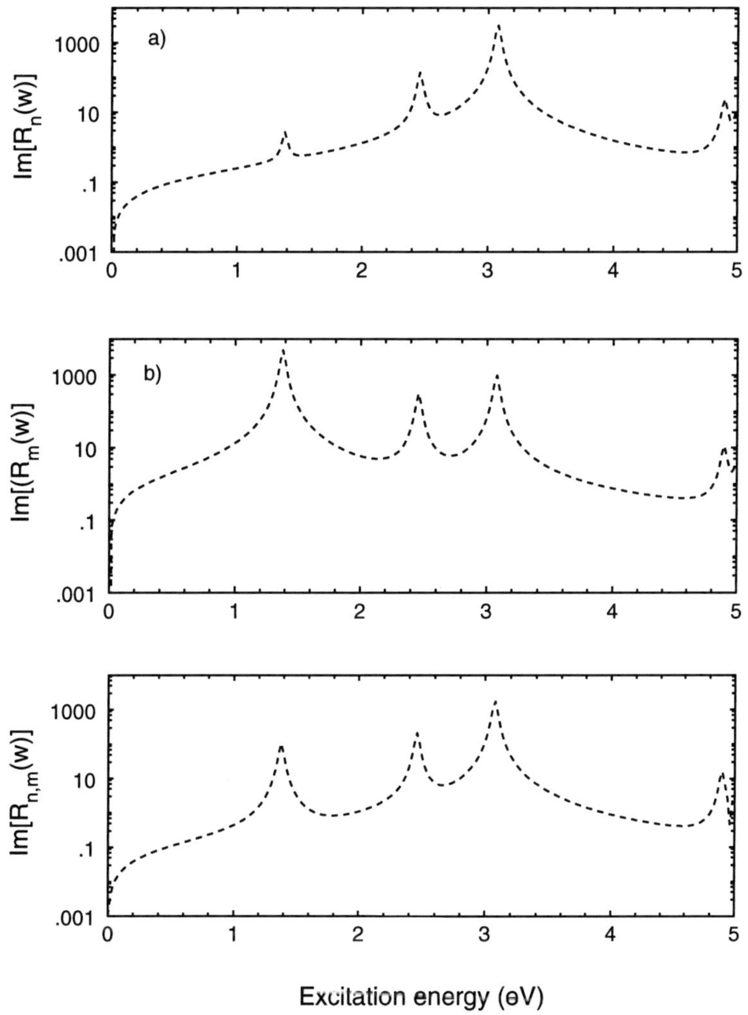


Fig. 3. The imaginary part of the response functions of $Na_6^+$ is plotted versus the excitation energy. a) The full response of the total density to an electric dipole field; b) the full response of the spin density to a spin dipole field; c) the full response of the total density (spin density) to a spin (electric) dipole field.

transition, neither the dipole nor the spin dipole modes appear. When the $1p^{\uparrow} \rightarrow 2s^{\uparrow}$transition is suppressed the peak at about 2.4 eV shown in Fig. 3 does not appear. However, the plasmon and the spin mode are shifted with respect to their original positions in the full spectra. The p-h transition $1s^{\downarrow} \rightarrow 1p^{\downarrow}$ is the origin of the interaction between the two channels, dipole and spin-dipole, and when it is eliminated the dipole spectrum looks like the full spectrum of $Na_6^+$ (see Fig. 3). We may conclude that the $1s^{\downarrow} \rightarrow 1p^{\downarrow}$ transition is responsible for the correspondence between the resonances observed in the full dipole and spin-dipole spectra and the cross talk between the two modes.

## *3.2. TDDFT results using the FAPG xc potential*

The experimental dipole plasmon peaks in $Na_9^+$, $Na_8$ and $Na_{20}$ are 2.62 eV, 2.53 eV and 2.42–2.78 eV, respectively, which are about 10% red-shifted with respect to our TDLSDA calculations in the previous

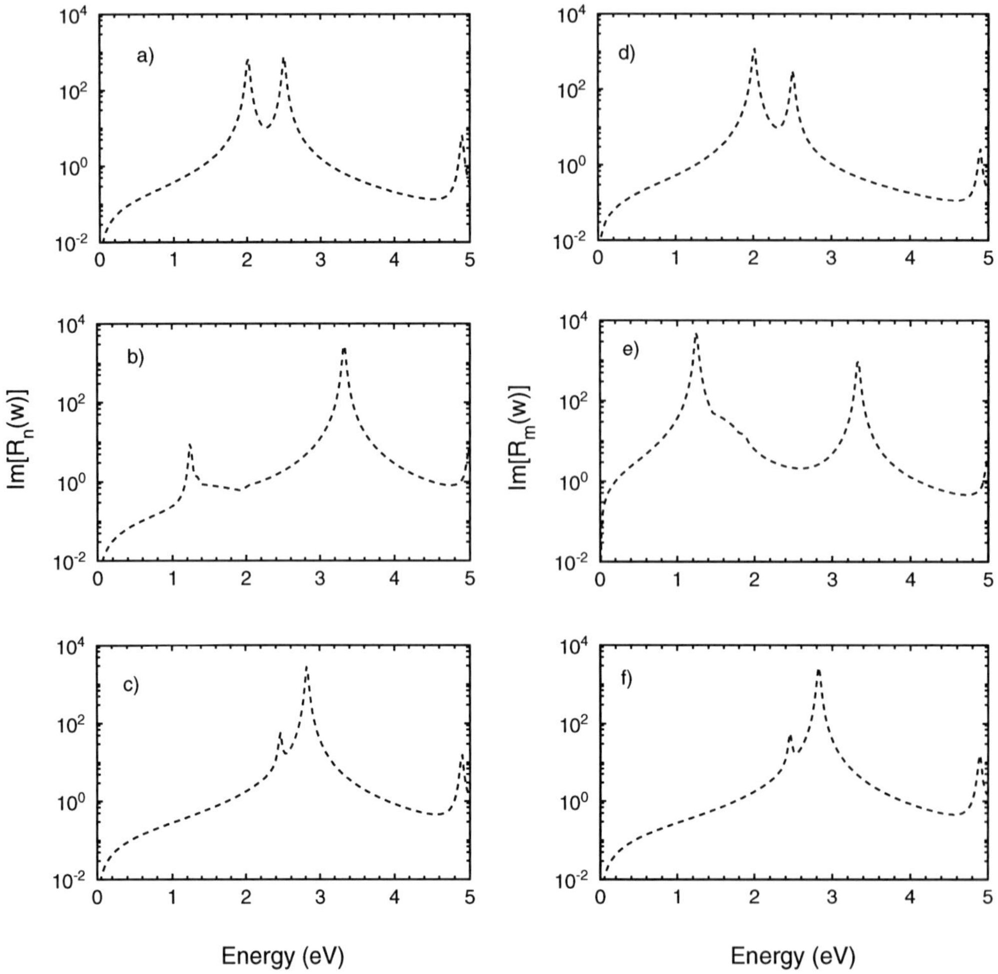


Fig. 4. The imaginary part of the response functions of $Na_6^+$, when some transitions are suppressed is plotted versus the excitation energy. a) The response, $R_n$, with the $1p^{\uparrow} \longrightarrow 1d^{\uparrow}$ transition eliminated. b) The response, $R_n$, with the $1p^{\uparrow} \longrightarrow 2s^{\uparrow}$ transition eliminated. c) The response, $R_n$, with the $1s^{\downarrow} \longrightarrow 1p^{\downarrow}$ transition eliminated. d) The response, $R_m$, with the $1p^{\uparrow} \longrightarrow 1d^{\uparrow}$ transition eliminated. e) The response, $R_m$, with the $1p^{\uparrow} \longrightarrow 2s^{\uparrow}$ transition eliminated. f) The response, $R_m$, with the $1s^{\downarrow} \longrightarrow 1p^{\downarrow}$ transition eliminated.

section. In this section we improve the agreement with experimental plasmon resonances and with static polarizabilities of sodium clusters by introducing in our TDDFT calculations the correct long-range behaviour of the xc potential via the Parr-Ghosh approximation outlined in Section 2.

### *3.2.1. Ground state properties and static polarizabilities of alkali-metal clusters*

We consider first the electronic ground state of some neutral ($Na_8$, $Na_{20}$, $Na_{40}$) and charged ($Na_9^+$, $Na_{21}^+$ and $Na_{19}^-$) closed-shell jellium-like sodium clusters. Our goal is a systematic comparison of the electronic properties obtained using LDA versus FAPG potentials. We will see that the FAPG electronic ground state shows a different energy level structure than the LDA one. On one hand, the eigenstates are more tightly bound and the grouping and ordering of levels are modified. On the other hand, there is a larger number of bound states and the gap between the HOMO and LUMO becomes wider. These two effects, as we will discuss in Section 3.2.2, lead to higher static dipole polarizabilities and to smaller

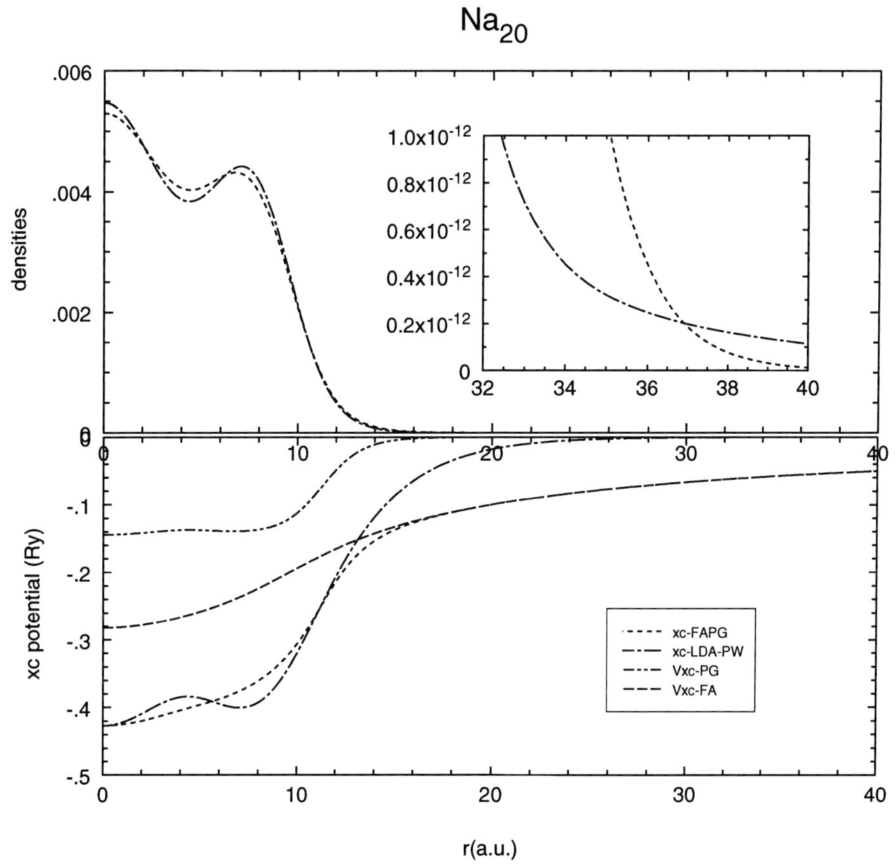


Fig. 5. Self-consistent electron densities and KS-potentials of $Na_{20}$ in the spherical jellium model, calculated within the LDA (dot-dashed) and FAPG (dotted) approximations. The two components of FAPG are also shown: FA (dashed) and PG (dash-dot-dot).

plasmon resonance frequencies, as well as to increase fragmentation of the plasmon.

A first insight is gained by comparing directly the behavior at different distances from the center of the cluster of the various xc potentials. In Fig. 5 we plot the KS-potentials and self-consistent densities of $Na_{20}$ obtained in this work for two cases, the LDA-PW of Perdew-Wang [39] that falls off exponentially as $r \to \infty$, and the FAPG one, which behaves as $-1/r$ for large $r$. We show also the two components of the FAPG potential, namely the Fermi-Amaldi term, $V_{xc}^{FA}$, which reproduces the correct asymptotic behavior, and the short-ranged Parr-Ghosh component, $V_{xc}^{PG}$, reflecting the shell structure of the system. The function of the Fermi-Amaldi term is mainly to subtract out the self-repulsion in the Hartree potential, $V_H$, while that of the Parr-Ghosh component represents correlation plus interorbital exchange. Note in the inset of Fig. 5 that the FAPG density is larger than the LDA one from the edge of jellium sphere, ($R_{jell} = 10.667$ a.u.) to a radial distance of 37 a.u., that is, the electron spill-out is larger for FAPG than for LDA. For $r > 37$ a.u. we have the asymptotic regime where the density decay exponentially as $\sim \quad e^{-2ar}$ with $a = (2\varepsilon_{HOMO})^{1/2}$, that is, $\varepsilon_{HOMO}^{FAPG} > \varepsilon_{HOMO}^{LDA}$.

In Table 1 we present the ratios of the negative of the HOMO eigenvalues, calculated using the LDA-PW, FAPG and PB xc potentials, to the experimental ionization potentials [52,53], for neutral clusters with 8, 20 and 40 valence electrons. These ratios are closer to unity, as they should be within exact

Table 1
The ratios of the negative of the HOMO eigenvalues to the experimental ionization potentials for sodium clusters, using different xc potentials. The PB results are taken from Ref. [22,49]. The experimental ionization potentials, $I_{exp}$, taken from Ref. [52,53] are (in eV) 4.22, 3.76 and 3.58 for the 8, 20 and 40 atom clusters

| | $-\varepsilon_{HOMO}/I_{\rm exp}$ | | |
|---|---|---|---|
| $N^o$ atoms$^{+/-}$ | 8 | 20 | 40 |
| LDA | 0.77 | 0.73 | 0.76 |
| FAPG | 0.92 | 0.94 | 0.95 |
| PB | 0.96 | 0.99 | 1.04 |

Table 2
Comparison of the occupied Kohn-Sham eigenvalues for jellium like $Na_{20}$ cluster obtained within different xc potentials. LDA-PW and FAPG: this work; PB from Ref. [55]; GW from Ref. [56]

| | EIGENVALUES (eV) | | | |
|---|---|---|---|---|
| Orbital | LDA-PW | FAPG | PB | GW |
| 1s | −5.1 | −5.9 | −5.8 | −5.8 |
| 1p | −4.4 | −5.2 | −5.2 | −5.2 |
| 1d | −3.4 | −4.2 | −4.2 | −4.4 |
| 2s | −2.7 | −3.6 | −3.7 | −3.8 |

Table 3
The Kohn-Sham eigenvalues (in eV) of the occupied orbitals for $Na_8$ cluster in the jellium model, calculated by using different xc approximations. LDA-PW and FAPG: this work; BLYP, KLI and GAM from Ref. [16]; PB from Ref. [55]

| | LDA-PW | BLYP | FAPG | PB | KLI | GAM |
|---|---|---|---|---|---|---|
| 1s | −4.53 | −4.62 | −5.19 | −5.30 | −5.66 | −5.90 |
| 1p | −3.27 | −3.44 | −3.88 | −4.04 | −4.39 | −4.59 |

DFT [54], by using the asymptotically correct potentials FAPG and PB than using the LDA one. The orbital energy eigenvalues are a sensitive indicator of the differences between various xc functionals. The Kohn-Sham eigenvalues and eigenvalue differences resulting from LDA, FAPG, PB and GW, KLI and GAM calculations will affect drastically the linear response of the clusters in the optical region. In Table 2 we compare the eigenvalues obtained using the LDA and the FAPG potentials for $Na_{20}$ cluster with the PB eigenvalues [55] and the GW quasi-particles energies obtained in ref. [56]. We see that the xc potentials, FAPG and PB, lead to eigenvalues very close to the GW energies, which are theoretically comparable to the experimental electron binding energies. Similarly, in Table 3 we compare our results for LDA-PW and FAPG eigenvalues $1s$ and $1p$ of $Na_8$ with those resulting from other asymptotically correct xc potentials (KLI [57], GAM [16,58], and PB [55]), and from the gradient corrected xc potential BLYP [42]. Gradient corrections have a much smaller effect on the orbital eigenvalues than the other non-local approaches. It turns out that the FAPG and PB eigenvalues are shifted down by $\sim 0.6$ eV and $\sim 0.8$ eV, respectively, compared to LDA; whereas, the KLI and GAM eigenvalues are more tight bound, leading to a $1p$ eigenvalue for $Na_8$, which is larger than the experimental ionization potential, $I_{exp} =$ 4.22 eV.

In Table 4 different calculations of the static polarizabilities for neutral and charged clusters with 8, 20 and 40 electrons are compared with experimental values. The FAPG, WDA and SIC results show a systematic improvement with respect to LDA, which is mainly due to a better description of the external part of the induced spill-out density. The WDA-PB and FULL-SIC calculations yield polarizabilities

Table 4
The static polarizability in units of the classical Mie polarizability, $\alpha_{Mie} = r_s^3 N$, for neutral and charged sodium clusters with $N = 8$, 20 and 40 electrons and $r_s = 3.93$ a.u. LDA-PW and FAPG (this work); PB from Ref [55]; SIC from Ref [24]; experimental values: a) Ref. [59], and b) Ref. [60]

| N° atoms$^{+/-}$ | 8 | 20 | 40 | $9^+$ | $19^-$ | $21^+$ |
|---|---|---|---|---|---|---|
| LDA-PW | 1.16 | 1.15 | 1.04 | 1.01 | 1.31 | 1.08 |
| FAPG | 1.58 | 1.40 | 1.30 | 1.26 | 1.73 | 1.27 |
| WDA | 1.49 | 1.42 | 1.37 | 1.27 | 1.72 | 1.29 |
| WDA-PB | 1.81 | 1.63 | 1.53 | 1.48 | 1.99 | 1.44 |
| SIC | 1.57 | 1.46 | 1.41 | 1.31 | 1.80 | 1.33 |
| FULL-SIC | 1.70 | 1.61 | 1.51 | 1.46 | 1.80 | 1.45 |
| Exp.$^{a)}$ | 1.72 | 1.58 | 1.56 | | | |
| Exp.$^{b)}$ | 1.86 | 1.77 | 1.55 | | | |

closer to experimental values. Since all calculations reported in Table 4 have been performed within the spherical jellium background model, at least a part of the remaining discrepancy with respect to experiment can be ascribed to geometric effects. For a recent discussion of the $Na_8$ case (and smaller sodium and lithium clusters) see Ref [20].

We would like to point out, that for each group of clusters with a given number of valence electrons, the smallest polarizability is that of the cations. This is qualitatively understood in terms of the net confining force with which the ionic background attracts the valence electrons. Because this force is largest for the cations, the electronic density is most compressed, leading to the smallest polarizability.

*3.2.2. Photoabsorption cross sections of neutral and charged sodium clusters*

A comparison of the linear TDDFT response to light of interacting valence electron of $Na_{40}$ calculated within LDA-PW (upper panel) and FAPG (lower panel) xc approaches is shown in Fig. 6. Dotted lines represent the results for the independent-particle-unscreened response and dot-dashed lines represent those of the full response of interacting electrons. The peaks are not delta functions because we have used complex photon energy $\omega + i\varepsilon$ with $\varepsilon = 0.02$ eV. The vertical arrow indicates for each case the position of the HOMO eigenvalue in the one-electron spectra. The multipeaked structure below the ionization threshold appearing in the FAPG response is similar to the one obtained previously within the WDA-PB xc approximation [49]. This increase of strength in the Ultra-Violet (UV) region is due to the coupling of the collective surface plasmon with one-electron excitations to the loosely bound Rydberg states, which are now properly, incorporated in the FAPG spectra. The increase of bound levels in FAPG with respect to LDA can be seen in Fig. 6 by simple inspection of the non-interacting spectra, which are composed by simple poles at the electron-hole excitation energies. The strong fragmentation of the FAPG response indicates that no particular transition dominates over the others, implying a reduction of the total strength in the visible region, which occurs without any significant change in the fragmentation pattern of the collective mode.

A direct comparison of the photoabsorption cross section calculated in the present work with the line shape and line width experimentally observed is not adequate, because of the existence of relaxation mechanisms of plasmon resonance not accounted for at any level of the linear response formalisms considered here. These relaxation mechanisms are responsible for the lifetime of the plasmon, as well as a sizeable line width associated to the geometry of the ions. The effect of these additional mechanisms can be simulated, on the average, by folding the calculated cross sections with normalized Lorentzian functions, including damping ratios. Using a Lorentzian width of 0.2 eV (ten times larger than the one

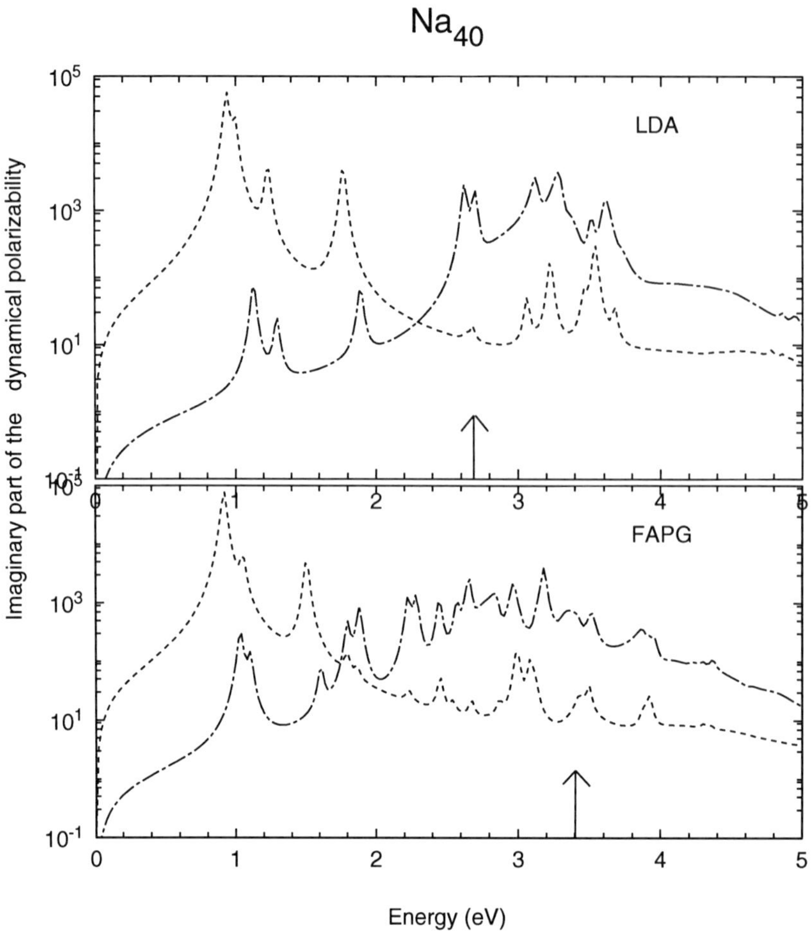


Fig. 6. Imaginary part of the dynamical polarizability per electron (in logarithmic scale and arbitrary units) of $Na_{40}$, versus excitation energy (in eV). The arrow indicates the HOMO eigenvalue. Complex photon energy $\omega + i\varepsilon$ with $\varepsilon = 0.02$ eV has been used.

used in Fig. 6), we have obtained two broad peaks (see Fig. 7 for $Na_{20}$), which are red-shifted in going from LDA-PW to FAPG, giving a better agreement with the experimental positions [59] (2.46 eV and 2.74 eV for $Na_{20}$, and 2.40 eV an 2.65 eV for $Na_{40}$). This qualitative behavior is the one expected from the results of Table 4, that is, the higher the static polarizability, the lower the plasmon frequency. Thus, the FAPG potential leads to an improvement over the LDA predictions of the photoabsorption spectra of sodium clusters.

The mechanism of the fragmentation of the plasmon line is not correctly described by the LDA. As an example, in $Na_{20}$ it occurs due to the proximity of the plasmon peak to the particle-hole transition $2s \rightarrow 3p$ at 2.8 eV. As the 3 p level and the vacuum level are practically degenerate with, the fragmentation line is broadened due to the proximity of transitions from the 2s level to scattering states. However, the experimental ionization threshold is 3.76 eV (see Section 3.2.1), and this mean that the loss of Rydberg states in the LDA is compensated by the reduction of the ionization threshold. This shortcoming of the LDA becomes corrected in the FAPG case, whose optical spectrum includes the effect of the Rydberg states.

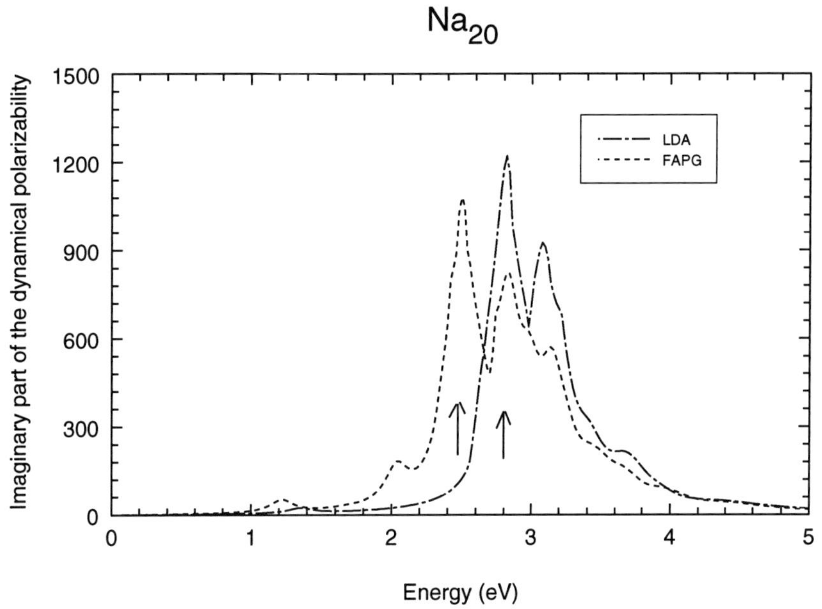


Fig. 7. Dipolar response to light of $Na_{20}$ cluster calculated within FAPG and LDA descriptions of xc effects. A Lorentzian width of 0.2 eV (ten times larger than in Fig. 6) is used. This width masks the many peaks in the FAPG spectrum resulting from transitions to loosely bound Rydberg states, which are not present in the TDLDA spectrum. Observed peaks are indicated with vertical arrows.

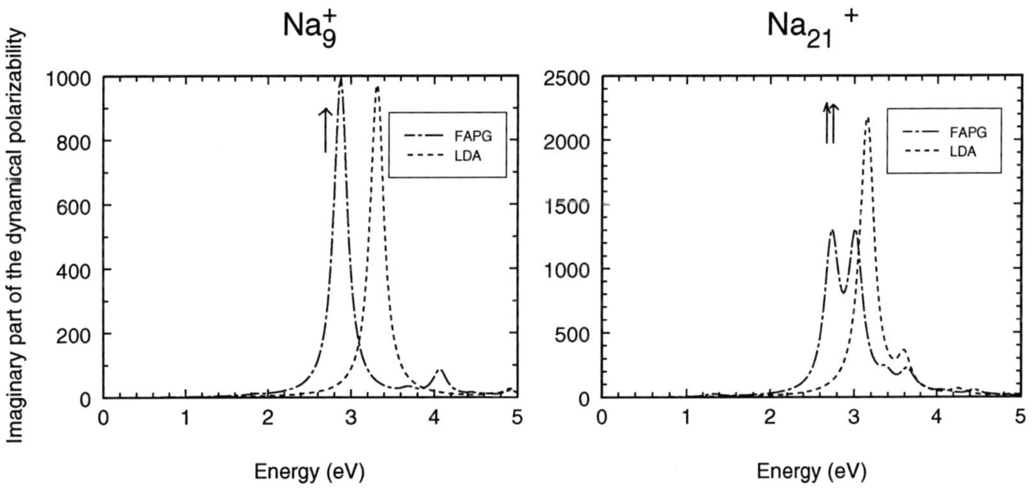


Fig. 8. Imaginary part of the dynamical polarizability per electron for $Na_9^+$ and $Na_{21}^+$, calculated with LDA and FAPG potentials. The arrows indicate the position of the observed peaks [61].

In Fig. 8 we compare the LDA and FAPG full linear responses (with a Lorentzian broadening of 0.2 eV) of $Na_9^+$ and $Na_{21}^+$. The position of experimental peaks [61] are indicated with arrows. The accumulation of strength in the UV region is smaller for the cations than for neutral clusters because the xc potential is deeper and the overlap between the bound and Rydberg states is smaller. For $Na_9^+$ we obtain only one FAPG broad peak at an energy very close to the experimental value, 2.62 eV. For $Na_{21}^+$ a fine structure near the plasmon peak results in the FAPG spectrum, as observed in the experiments [61]. This double peak is absent in both, the LDA and SIC [24] spectra, and consequently it lies beyond the consideration of self-interaction corrections in the xc potential.

## 4. Dipolar response properties of doped metal-clusters

In this section we consider explicitly, in an approximate manner, by means of the SAPS method [32], the granularity of the ionic distribution. The latter is quite successful in predicting the stabilizing of alkali metal clusters produced by the closing of spherical electronic shells, giving in addition the corresponding optimized skeleton of atomic cores. For $Na_nPb$ ($n = 4$, 6, 16) the total density TDDFT response to an external electrical field was studied in reference [21] within the LDA and FAPG xc potentials and kernels. Here we study in Subsection 4.1, within the TDLDA, the spin-density dipole response of $Na_nPb$ doped clusters, starting with their SAPS ground state, for closed shell ($n = 4$, 5) and open shell ($n = 5$) systems. In Subsection 4.2 we study the ground state electronic and ionic structure of doped $Al_nFe$ clusters as given by SAPS method, and the corresponding photo-absorption spectra and static polarizabilities.

### 4.1. Electric and spin responses of doped $Na_nPb$ clusters with $n = 4$, 5, 6

The ground state geometries of $Na_4Pb$, $Na_5Pb$ and $Na_6Pb$ resulting from our SAPS optimizations, using the atomic pseudo-potentials of Fiolhais et al. [33] for Na and Pb, contain a central Pb atom and an atomic shell of Na atoms forming a pyramid and an octahedron. The Pb atom contributes four valence electrons and each Na atom contributes one valence electron. The electronic structure is dominated by the strong attractive Pb potential [38], resulting in an ordering of electronic levels 1s 1p 2s 1d, that is, changing the order of the 2s and 1d levels with respect to pure sodium clusters. Thus, $Na_4Pb$ and $Na_6Pb$ are closed shell clusters with 8 and 10 electrons in the configurations 1s 1p and 1s 1p 2s, respectively. On the other hand, $Na_5Pb$ is an open shell cluster with a half-filled 2s HOMO level. In Fig. 9 the dipole and spin dipole responses of $Na_4Pb$ (upper panel), $Na_5Pb$ (middle panel), and $Na_6Pb$ (lower panel) are represented against the excitation energy. The electric dipole response of $Na_4Pb$ is similar to that of the isoelectronic $Na_8$ pure sodium cluster, as shown in reference [21]. For $Na_6Pb$ a similar spectrum as that of pure $Na_8$ is also observed because the strongly bound 1s level does not participate in the optical dipole response. However, the configuration for the eight active electrons in the optical region is 1p 2s for the doped clusters instead of 1s 2p for pure $Na_8$ clusters. On one hand, the plasmon resonance of $Na_6Pb$ appears close to the 2s eigenvalue (ionization threshold), causing the broadening of the plasmon peak. On the other hand, the energies for the p-h transitions $1p \rightarrow 2s$ and $1p \rightarrow 1d$in $Na_4Pb$, as well as for $2s \rightarrow 2p$ and $1p \rightarrow 1d$ in $Na_6Pb$, appear in reverse order than those for pure eight electron pure clusters, $Na_8$ and $Na_9^+$. This is not very important for the electric response because the plasmon peak is due mainly to the transition $1p \rightarrow 1d$ (see the discussion for $Na_8$ and $Na_9^+$ above), but the spin response of $Na_4Pb$ and $Na_6Pb$ suffer large changes with respect to that of pure eight electron clusters. For example, the energies for the transitions $2s \rightarrow 2p$ and $1p \rightarrow 1d$ in $Na_6Pb$ are 1.38 eV and 2.15 eV respectively, and, considering the screening the full spectrum becomes slightly red-shifted, showing two main peaks at 1.2 eV and 2.0 eV. The first resonance is due mainly to the $2s \rightarrow 2p$ p-h transition, and the other peak at $\sim 2$ eV is due mainly to $1p \rightarrow 1d$. The latter resonance appears at larger energy than for pure eight electron clusters (about 1.16 eV, see Fig. 2). For the dipole and spin dipole responses of $Na_5Pb$ cluster, similar facts as for the closed shell $Na_4Pb$ and $Na_5Pb$ are found, and the mixed response (not shown in the figure) is similar to that the half-filled shell $Na_6^+$ cluster.

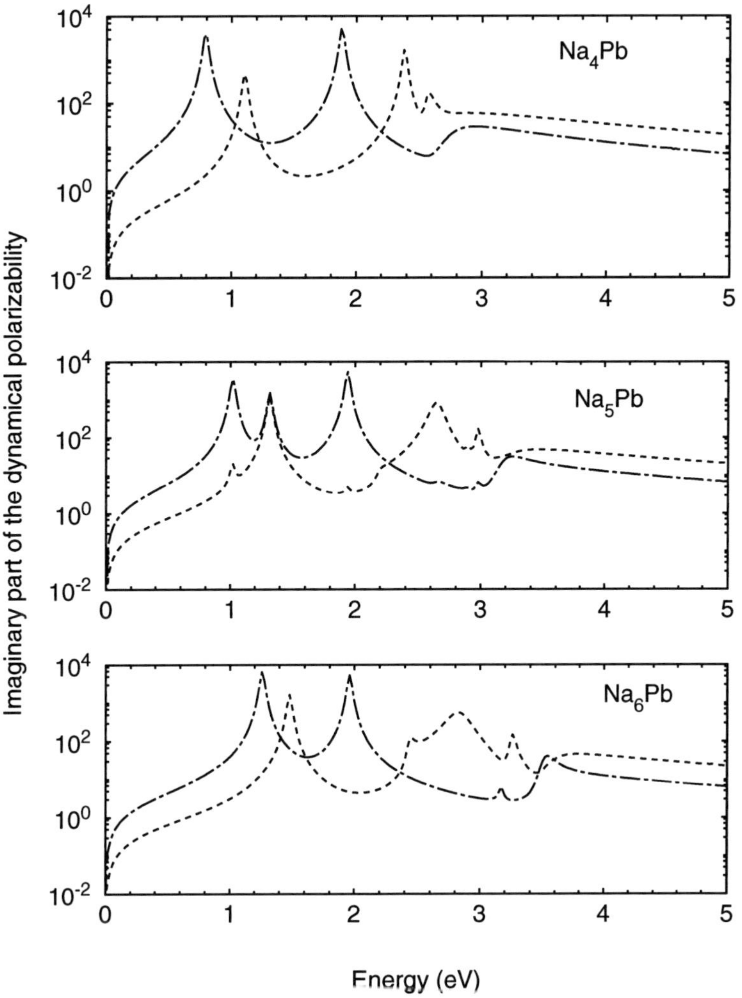


Fig. 9. The imaginary part of the dipole (dotted lines) and spin-dipole (dash-dotted lines) response functions within TDLDA is plotted versus the excitation energy for $Na_4Pb$ (upper panel), $Na_5Pb$ (middle panel) and $Na_6Pb$ (lower panel).

### 4.2. *Study of ground state and response properties of doped $Al_nFe$ clusters*

In this section we present the results obtained for clusters consisting of 4 to 12 and 18 aluminum atoms and one impurity atom (iron) in the center. Firstly, we have studied the ground-state electronic and geometric structure and, secondly, we have obtained the photoabsorption cross sections using the TDLSDA in two energy ranges: near the $3p$-ionization threshold of the impurity atom and at lower energies (0–5 eV). For comparison, we have computed the cross sections of pure aluminum clusters.

#### 4.2.1. *Electronic structure*

To obtain the electronic structure of $Al_nFe$ doped clusters, we model the ionic potential as

$$V_{ion} = -\frac{Z}{r} + V_{ion}^{Al} \tag{43}$$

Table 5
Electronic structure of the ground state (for the (26 +$3n$) electrons) of $Al_nFe$ clusters within the Jellium and SAPS models. The first row gives the single particle states common to both models. The states between square brackets contain the 18 electron (Argon like) core of Fe

| Cluster | Jellium | SAPS |
|---|---|---|
| Electronic structure | $[1s^2 2s^2\ 1p^6 3s^2 2p^6] 4s^2 3p^6 1d^{10}$ ... | |
| $Al_4Fe$ | $5s^1 2d^1$ | $5s^2 2d^0$ |
| $Al_5Fe$ | $5s^0 2d^5$ | $5s^1 2d^4$ |
| $Al_6Fe$ | $5s^2 2d^6$ | $5s^2 2d^6$ |
| $Al_7Fe$ | $5s^2 2d^9$ | $5s^2 2d^9$ |
| $Al_8Fe$ | $5s^2 2d^{10} 1f^2$ | $5s^2 2d^{10} 1f^2$ |
| $Al_9Fe$ | $5s^2 2d^{10} 1f^5$ | $5s^2 2d^9 1f^6$ |
| $Al_{10}Fe$ | $5s^2 2d^{10} 1f^8$ | $5s^2 2d^8 1f^{10}$ |
| $Al_{11}Fe$ | $5s^2 2d^{10} 1f^{11}$ | $5s^2 2d^7 1f^{14}$ |
| $Al_{12}Fe$ | $5s^2 2d^{10} 1f^{14}$ | $5s^2 2d^{10} 1f^{14}$ |
| $Al_{18}Fe$ | $5s^2 2d^{10} 1f^2 4p^6 1g^{12}$ | |

where $Z$ is the nuclear charge of the impurity atom ($Z = 26$ for Fe) and $V_{ion}^{Al}$ is the ionic pseudo-potential created by the Al atomic cores, where each Al atom contributes three valence electrons. In order to study the influence of the ionic granularity we compare the results using two methods to construct $V_{ion}^{Al}$, the jellium spherical model and the SAPS model, both described in Section 2.1. The study of the photoabsorption spectra of $Al_nFe$ clusters within the jellium model was first reported by Kurkina and collaborators [62]. The effective potential of Eq. (5) shows, for the different orbital states, two minima: the inner and deeper one, due to the iron atom, and the other, similar to the one of the pure aluminum cluster. These features explain most of the relevant properties discussed below. The ground state configurations (for the (26 +$3n$) electrons) of the $Al_nFe$ clusters that minimize the cluster total energy are given in Table 5 for the jellium and SAPS models. The two ionic models show different configurations for $n = 4, 5, 9, 10, 11$, which lead to different photoabsorption cross sections (as we will discuss below).

The 18 electrons occupying the five inner single particle orbitals have eigenvalues that correspond very closely to those of the argon-like core of pure atomic Fe [36]. Apart of these 18 core electrons from the $Al_nFe$ clusters, the rest of the electrons occupy analogous *delocalized* orbitals corresponding to the $Al_n$ clusters, and the remaining 8 electrons always occupy *localized* $d$ orbitals. These facts are illustrated in Table 6 and Fig. 10. When a new $d$ state begins to be filled it becomes progressively localized, and simultaneously the previous $d$ state becomes delocalized. This phenomenon is shown in Fig. 10. When the $2d$ state is empty, the $1d$ electronic density is mostly localized in the inner region where the iron atom dominates the potential. During the filling of the $2d$ shell, the $1d$ density spreads over the two regions of the potential, and when its is filled the charge goes to the outer region, which belongs to the aluminum atoms. Simultaneously, the $2d$ electronic charge gradually becomes localized in the inner region. The same process occurs during the filling of the $3d$ orbital and so on.

As seen in Table 5, the number of electrons in the last occupied $d$ orbital for the $n = 8$ to 11 $Al_nFe$ clusters decreases within the SAPS model, whereas for the jellium model it is always fully occupied. This fact, together with the discussion above, establishes important differences between the localization character of the $d$ valence electrons in doped $Al_nFe$ clusters described by a jellium impurity atom or by a discrete ion model like SAPS. As an example, we compare in Fig. 11 the photoabsorption spectrum of $Al_nFe$ in the region of several tens of eV, as calculated within the TDLDA using the jellium or the

Table 6
Comparison of the external orbitals and eigenvalues of $Al_nFe$ (within SAPS method) and $Al_n$ clusters for $n$ = 4, 7 and 12. The additional $d$ level in the doped clusters is always localized whereas the other $d$ orbitals are delocalized

| | $Al_4Fe$ | | | $Al_4$ | |
|---|---|---|---|---|---|
| *Orbital* | $\varepsilon^{\uparrow}$ | $\varepsilon^{\downarrow}$ | *orbital* | $\varepsilon^{\uparrow}$ | $\varepsilon^{\downarrow}$ |
| $4s^2$ | −9.48 | −9.48 | $1s^2$ | −12.51 | −12.25 |
| $3p^6$ | −6.40 | −6.40 | $1p^6$ | −7.70 | −7.29 |
| $1d^{10}$ | −5.58 | −5.58 | — | — | — |
| $5s^2$ | −2.64 | −2.64 | $2s^2$ | −3.99 | −3.68 |
| $2d^0$ | −2.23 | −2.23 | $1d^2$ | −3.38 | −2.95 |
| | $Al_7Fe$ | | | $Al_7$ | |
| $4s^2$ | −12.23 | −12.18 | $1s^2$ | −12.57 | −12.45 |
| $3p^6$ | −9.43 | −9.39 | $1p^6$ | −9.36 | −9.27 |
| $1d^{10}$ | −6.96 | −6.82 | $1d^{10}$ | −5.67 | −5.60 |
| $5s^2$ | −4.65 | −4.57 | $2s^2$ | −4.93 | −4.80 |
| $2d^9$ | −3.52 | −3.19 | — | — | — |
| — | — | — | $2p^1$ | −2.28 | −2.14 |
| | $Al_{12}Fe$ | | | $Al_{12}$ | |
| $4s^2$ | −12.84 | −12.84 | $1s^2$ | −12.40 | −12.23 |
| $3p^6$ | −10.85 | −10.85 | $1p^6$ | −10.64 | −10.50 |
| $1d^{10}$ | −8.34 | −8.34 | $1d^{10}$ | −8.01 | −7.90 |
| $5s^2$ | −5.86 | −5.86 | $2s^2$ | −5.44 | −5.21 |
| $1f^{14}$ | −4.97 | −4.97 | $1f^{14}$ | −4.96 | −4.86 |
| $2d^{10}$ | −3.85 | −3.85 | — | — | — |
| — | — | — | $2p^2$ | −3.57 | −3.36 |

SAPS model for the ionic distribution. That difference should occur when the filling of the outermost $d$ orbital is different in both models, independent of the number of Al atoms. On the other hand, the TDLDA response in the optical region is similar for both ionic models, and very close to the response of the corresponding $Al_n$ clusters, as illustrated in Fig. 12 for the case of $Al_{18}Fe$ versus $Al_{18}$.

### 4.2.2. *Geometric structure and static polarizabilities*

In Fig. 13, taken from Figs 4–7 of ref [36], it is shown the geometrical arrangements of the studied doped clusters as optimized by the SAPS model. In all cases the Fe atom is at the center of the cluster. Up to $n = 9$, the average size of $Al_n$ clusters is smaller that those of $Al_nFe$. Starting with $n = 9$, when a second shell of atoms begin to develop, the sizes of $Al_nFe$ and $Al_n$ are similar. This means that the static polarizabilities, which are mainly controlled by the size of the cluster, will be similar in both types of systems when the sizes are large. In Fig. 14 are represented the static dipolar polarizability per aluminum atom of $Al_nFe$ clusters divided by the cube of the equivalent radius of the cluster, calculated within the SAPS and within the jellium models. The equivalent radius for jellium clusters is given by $r_s(3N)^{1/3}$, being $N$ the number of Al atoms, and for SAPS clusters it is determined by the radius of the external ionic shell. We see in the figure that the effect of Fe atom is substantial for clusters up to 9 atoms. The jellium and SAPS static polarizabilities of $Al_nFe$ clusters are similar one to the other, but the dynamical polarizability, in the energy region between 50–60 eV will be very different, as we show below.

### 4.2.3. *Photoabsorption spectra of $Al_nFe$ clusters*

Firstly, we have investigated the cross sections of $Al_nFe$ clusters in the energy region near the $3p$ ionization threshold of the impurity atom. There are two varieties of cluster cross sections: spectra

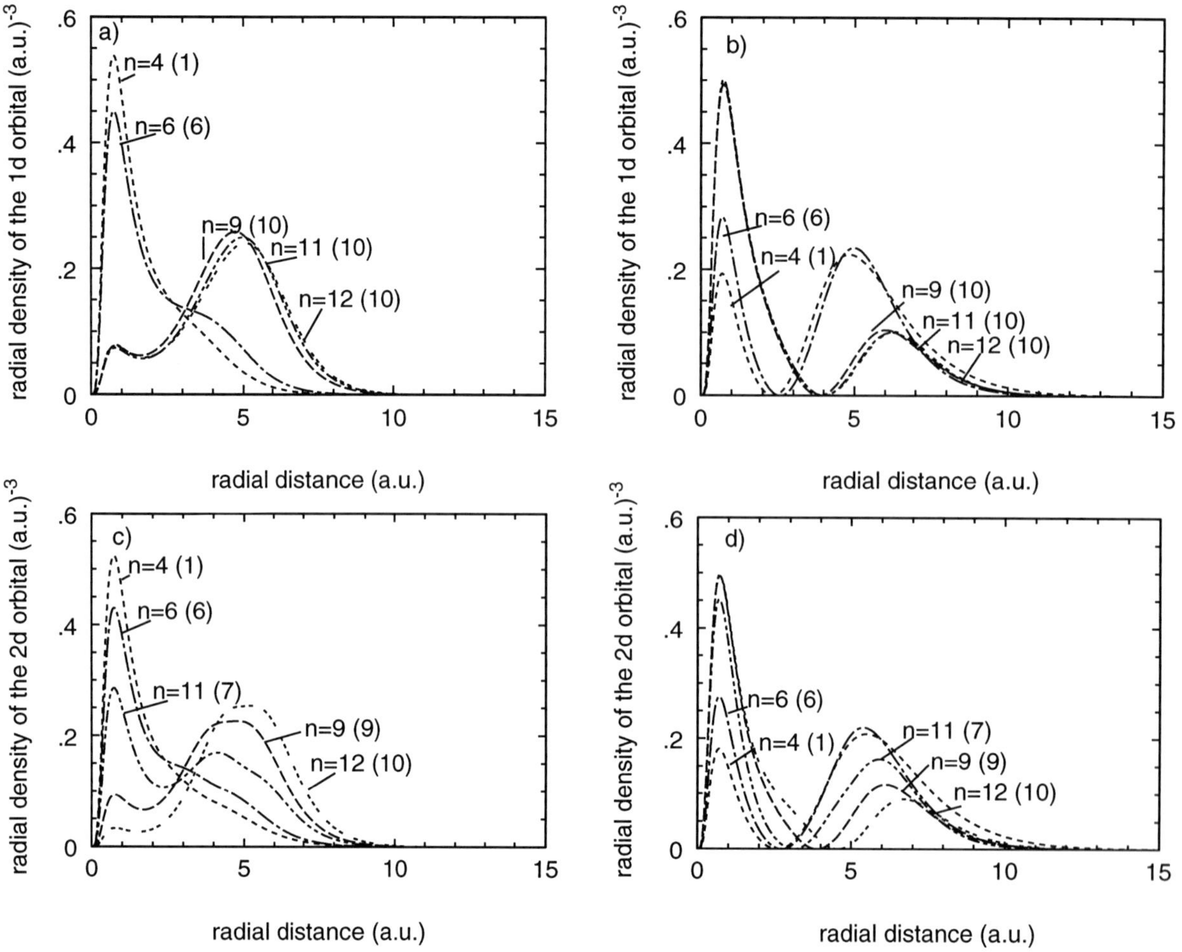


Fig. 10. The radial density of the 1d and 2d orbitals are represented for the ground state of $Al_nFe$ clusters. Each curve is identified with two numbers, $n$ $(m)$, corresponding to the number of aluminium atoms ($n$) and to the occupancy of the orbital ($m$). Panels a) and b) correspond to the jellium type of calculation, and panels c) and d) correspond to the SAPS model calculation. Note that when the 2d orbital begins to be filled it becomes progressively localized, and, simultaneously, the 1d orbital becomes localized. See text for details.

containing a double asymmetric resonance and those consisting of a resonance group. The first case is realized for clusters with completely occupied $d$ shells. The second case occurs for the systems with a $d$ shell being filled. In both cases the resonances lie in the energy region of the giant resonance ($3p^{\downarrow} \rightarrow 3d^{\downarrow}$) in $3d$ atoms. Our calculations show that in the ground state there is at least one empty $d$ level (which can be spin-split) above the occupied energy region. For clusters with closed $d$ shells the single resonance is connected with the excitation of $2p$ electrons to the empty $d$ level. (We note that the $2p$ state in the $Al_nFe$ doped clusters corresponds to the $3p$ core state of the impurity atom in the usual atomic notation). In clusters with a $d$ shell being filled discrete transitions of $2p$ electrons are possible not only to the empty $d$ level but also to the partially occupied one.

For illustration we consider in the spectrum of $Al_{11}Fe$ (Fig. 11). In the energy range 45–60 eV the spectrum consists of three asymmetric peaks at 50.65 eV ($2p^{\downarrow} \rightarrow 2d^{\downarrow}$), 56.82 eV ($2p^{\uparrow} \rightarrow 3d^{\uparrow}$) and 54.69 eV ($2p^{\downarrow} \rightarrow 3d^{\downarrow}$) within the SAPS model where we obtain a double resonance at 5.97–54.97 eV ($2p^{\downarrow} \rightarrow 3d^{\downarrow}, 2p^{\uparrow} \rightarrow 3d^{\uparrow}$) using the jellium model.

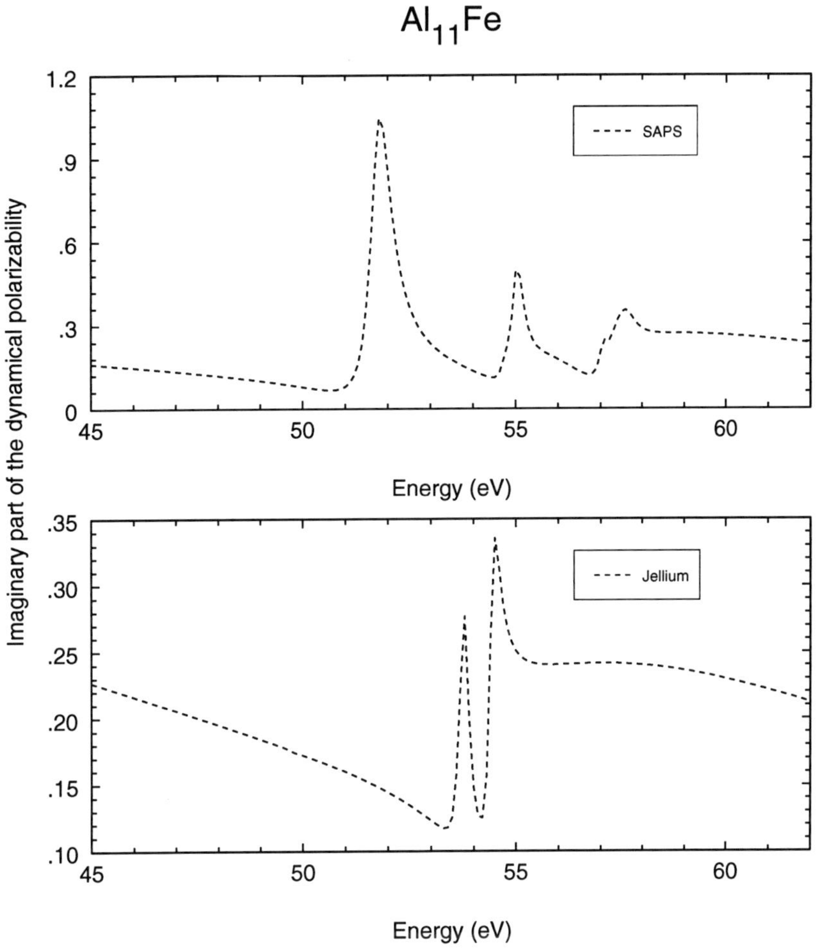


Fig. 11. The imaginary part of the TDLDA response function of $Al_{11}Fe$ doped cluster in the energy range 45–60 eV within the Jellium (lower panel) and SAPS (upper panel) models.

## 5. Summary and conclusions

By means of methods based on time dependent density functional theory we have investigated the behaviour (dynamical polarizability) of the electronic cloud of spin saturated and spin polarized states of pure and doped metal clusters in response to electric as well as spin-dipole external time dependent perturbations in the linear regime. Thus, we have analyzed the cross talk between dipole and spin-dipole modes. As test cases, we have studied in this paper the photoabsorption cross section of i) the spin saturated clusters $Na_9^+$, $Na_8$, $Na_{21}^+$, $Na_{20}$, $Na_{40}$, $Na_4Pb$ and $Na_6Pb$, and ii) the spin polarized clusters $Na_6^+$, $Na_5Pb$, $Al_{11}Fe$, $Al_{18}$ and $Al_{18}Fe$. The static polarizability is obtained as the zero frequency limit of the dynamical polarizability, $R_n(\omega)$, where omega is the frequency of the perturbing external field.

In addition to the well-known Mie Plasmon resonance, which dominates the optical response when a no-spin dependent dipole field is applied, we obtain, under the influence of a spin dependent dipole field, another collective mode that lies in the low energy part of the spectrum (about 1 eV). Contrary to the plasmon resonance, which is blue-shifted with respect to the unperturbed particle-hole component lines due to the induced Hartree repulsion, the collective spin resonance becomes red shifted as a consequence

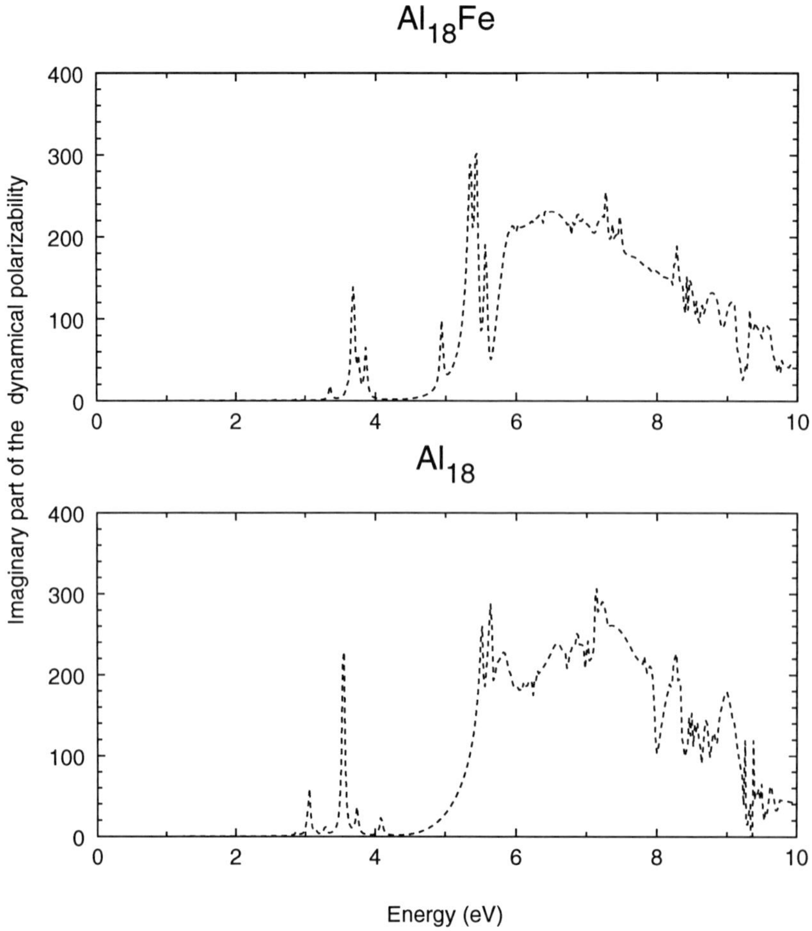


Fig. 12. The imaginary part of the TDLDA response functions of $Al_{18}Fe$ and $Al_{18}$ (both within the Jellium model) at energies of excitation below to ten eV.

of the attractive part (xc) of the residual interaction. That spin-mode, appearing at decreasing energy when the size of the cluster increases, can be visualized as oscillations of the spin-up electronic cloud against the spin-down electronic cloud, and may be regarded as the counterpart, in the spin channel, of the dipole surface oscillations between electrons and positive background appearing in the electrical dipole channel. Both, dipole and spin-dipole, modes are well separated for cluster in a saturated state, but they intertwine for spin-polarized clusters, independently of the spin or no-spin dependence of the external fields. This happen because a global motion of the whole electron cloud always excites the spin mode when up and down densities are different. This crosses talks means that there is a real chance to observe the spin modes by optical absorption, at least indirectly.

While the jellium model is presumably sufficient for exhibiting qualitative trends, access to details at spectroscopy accuracy requires a full account of the ionic structure. The fact to take into account is that a detailed ionic background produces a departure from spherical symmetry for ionic and electronic charge distributions leading to splitting of the plasmon into different frequencies. This splitting is one of the basic features of the dipole plasmon which was noted very early [8] and which has been exploited extensively to explore the cluster geometry by analyzing optical response. We expect also a fragmentation

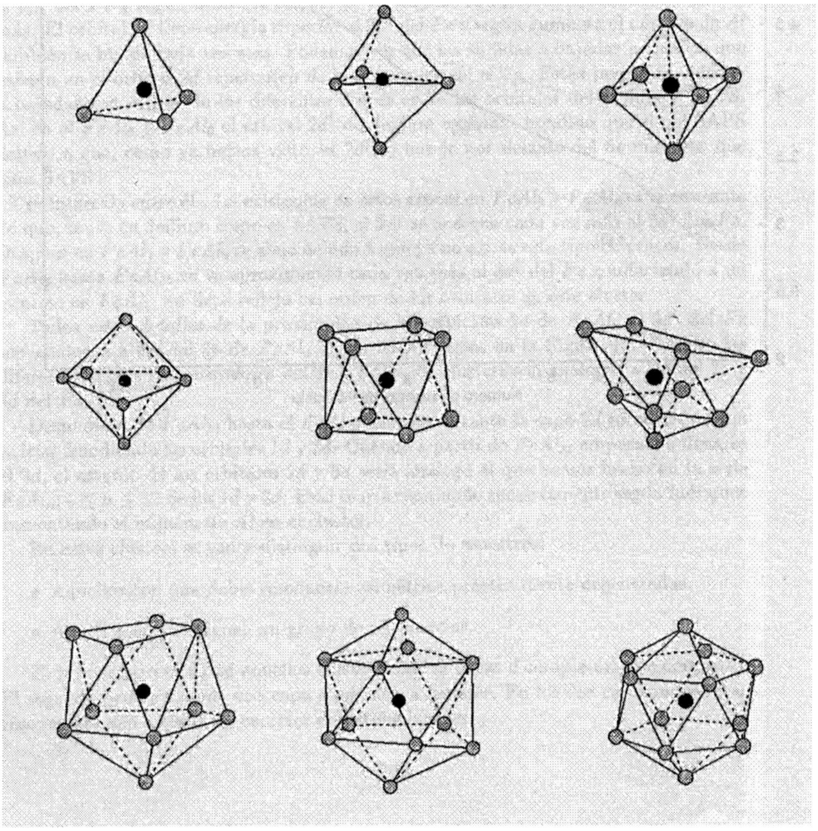

Fig. 13. Geometries of $Al_nFe$ clusters with $3 < n < 13$ resulting from SAPS model optimization within the LSDA for exchange-correlation effects.

of the spin dipole spectra in the same way, negligible for the unpolarized ground state but large for the polarized clusters. But, the coupling between dipole and spin-dipole modes that we have studied here will remain as a basic feature, and moreover, the splitting effect will be smaller for large clusters where the electronic fluctuations seem to dominate the spatial structure.

Another source of inaccuracy arises when theoretical TDLDA photoabsorption spectra are compared to experimental photodepletion spectra. The reason is that the predicted plasmon peaks are systematically blue-shifted, whereas the static polarizability results are systematically underestimated. In view of the fact that these failures occur also for spherically symmetric magic clusters (with closed electronic shells), even when their ionic structure is fully taken into account, we have explored the effects of an improved new xc potential due to Parr and Ghosh [45], which exhibits the correct asymptotic behavior. The use of this new potential in the context of TDDFT provides a simple and convenient framework, which does not require extra computational effort and yet yields results that remedy many of the unpleasant features intrinsic to LDA. The FAPG potential of Parr and Ghosh [45] eschews the traditional resolution of exchange-correlation into exchange plus correlation, in favor of a resolution into a Fermi-Amaldi term plus a constraint term (Eq. (12)). The first provides the correct normalization of the xc hole and the correct long-range behavior of the potential, which leads to an increase in the number of bound states with respect to the LDA case. The second is obtained as a function of the density by fitting to ab-initio data for the atoms He to Ar following the procedure of Zhao, Morrison and Parr [46]. For the present

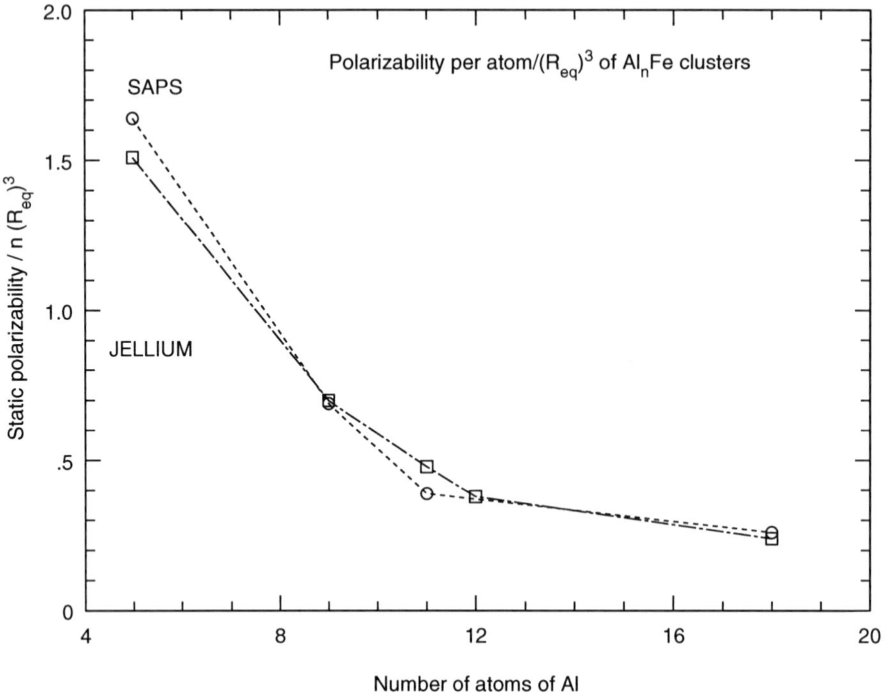


Fig. 14. The static polarizability per atom, divided by the cube of the equivalent cluster radius, for $Al_n Fe$ clusters calculated within the jellium model (dashed lines) and within the SAPS model (dotted lines).

applications to metal clusters we choose the value of the FAPG potential at $r = 0$ to be the same one as that of the LDA-PW potential, which we have used here for comparative purposes.

We have first considered the ground state of sodium clusters in the spherical jellium model. The HOMO eigenvalue from the FAPG calculations gives a very good approximation to the ionization threshold. Other eigenvalues are closer to reported GW quasi-particles energies [56] than those obtained from KLI or GAM potentials [16], which are considerably deeper than the FAPG potential inside the cluster. Moreover, the FAPG approximation yields a better description of the local-field correction entering into the response function, accounting well for the polarizability, both qualitative and quantitatively. As for the dynamical response properties of sodium clusters, we observe that the Landau damping is more noticeable in the FAPG calculation than in the LDA one, due to the larger number of loosely bound states. In the case of $Na_{20}$ and $Na_{40}$ we predict accurately the position of the surface plasmon. The observed fragmentation into two main peaks of these plasmons is obtained in both LDA and FAPG calculations, although the underlying mechanism is different. The fragmentation peak in the LDA is broadened by transitions, promoting the electron into the continuum. In contrast, the fragmentation of the plasmon using the FAPG calculation is due to particle-hole transitions in the discrete spectrum, which contain now much more loosely bound states than LDA.

For the positively charged clusters, $Na_9^+$ and $Na_{21}^+$, the FAPG calculation provides a red shift of the main features of the spectra with respect to those obtained in the LDA calculation. For $Na_9^+$ we obtain only one peak within both FAPG and LDA calculations, but the FAPG one is closer to the experimental plasmon. In the $Na_{21}^+$ case, only the FAPG calculation reproduces the experimental splitting of the plasmon in to two peaks [61]. This is a result that has not been obtained using other non-local functionals such as WDA-PB [22] and SIC [24]. Moreover, we are now able to provide non-trivial predictions for the photoabsorption cross section of negatively charged clusters. Because the surface plasmon is embedded in a continuum of single particle states to which it couples within the FAPG approach, we were able to single out the contribution to the total line width arising from the probability for electron detachment. This opens the possibility to study previously intractable systems (at the level of LDA), such as the dynamical response of small anions and anionic clusters.

To finish off, in Section IV we have studied the linear response of metal clusters doped with an impurity atom. For these calculations we take into account the ionic structure as described by the SAPS model [32]. Firstly, we have calculated and analyzed the dynamical response of spin saturated $Na_4Pb$ and $Na_6Pb$ clusters, and of the spin-polarized $Na_5Pb$ one. Both, FAPG and LDA predict that the 1p level is responsible of the broad resonance near the ionization threshold whereas the deeply bound 1s level has little effect on the full response. In both calculations the spectra of $Na_4Pb$ and $Na_6Pb$ are similar to those of the isoelectronic pure sodium clusters $Na_8$, apart from a global blue shift (mainly in $Na_4Pb$) due to the strong Pb pseudo-potential. For the responses of $Na_5Pb$ cluster similar facts as for the half-filled shell cluster $Na_6^+$ are found.

Finally we compare the TDLDA static and dynamical polarizabilities of $Al_nFe$ clusters, within the jellium and within the SAPS models for the ionic structure of aluminum atoms and taking the atomic impurity with all its electrons. A remarkable property in these clusters as compared to the pure $Al_n$ clusters is that always contain an additional localized d orbital, whereas for the pure cluster all orbitals are delocalized. This lead to qualitative differences in the response to excitations of several tens eV but the dynamical polarizabilities in the optical region (a few eV) are similar for pure $Al_n$ and doped $Al_nFe$ aluminum clusters. Correspondingly, the static polarizabilities (the limit $\omega \to 0$ of the dynamical polarizability) are also similar. This conclusion is valid independently of the model for the ionic structure, jellium or SAPS, used in our calculations.

## Acknowledgments

Work supported by JCyL (grant VA073/02) and DGICYT (grant MAT2002-04393-C02-01)

## References


[1] W.D. Knight, K. Clemenger, W.A. de Heer, W.A. Saunders, M.Y. Chou and M.L. Cohen, *Phys. Rev. Lett.* **52** (1984), 2141.
[2] W.D. Knight, K. Clemenger, W.A. de Heer and W.A. Saunders, *Phys. Rev. B* **31** (1985), 2539.
[3] W.A. de Heer, K. Selby, V. Kresin, J. Masui, M. Vollmer, A. Châtelain and W.D. Knight, *Phys. Rev. Lett.* **59** (1987), 1805; K. Selby et al, *Phys. Rev. B* **43** (1991), 4565.
[4] W.A. de Heer, *Rev. Mod. Phys.* **65** (1993), 611.
[5] M. Brack, *Rev. Mod. Phys.* **65** (1993), 677.
[6] Metal Clusters, *Wiley Series in Theoretical Chemistry*, Vol X, W. Ekardt, Ed., Wiley, New York, 1999,
[7] K.D. Bonin and V.V. Kresin, *Electric-Dipole Polarizabilities of Atoms, Molecules and Clusters*, World Scientific, Singapore, 1997.
[8] W. Ekardt, *Phys. Rev. Lett.* **52** (1984), 1925; W. Ekardt, *Phys. Rev. B* **31** (1985), 6360; W. Ekardt and Z. Penzar, *Phys. Rev. B* **43** (1991), 1322.
[9] M. Brack, *Phys. Rev. B* **39** (1989), 3533.
[10] L. Serra, F. García, M. Barranco, J. Navarro, L.C. Balbás and A. Mañanes, *Phys. Rev. B* **39** (1989), 8247; *J. Phys. Condens. Matter* **1** (1989), 10391.
[11] P.-G. Reinhard, M. Brack and O. Genzken, *Phys. Rev. A* **41** (1990), 5568.
[12] A. Rubio, L.C. Balbás and J.A. Alonso, *Z. Phys. D* **19** (1991), 93.
[13] M. Madjet, C. Guet and W.R. Johnson, *Phys. Rev. A* **51** (1995), 1327.

[14] A. Rubio, J.A. Alonso, X. Blase, L.C. Balbás and S.G. Louie, *Phys. Rev. Lett.* **77** (1996), 247; M.A.L. Marques, A. Castro and A. Rubio, *J. Chem. Phys.* **115** (2001), 3006.
[15] V. Bonacic-Koutecky, J. Pittner, C. Fuchs, P. Fantucci, M.F. Guest and J. Koutecký, *J. Chem. Phys.* **104** (1996), 1427.
[16] C.A. Ullrich, P.-G. Reinhard and E. Suraud, *Phys. Rev. A* **62** (2000), 053202.
[17] I. Vasiliev, S. Ögüt and J.R. Chelikowsky, *Phys. Rev. Lett.* **82** (1999), 1919.
[18] S. Kümmel, M. Brack and P.-G. Reinhard, *Phys. Rev. B* **62** (2000), 7602.
[19] M.B. Torres and L.C. Balbás, *J. Phys. Condens. Matter* **12** (2000), 4365.
[20] S.J.A. Van Gisbergen, J.M. Pacheco and E.J. Baerends, *Phys. Rev. A* **63** (2001), 063201.
[21] M.B. Torres and L.C. Balbás, *J. Phys. Condens. Matter* **14** (2002), 5795.
[22] A. Rubio, L.C. Balbás, L. Serra and M. Barranco, *Phys. Rev. B* **42** (1990), 10959.
[23] S. Kümmel, J. Akola and M. Manninen, *Phys. Rev. Lett.* **84** (2000), 3827.
[24] J.M. Pacheco and W. Ekardt, *Ann. Physik.* **1** (1992), 254; Z. Phys. D **24** (1992), 65.
[25] S. Kümmel, K. Andrae and P.G. Reinhard, *Applied Phys. B* **73** (2003), 293.
[26] W. Ekardt, *Phys. Rev.* **29** (1984), 1558.
[27] A. Zangwill and P. Soven, *Phys. Rev. A* **71** (1980), 1561.
[28] E.K.U. Gross, J.F. Dobson and M. Petersilka, in: *Density Functional Theory*, (Vol. 181), R.F. Nalewajski, ed., Topics in Current Chemistry, Springer, Berlin, 1996.
[29] P. Hohenberg and W. Kohn, *Phys. Rev.* **136** (1964), B864.
[30] W. Kohn and L.J. Sham, *Phys. Rev.* **140** (1965), A1133.
[31] U.V. Barth and L. Hedin, *J. Phys. C* **5** (1972), 1629.
[32] M.P. Iñiguez, M.J. López, J.A. Alonso and J.M. Soler, *Z. Phys. D* **11** (1989), 163.
[33] C. Fiolhais, J.P. Perdew, S.Q. Armster, J.M. Maclaren and M. Brajczewska, *Phys. Rev. B* **51** (1995), 14001.
[34] A.R. Williams et, Von Barth U, 1983 in: *Theory of the Inhomogeneous Electron Gas*, S. Lundqvist and N.H. March, eds, Plenum, New York, See also K.L. Liu and S.H. Vosko, *Can. J. Phys.* **67** (1989), 1015.
[35] Ll. Serra, R.A. Broglia, M. Barranco and J. Navarro, *Phys. Rev. A* **47** (1993), R 1601.
[36] M.B. Torres, PhD Thesis, Valladolid University, 1997, ISBN: 0-493-9835-1.
[37] C. Yeretzian, U. Röthlisberger and E. Schumacher, *Chem. Phys. Lett.* **237** (1995), 334.
[38] L.C. Balbás and J.L. Martins, *Phys. Rev. B* **54** (1996), 2937.
[39] J.P. Perdew and Y. Wang, *Phys. Rev. B* **45** (1992), 13244.
[40] S.H. Vosko, L. Wilk and M. Nusair, *Can. J. Phys. B* **58** (1980), 1200.
[41] J.P. Perdew and A. Zunger, *Phys. Rev. B* **23** (1981), 5048; see also, A. Svane and O. Gunnarsson, *Phys. Rev. Lett.* **65** (1990), 1148; K.D. Sen and M.K. Harbola, *Chem. Phy. Lett.* **178** (1991), 347.
[42] M. Ernzerhof, J.P. Perdew and K. Burke, in: *Density Functional Theory I, p.1, Topics in Current Chemistry*,(Vol. 180), R.F. Nalewajski, ed., Springer, 1996.
[43] J.A. Alonso and L.A. Girifalco, *Phys. Rev. B* **17** (1978), 3735; O. Gunnarsson, M. Jonson and B.I. Lundqvist, *Phys. Rev. B* **20** (1979), 3136.
[44] H. Przybylski and G. Borstel, *Solid State Commun.* **49** (1984), 317; *Solid State Commun.* **52** (1984), 713.
[45] R.G. Parr and S.K. Ghosh, *Phys. Rev. A* **51** (1995), 3564.
[46] Q. Zhao and R.G. Parr, *Phys. Rev. A* **50** (1994), 2138; R.C. Morrison and Q. Zhao, *Phys. Rev A* **51** (1995), 1980; Q. Zhao and R.G. Parr, *Phys. Rev A* **46** (1992), 2337; *J. Chem. Phys.* **98** (1993), 543.
[47] M. Levy, *Proc. Natt. Acad. Sci. USA* **76** (1979), 6062.
[48] E. Fermi and E. Amaldi, *Mem. R. Accad. Italia* **6** (1934), 117.
[49] A. Rubio, L.C. Balbás and J.A. Alonso, *Phys. Rev. B* **46** (1992), 4891.
[50] C.R.C. Wang et al, *Chem. Phys. Lett.* **166** (1990), 26; *J. Chem. Phys.* **94** (1991), 2496; C.R.C. Wang et al., *J. Chem. Phys.* **93** (1990), 3787; C.R.C. Wang et al., *Z. Phys. D.* **19** (1991), 13; K. Selby, V. Kresin, J. Masui, M. Vollmer, W.A. de Heer, A. Scheidemann and W. Knight, *Phys. Rev. B* **43** (1991), 4565.
[51] C. Kohl, B. Montag and P.G. Reinhard, *Z. Phys. D* **35** (1995), 57.
[52] W.A. de Heer, W.D. Khight, M.Y. Chou and M.L. Cohen, in: *Solid State Physics*, (Vol. 40), H. Ehrenreich, F. Seitz and D. Turnbull, eds, Academic Press, New York, 1987, p. 93.
[53] M.M. Kappes, M. Schär, U. Röthlisberger, Ch. Yeretzian And E. Schumacher, *Chem Phys. Lett* **143** (1988), 25.
[54] For a recent discussion of the ionization potential theorem see J.P. Perdew and M. Levy, *Phys. Rev.* **B** 56 (1997), 16021; L. Kleinman, *Phys. Rev.* **B** 56 (1997), 12042; *ibid Phys. Rev.* **B** 56 (1997), 16029. For a discussion of the eigenvalues of unoccupied orbitals see, for example, A.I. Al-Sharif, R. Resta and L.J. Umrigar, *Phys. Rev. A* **57** (1998), 2466. For a discussion of the eigenvalues of occupied orbitals see, for example, E.J. Baerends, O.V. Gritsenko and R. Van Leeuwen in: *Chemical Applications of DFT*, (Vol. 629), B.B. Loird, R.B. Ross and T. Ziegler, eds, of American Chemical Society Symposia Series, Ameriacn Chemical Society, Washington, D.C. 1996.
[55] L.C. Balbás, J.A. Alonso and A. Rubio, *Europhys. Lett.* **14** (1991), 323.

[56] S. Saito, S.B. Zhang, S.G. Louie and M.L. Cohen, *Phys. Rev. B* **40** (1989), 3643; *ibid, J. Phys.: Condens. Matter* **2** (1990), 9041.
[57] J.B. Krieger, Y. Li and G.J. Iafrate, *Phys. Rev. A* **46** (1992), 5453.
[58] M.F. Politis, P.A. Hervieux, J. Hanssen, M.E. Madjet and F. Martin, *Phys. Rev. A* **58** (1998), 367.
[59] K. Selby, M. Vollmer, J. Masui, V. Kresin, W.A. de Heer and W.D. Knight, *Phys. Rev. B* **40** (1989), 5417.
[60] G. Tikhonov, V. Kasperovich, K. Wong and V.V. Kresin, *Phys. Rev. A* **64** (2001), 063202.
[61] C. Bréchignac et al., *Chem. Phys. Lett.* **189**(1992), 28; C. Ellert et al., *Phys. Rev. Lett.* **75** (1995), 1731.
[62] L.I. Kurkina and O.V. Farberovich, *Z. Phys. D* **28** (1993), 215.

Journal of Computational Methods in Sciences and Engineering 4 (2004) 551–558
IOS Press


# The influence of solvent on the two-photon absorption cross section and hyperpolarizability of molecules exhibiting large solvatochromic shifts

Wojciech Bartkowiak[a,*], Bartłomiej Skwara[a] and Robert Zaleśny[a,b]
[a]*Institute of Physical and Theoretical Chemistry, Wroclaw University of Technology, Wyb. Wyspiańskiego 27, 50-370 Wroclaw, Poland*
[b]*Department of Quantum Chemistry, Institute of Chemistry, Nicolaus Copernicus University, Gagarina 7, 87-100, Toruń, Poland*

**Abstract.** A theoretical approach to the effect of influence of solute/solvent interaction on the electronic contributions to the two-photon absorption cross section ($\delta$) is presented. The solvent effect was included via a discrete quantum-mechanical Langevin dipoles/Monte Carlo method. The calculations of $\delta$ were performed for the first excited singlet state connected with the intramolecular charge transfer (CT). The results of calculations demonstrate the existence of the significant solvent effect on the two-photon absorption cross section ($\delta$) of the charge-transfer chromophore (p-nitroaniline) exhibiting large solvatochromic shifts. The results of quantum-chemical calculations are analyzed based on the simple two-state model.

Keywords: Two-photon absorption, solvent effect, hyperpolarizability, p-nitroaniline

## 1. Introduction

The construction of computational methods, which can account accurately for the solvent effects is one of the largest challenges in the contemporary quantum chemistry. It is related to the fact that the nature of intermolecular interactions in the condensed phases is extremely complicated [1–5]. Many authors, in their theoretical and experimental works, have shown that the solvent effects on the values of the first-order hyperpolarizability ($\beta$) and second-order hyperpolarizability ($\gamma$) of donor–acceptor $\pi$-conjugated organic compounds is significant [6–44]. For example, the calculated static values of $\beta$ for p-nitroaniline (PNA) in polar solvents can be larger by a factor of 1.4–2.2 than the corresponding values in the gas phase [32,33]. The UV-Vis absorption spectra for this type of molecules are characterized by a strong low-lying ($\pi \rightarrow \pi^*$) transition which is assigned to the intramolecular charge-transfer (CT) transition [4,5]. This lowest excited state gives significant contribution to the value of $\beta$ and $\gamma$. Recently, it has been shown, in the case of static vector component of $\beta$ and scalar part of $\gamma$, that the calculated ratio of $\beta^{sol}/\beta^{gas}$ is approximately equal to the $\gamma^{sol}/\gamma^{gas}$ ratio for many donor-acceptor chromophores [36].

*Corresponding author. Tel.: +48 71 320-26-75; Fax: +48 71 320 33 64; E-mail: bartkowiak@kchf.ch.pwr.wroc.pl.

The solvent effect on $\beta$ and $\gamma$ can be discussed based on the simple two-state model [45–48]. According to the two-level picture, the expression for first-order hyperpolarizability (in atomic units) reads:

$$\beta_\mu \approx \beta_{xxx} = 6\frac{\langle g|r_x|CT\rangle^2 \Delta\mu}{(\omega_{gCT})^2} \tag{1}$$

Where $\Delta\mu$ is the difference of dipole moments between the ground state and the CT excited state, $\hbar\omega_{gCT}$ is the transition energy to the CT state, and $\langle g|r_x|CT\rangle$ is transition dipole moment. Similarly, the two-state model can be developed for $\gamma$ as

$$\langle\gamma\rangle \approx \gamma_{xxxx} = 24\frac{\langle g|r_x|CT\rangle^2(\Delta\mu)^2 - \langle g|r_x|CT\rangle^4}{(\omega_{gCT})^3} \tag{2}$$

It is worth noting that the concept of two-state models are derived from the sum-over-states (SOS) expression for $\beta$ and $\gamma$. For example, the relationship between the electronic structure of the molecule and the first molecular hyperpolarizability ($\beta_{ijk}$) tensor, as derived from time-dependent perturbation theory, is given by [49,50]

$$\beta_{ijk}(-\omega_\sigma;\omega_1,\omega_2) = \frac{1}{\hbar^2}P(i,j,k;-\omega_\sigma,\omega_1,\omega_2)\sum_{i\neq 0}\sum_{m\neq 0}\frac{\langle 0|\mu_i|l\rangle\langle l|\overline{\mu}_j|m\rangle\langle m|\mu_k|0\rangle}{(\omega_{0l}-\omega_\sigma)(\omega_{0m}-\omega_2)} \tag{3}$$

In Eq. (3), the matrix elements $\langle 0|\mu_i|l\rangle$ and $\langle l|\overline{\mu}_j|m\rangle = \langle l|\mu_j|m\rangle - \langle 0|\mu_j|0\rangle\delta_{lm}$ are the electronic transition moments, $\omega_{0l}$ (times $\hbar$) is the energy difference between the electronic ground and excited state $l$, and $\omega_\sigma = \omega_1 + \omega_2$ is the polarization frequency (below electronic resonances). The superscripts $i, j$, and $k$ refer to the molecular Cartesian coordinates $x, y$, and $z$. $P$ is a permutation operator and indicates a summation over six terms obtained by permuting frequencies. The summations over excited states used in the SOS expressions obtained from the different quantum-mechanical methods are in general infinite. The two-state models are the approximation to the SOS method. In this case only one excited state in the summation is taken into consideration.

The solvent dependence of molecular hyperpolarizabilities can be understood in terms of the solvatochromic shift direction of the strong $\pi \longrightarrow \pi^*$ transition with increasing polarity of the medium. In the case of molecules exhibiting positive solvatochromism, the increase of $\beta$ and $\gamma$ is connected with decrease of $\hbar\omega_{gCT}$ and increase of $\Delta\mu$ (see Eqs (1) and (2)), while increasing solvent polarity. In the case of molecules exhibiting negative solvatochromism (blue-shift of the CT absorption band), the solvent dependence of $\beta$ and $\gamma$ shows opposite trends (NLO responses are significantly reduced).

In contrast to the extensive investigations on the solvent influence on the molecular (hyper)polarizabilities, there are only several theoretical works on the two-photon absorption (TPA) cross sections ($\delta$) of molecules in the condensed phases [51–53]. TPA is one of the nonlinear optical processes. It can be described by the imaginary part of $\gamma(Im\gamma(-\omega;\omega,\omega,-\omega))$ [51]. Luo et al. have investigated the solvent polarity influence on the nonlinear optical properties of simple D/A polyene at an *ab initio* level of theory applying continuum model of solvent [51]. These authors have noticed that the solvent effect on the TPA (for the CT excited state) in typical D/A chromophore with positive solvatochromism is smaller as compared to $\beta$. On the other hand, Das and Dudis have found that the second-order transition moments Eq. (6) are quite strongly influenced by the presence of solvents for PNA molecule [52]. This result is supported by our recent investigations. We have observed significantly decreased values of $\delta$, for 4-(1-pyridinium-1-yl)phenolate betaine dye in aqueous solution as compared to the gas phase [53]. This molecule exhibits large blue-shift of the longest-wavelength absorption band. Moreover, it has been shown that the bond length alternation (BLA) and bond order alternation (BOA) as parameters related to the molecular structure strongly influence the values of $\delta$ [54,55]. These structural parameters can be controlled by the polarity of solvent or external electric field.

The aim of this work is a qualitative and quantitative analysis of the influence of solvent on the two-photon absorption (TPA) and nonlinear optical properties of donor-acceptor molecule based on a recently proposed quantum-mechanical Langevin dipoles/Monte Carlo (QM/LD/MC) method [56]. In this paper a theoretical study on the influence of the solvent effect on the TPA of the prototypical $\pi$-conjugated donor-acceptor molecule: p-nitroaniline (PNA) is presented. One of the aims of this work is to investigate the solvent dependence of $\delta$ using (similarly as in the case of $\beta$ and $\gamma$) the two-state model.

## 2. Methodology and computational aspects

The reaction field contribution of the solvent was evaluated using, recently developed, non-cavity quantum-mechanical Langevin dipoles/Monte Carlo (QM/LD/MC) method [56]. In the QM/LD/MC model the solvent molecules are represented by an ensemble of polarizable Langevin point dipoles, distributed amongst the nodes of the regular, cubic lattice. The local field on the i-th node results from the charge density distribution (set of charges, dipoles and quadrupoles) of the solute molecules as well as from the electrostatic field generated by other dipoles. In the iterative calculations there are two competitive processes taken into account. One of them is the orienting effect of the electric field of the solute molecule on the permanent dipole moments of solvent. On the other hand these in turn interact among themselves trying to assume anti-parallel arrangements corresponding to their maximum stabilization. The optimum position and orientation of the solute molecule, placed in a cubic grid of polarizable solvent molecules, was determined using the Monte Carlo (MC) method. The discussed model realistically describes the actual interactions among the dissolved molecule and solvent molecules because of the inclusion of anisotropy of intermolecular interactions. The effect of the reaction field, produced by solvent, is introduced into the solute Hamiltonian by means of a perturbation operator $\mathcal{R}$ according to the equation

$$H = H^o + \mathcal{R}, \tag{4}$$

where $H^o$ is the unperturbed Hamiltonian of isolated molecule. The total potential $\mathcal{R}$ acting on the solute atoms is a sum of averaged (in the meaning of the MC method) potential due to the permanent ($\mathcal{R}_{\text{perm}}$) and induced ($\mathcal{R}_{\text{ind}}$) dipole moments of solvent molecules. It should be noted that the QM/LD/MC technique is less expensive in comparison with the classical dynamics simulations. Details of the QM/LD/MC method are given in Ref. [56]. The calculations of NLO properties were carried out according to description given in Ref. [36].

The electronic contributions to the two-photon absorption were studied based on the formalism described by Monson and McClain [57,58]. The two-photon absorption cross section ($\delta$) from a ground ($g$) to a final state ($f$) for molecules in isotropic media is defined as

$$\langle \delta_{gf} \rangle = \langle |S_{gf}(\overline{\mu}_1, \overline{\mu}_2)|^2 \rangle = \frac{1}{30} \sum_{ij} [S_{ii} S_{jj}^* F + S_{ij} S_{ij}^* G + S_{ij} S_{ji}^* H], \tag{5}$$

where $F(\overline{\mu}_1, \overline{\mu}_2), G(\overline{\mu}_1, \overline{\mu}_2), H(\overline{\mu}_1, \overline{\mu}_2)$ are polarization variables and $S_{ij}$ is the two-photon transition moment. The equation for the two-photon matrix elements:

$$S_{ij}^{gf} = \sum_k \left( \frac{\langle g|\vec{u}_1 \vec{r}|k\rangle \langle k|\vec{\mu}_2 \vec{r}|f\rangle}{\omega_{kg} - \omega_1} + \frac{\langle g|\vec{u}_2 \vec{r}|k\rangle \langle k|\vec{\mu}_1 \vec{r}|f\rangle}{\omega_{kg} - \omega_2} \right) \tag{6}$$

can be directly obtained from time-dependent perturbation theory, treating an interaction of a molecule with an electromagnetic field as a perturbation [58]. Labels $i, j$ stand for Cartesian co-ordinates and $\vec{\mu}_1, \vec{\mu}_2$ denote polarization of photons with energies $\hbar\omega_1, \hbar\omega_2$. Summation in Eq. (6) goes over all the intermediate states ($k$) including the ground ($g$) and the final state ($f$). The energies of both photons are equal (i.e. $\hbar\omega_1 = \hbar\omega_2 = \frac{1}{2}\hbar\omega_{gf}$). In the case of both photons polarized linearly with parallel polarization Eq. (5) can be written in simpler form

$$\langle\delta_{gf}\rangle = \frac{1}{15}\sum_{ij}[S_{ii}S^*_{jj} + 2S_{ij}S^*_{ij}]. \tag{7}$$

Direct comparison between theoretically estimated $\langle\delta_{gf}\rangle$ (in atomic units) from Eq. (7) and experimental data requires knowledge of the line shape function $g(\omega)$ for investigated molecule. It should be noted that $g(\omega)$ depends also on the environment in which measurement is carried out. Experimental TPA cross section $\sigma^{(2)}_{gf}$ (in the conventional cm$^4$ s/photon units) is connected with the $\langle\delta_{gf}\rangle$ by following equation [59,60]:

$$\sigma^{(2)}_{gf} = \frac{8\pi^3\alpha^2\hbar^3}{e^4}\frac{\omega^2 g(\omega)}{\Gamma_f/2}\langle\delta_{gf}\rangle, \tag{8}$$

where $\alpha$ is a fine-structure constant, $\Gamma_f$ is the lifetime broadening of final state (set to 0.1 eV), $\hbar\omega$ is the photon energy. $g(\omega)$ is assumed to be a $\delta$-function. In the present work the solute properties were not corrected by local-field factors [51,61].

The calculations of the two-photon transition matrix elements require knowledge of the transition dipole moments and the transition energies in the SOS expression Eq. (6). In the present study, the single-determinant, molecular orbital approximation was used, and the monoexcited configuration interaction (SCI) was employed to describe the excited states (all singly excited configurations were included for PNA molecule). Our experience indicate that this level of theory is sufficient for description of first-order hyperpolarizabilities ($\beta$) and TPA cross section ($\langle\delta\rangle$) of donor-acceptor organic molecules with the low-laying CT electronic excited state [36,53,55]. However, similarly as in the case of $\beta$, the absolute values of $\langle\delta\rangle$ can be slightly overestimated [36].

It should be noted that the solvent effect on the one- and two-photon absorption should be described as nonequilibrium solvation process [1,2]. It is related to the fact that the vertical transitions are faster that the relaxation time of the solvent. At the QM/LD/MC method level, the solvation energy of the excited states is evaluated using the ground-state solvent configurations, since the absorption of light is faster than the reorientation time of the permanent dipoles of the solvent and allowing only the induced dipoles to be reoriented. Thus, in the calculation of the electronic transition of a solute molecule in a field of polarizable solvent molecules, we must include the change of the energy necessary to polarize the solvent in the excited and ground state of the solute molecule by the solvent induced dipoles (see discussion in Ref [56]). This procedure is only included for calculations of the transition energies between the ground and the CT excited states.

All methods are implemented in the GRINDOL code based on all-valence INDO-like approximation [62]. This semiempirical method has a less degree of parametrization in the evaluation of the one- and two-electron integrals as compared with other semiempirical approaches. The GRINDOL method combined with QM/LD/MC, finite-field (FF) and SOS methods is of proven reliability in the description of molecular nonlinear optical and spectroscopic properties in the gas phase and in solutions [32–36,53, 55,56].

In this work, TPA cross section from the ground to the first excited (CT) state is the only considered. Including only the ground and CT excited state in the expression of Eq. (6), one obtains the two-state approximate equation for dominate component along the molecular axis (in this case x axis):

$$S_{xx}^{gCT} = \frac{4\langle g|r_x|CT\rangle\Delta\mu}{\omega_{gCT}}. \tag{9}$$

In order to obtain the expression of $\langle\delta\rangle$, one can combine this equation with Eq. (7). Finally, the expression for TPA cross section becomes:

$$\langle\delta_{gCT}\rangle \approx \langle\delta_{gCT}^{xx}\rangle = \frac{16}{5}S_{xx}^2 = \frac{16}{5}\frac{\langle g|r_x|CT\rangle^2(\Delta\mu)^2}{\omega_{gCT}^2} \tag{10}$$

It should be noted that meaning of the spectroscopic parameters in above equations is the same as in the case of Eq. (1). $\langle g|r_x|CT\rangle^2$ is proportional to the oscillator strength ($f$) [63]. In the resulting two-state approximation the electronic $\langle\delta_{gCT}\rangle$, in atomic units, can be given as

$$\langle\delta_{gCT}\rangle = \frac{24}{5}\frac{f(\Delta\mu)^2}{\omega_{gCT}^3}. \tag{11}$$

It has been shown that for the CT state of donor-acceptor molecules possessing the largest cross section, a two state model, including only ground state and final state, provides essential features of the TPA process [55,63]. A comparison between Eq. (1) and Eq. (11) show that the TPA cross section ($\langle\delta_{gCT}\rangle \approx f(\Delta\mu)^2$) is a quadratic function of $\Delta\mu$ while $\beta$ being linear ($\beta \approx f\Delta\mu$). This observation suggests that the solvent effect on the TPA cross section of donor-acceptor molecules should be larger than that for the first – order hyperpolarizability. In the next section it will be shown that this assumption is correct.

## 3. Results and discussion

In order to study the electronic linear and nonlinear transition processes and NLO properties, the PNA molecule was selected as an appropriate model system. The choice is connected with the fact that PNA represents one of the simplest push-pull chromophores with large positive solvatochromism and substantial nonlinear optical response [25–34]. Moreover, results of investigations of the solvent effect on the spectroscopic properties and NLO activity of PNA have been presented in several papers [64–71]. We have already demonstrated the applicability of the QM/LD/MC technique in the calculations of the solvent effect on the electronic transitions of PNA molecule [32–34,56] (see Table 1). Our previous study of the absorption and NLO properties of PNA in various solvents leads to the following main conclusions: (1) The calculated red shift for maximum of the low-lying CT absorption band ($\Delta\tilde{\nu}_{\max}$) is equal to 5753 cm$^{-1}$ (going from the gas phase to the water solvent), which is in good agreement with measurements of Gorman et al. (5855 cm$^{-1}$) [65] and another experimental data [66]. (2) As expected, the dipolar CT state is stabilized by surrounding solvent. On going to polar solvents, the increase of the difference between dipole moment ($\Delta\mu$) in the ground and the CT state is observed. For example, the calculated value of $\Delta\mu$, in chloroform solvent, is equal to 9.2 D. This result is very closed to the electrooptical absorption measurements of $\Delta\mu = 9.3$ D (in dioxan solvent) [67]. (3) The calculations demonstrate the existence of the large solvent effect on the values of $\beta$ and $\gamma$ for PNA [32–34]. The calculated values of static $\beta$ and $\gamma$ in chloroform and water are larger by a factor 1.4 and 2.2 in comparison

Table 1
Calculated solvent effect on the absorption energies ($E_{CT}$), difference of the dipole moments between the electronic excited CT state and the ground state, oscillator strengths ($f$) and TPA cross sections ($\langle\delta\rangle$ and $\sigma^{(2)}$) for PNA molecule. The gas phase fully optimized geometry of PNA and PNA-water complex was obtained at the AM1 method level [76]

| Solvent | $E_{CT}$ [$10^3$ cm$^{-1}$] | $\Delta\mu$ [D] | $f$ | $\langle\delta\rangle$ [$10^3$ a.u.] | $\sigma^{(2)}$ [$10^{-48}$ cm$^4$ photon$^{-1}$ s] |
|---|---|---|---|---|---|
| Gas | 30.50 | 7.2 | 0.3560 | 4.81(5.36)[a] | 3.17(3.53) |
| $CHCl_3$ | 28.74 | 9.2 | 0.3538 | 8.33(10.2) | 4.87(5.96) |
| $H_2O$ | 24.75 | 10.4 | 0.3332 | 15.71(18.2) | 7.36(8.53) |
| PNA + 2 $H_2O$[b] | 25.58 | 10.0 | 0.4191 | 14.18(17.9) | 7.07(8.94) |

[a]In parentheses are given values obtained from the two-state model Eq. 11.
[b]The complex was solvated in bulk water using the QM/LD/MC model.

with the corresponding values in the gas phase. In our calculations influence of the solvent on the static $\alpha$ is smaller as compared with $\beta$ and $\gamma$. The calculated ratio of $\alpha$ ($H_2O$)/$\alpha$ (gas) is equal to $\sim$ 1.1.

It should be noted that PNA molecule can form H-bonds involving the NH groups and electron donor atoms of the solvent. This observation comes from experimental studies [64,65]. Huyskens et al. [31] have shown that the formation of specific solute/solvent interactions as H-bonds always increases the hyperpolarizability $\beta$. These authors based on statistical study of a large number of solvents (when H-bonds are formed) have found following relation for $\beta$:

$$\beta = \beta_{gas} + \gamma^0 a\sqrt{\mu_S/V_S} + (1-\gamma^0)b\sqrt{\mu_S}, \tag{12}$$

where $a$ and $b$ are a constants for a given solute molecule. $\mu_S$ is the gaseous dipole moment, and $V_S$ is its molar volume. $\gamma^0$ is the fraction of the time during which the solute is not involved in H-bonding with the solvent molecules. The presence of H-bonds probably strongly influences the intensity of the CT absorption band for PNA molecule. Kovalenko et al. [70] have observed that the width $\Gamma$ of this band decreases by going from a nonpolar (cyclohexan) to highly polar solvent (acetonitryle and DMSO). In the case of water solvent this trend is opposite. Kovalenko et al. have indicated that the break of this trend is due to extra-broadening from hydrogen bonding [70]. In this point our results of calculations do not confirm above observations [34]. The results of calculations of the oscillator strengths $f$ indicate that the following sequence holds for PNA: $f_{gas} > f_{CHCl3} > f_{H_2O}$. In order to explain this disagreement we also carried out calculations using a supermolecule (SM) approach combined with QM/LD/MC method (two H-bonds between oxygen atoms of water molecule and hydrogen atoms of –$NH_2$ group of PNA). As we can see from Table 1 (last row of this Table) such an approach gives correct picture of change of the intensity ($f_{H_2O/SM} > f_{gas} > f_{CHCl3}$). It should be noted that the direct inclusion of H-bonds in PNA molecule (in the supermolecular approach) does not change significantly remaining spectroscopic parameters of PNA in the water solvent. This result shows that effect of specific interactions that comes from H-bonds is very subtle in the case of PNA molecule.

The results of calculated values of the TPA cross section ($\delta$) are shown in Table 1. The following conclusion can be drown from our quantum chemical calculation: (1) The solvent dependence of $\langle\delta\rangle$ is very large; the $\langle\delta\rangle$ value (in a.u.) in the chloroform and water solvent is larger by a factor 1.7 and 3.3 than the corresponding calculated values in the gas phase. (2) The two-state model Eq. (11) gives correct description of behavior of $\langle\delta\rangle$ both in the gas phase and in solvents. Moreover, two-state model shows that the CT state gives the most important contribution to $\langle\delta\rangle$ value when compared with remaining excited states included in the summations Eq. (6). However, absolute values of $\langle\delta\rangle$ obtained from Eq. (11) are larger than calculated value from Eq. (7). (3) The SM approach shows that the explicit addition of water molecules (H-bond complex) does not change value of $\langle\delta\rangle$ for PNA molecule in the water solvent significantly.

In summary, the solvent effect on $\langle\delta\rangle$ can be substantial for donor-acceptor chromophores. The results of our calculations for PNA molecule show that the solvent dependence in the case of $\langle\delta\rangle$ is significantly larger than that for $\alpha, \beta$ and $\gamma$. This is due to the quadratic dependence of $\langle\delta\rangle$ on dipole moment difference between the ground and CT excited state. These results of calculations can be important for understanding of TPA of organic donor-acceptor molecules in condensed phases. It should be pointed out that in the present paper only electronic contributions to the two-photon absorptivities were taken into account. Several authors concluded that the vibronic contributions to $\langle\delta\rangle$ can be substantial [72, 73]. However, up to now no attempts have been made to investigate the environmental effects on the vibronic contribution to TPA cross section. We believe that similar to the molecular polarizabilities solvent influence on the vibronic contributions can be important as well [38,74,75].

## Acknowledgments

This work was supported by the Wroclaw University of Technology.

## References

[1] J. Tomasi and M. Persico, *Chem. Rev.* **94** (1994), 2027.
[2] C.J. Cramer and D.G. Truhlar, *Chem. Rev.* **99** (1999), 2161.
[3] A. Warshel, *Computer Modeling of Chemical Reactions in Enzyme and Solutions*, Wiley, New York, 1991.
[4] C. Reichardt, *Solvents and Solvent Effects in Organic Chemistry*, VCH, Weinheim, 1998.
[5] C. Reichardt, *Chem. Rev.* **94** (1994), 2319.
[6] D.M. Bishop and P. Norman, Calculations of dynamic hyperpolarizabilities for small and medium sized molecules, in: *Handbook of Advanced Electronic and Photonic Materials and Devices*, (Vol. 9), H.S. Nalwa, ed., Academic, San Diego, 2000, p. 1.
[7] B. Champagne and B. Kirtman, Theoretical approach to the design of organic molecular and polymeric nonlinear optical materials, in: *Handbook of Advanced Electronic and Photonic Materials and Devices*, (Vol. 9), H.S. Nalwa, ed., Academic, San Diego, 2000, p. 63.
[8] H.A. Kurtz and D.S. Dudis, Quantum mechanical methods for predicting nonlinear optical properties, in: *Review in Computational Chemistry*, (Vol. 12), K.B. Lipkowitz and D.B. Boyd, eds, VCH, New York, 1998, p. 241.
[9] B.F. Levine, C.G. Bethea, E. Wasserman and L. Leenders, *J. Chem. Phys.* **68** (1978), 5042.
[10] C.C. Teng and A.F. Garito, *Phys. Rev. Lett.* **50** (1983), 350.
[11] C.C. Teng and A.F. Garito, *Phys. Rev. B* **28** (1983), 6766.
[12] L.-T. Cheng, W. Tam, S.H. Stevenson, G.R. Meredith, G. Rikken and S.R. Marder, *J. Phys. Chem.* **95** (1991), 10631.
[13] L.-T. Cheng, W. Tam, S.R. Marder, A.E. Stiegman, G. Rikken and C.W. Spangler, *J. Phys. Chem.* **95** (1991), 10643.
[14] I.A. Maslianitsin, V.D. Shigorin and G.P. Shipulo, *Chem. Phys. Lett.* **194** (1992), 355.
[15] R. Sen, D. Majumdar and S.P. Bhattacharyya, *Chem. Phys. Lett.* **190** (1992), 443.
[16] S.R. Marder, D.N. Bretan and L.-T. Cheng, *Science* **252** (1991), 103.
[17] K. Clays, E. Hendrickx, M. Triest, T. Verbiest, A. Persoons, C. Dehu and J.L. Brdas, *Science* **262** (1993), 1419.
[18] S.R. Marder, C.B. Gorman, F. Meyers, J.W. Perry, G. Bourhill, J.L. Brdas and B.M. Pierce, *Science* **265** (1994), 632.
[19] G. Bourhill, J.L. Brdas, L.-T. Cheng, S.R. Marder, F. Meyers, J.W. Perry and B.G. Tiemann, *J. Am. Chem. Soc.* **116** (1994), 2619.
[20] M. Barzoukas, A. Fort, P. Boy, C. Combellas and A. Thibault, *Nonlinear Opt.* **7** (1994), 41.
[21] C. Runser, A. Fort, M. Barzoukas, C. Combellas, C. Suba, A. Thibault, R. Graff and J.P. Kintzinger, *Chem. Phys.* **193** (1995), 309.
[22] G. Chen., D. Lu and W.A. Goddard III., *J. Chem. Phys.* 101 (1994), 5860.
[23] S. Di Bella, T.J. Marks and M.A. Ratner, *J. Am. Chem. Soc.* **116** (1994), 4440.
[24] I.D.L. Albert, T.J. Marks and M.J. Ratner, *J. Phys. Chem.* **100** (1996), 9714.
[25] M. Sthelin, D.M. Burland and J.E. Rice, *Chem. Phys. Lett.* **191** (1992), 245.

[26] J. Yu and M.C. Zerner, *J. Chem. Phys.* **100** (1994), 7487.
[27] K.V. Mikkelsen, Y. Luo, H. Ågren and P. Jorgensen, *J. Chem. Phys.* **100** (1994), 8240.
[28] P.-E. Larsson, L.M. Kristensen and K.V. Mikkelsen, *Chem. Phys.* **75** (1999), 449.
[29] C. Dehu, F. Meyers, E. Hendrickx, K. Clays, A. Persoons, S.R. Marder and J.L. Brdas, *J. Am. Chem. Soc.* **117** (1995), 10127.
[30] J.N. Woodford, M.A. Pauley and C.H. Wang, *J. Phys. Chem. A* **101** (1997), 1989.
[31] F.L. Huyskens, P.L. Huyskens and A.P. Persoons, *J. Chem. Phys.* **108** (1998), 8161.
[32] W. Bartkowiak and J. Lipiński, *Computers and Chemistry* **1** (1998), 31.
[33] W. Bartkowiak and J. Lipiński, *Chem. Phys. Lett.* **292** (1988), 92.
[34] J. Lipiński and W. Bartkowiak, *Chem. Phys.* **245** (1999), 263.
[35] W. Bartkowiak and J. Lipiński, *J. Phys. Chem. A* **102** (1998), 5236.
[36] W. Bartkowiak, R. Zaleśny, W. Niewodniczański and J. Leszczynski, *J. Phys. Chem. A* **105** (2001), 10702.
[37] W. Bartkowiak, K. Strasburger and J. Leszczynski, *J. Mol. Struct.* (*THEOCHEM*) **549** (2001), 159.
[38] R. Cammi, B. Mennucci and J. Tomasi, *J. Am. Chem. Soc.* **120** (1998), 8834.
[39] R. Cammi, B. Mennucci and J. Tomasi, *J. Phys. Chem. A* **104** (2000), 4690.
[40] W.H. Thompson, M. Blanchard-Desce and J.T. Hynes, *J. Phys. Chem. A* **102** (1998), 7712.
[41] W.H. Thompson, M. Blanchard-Desce, V. Alain, J. Muller, A. Fort, Barzoukas and J.T. Hynes, *J. Phys. Chem. A* **103** (1999), 3766.
[42] A. Painelli and F. Terenziani, *Chem. Phys. Lett.* **312** (1999), 211.
[43] A. Painelli, *Chem. Phys.* **245** (1999), 185.
[44] A. Painelli and F. Terenziani, *J. Phys. Chem. A* **104** (2000), 11041.
[45] J.L. Oudar and D.S. Chemla, *J. Chem. Phys.* **66** (1977), 2664.
[46] C.W. Dirk, L.-T. Cheng and M.G. Kuzyk, *Int. J. Quantum Chem.* **43** (1992), 27.
[47] D.R. Kanis, M.A. Ratner and T.J. Marks, *Chem. Rev.* **94** (1994), 195.
[48] J.L. Brédas, C. Adant, P. Tackx, A. Persoons and B.M. Pierce, *Chem. Rev.* **94** (1994), 77.
[49] B.J. Orr and J.F. Ward, *Mol Phys.* **20** (1971), 513.
[50] A. Willetts, J.E. Rice, D.M. Burland and D.P Shelton, *J. Chem Phys.* **97** (1992), 7590.
[51] Y. Luo, P. Norman, P. Macak and H. Ågren, *J. Phys. Chem. A* **104** (2000), 4718.
[52] G.P. Das and D.S. Dudis, *J. Phys. Chem. A* **104** (2000), 4767.
[53] R. Zaleśny, W. Bartkowiak, S. Styrcz and J. Leszczynski, *J. Phys. Chem. A* **106** (2002), 4037.
[54] T. Kogej, D. Beljonne, F. Meyers, J.W. Perry, S.R. Marder and J.L. Brédas, *Chem. Phys. Lett.* **298** (1998), 1.
[55] W. Bartkowiak, R. Zaleśny and J. Leszczynski, *Chem. Phys.* **287** (2003), 103.
[56] J. Lipiński and W. Bartkowiak, *J. Phys. Chem. A* **101** (1997), 2159.
[57] P.R. Manson and W.M. McClain, *J. Phys. Chem.* **53** (1970), 29.
[58] W.M. McClain and R.A. Harris, Two-Photon Molecular Spectroscopy in Liquids and Gases, in: *Excited States*, (Vol. 3), E.C. Lim, ed., Academic, New York, 1977, p. 1.
[59] D.M. Bishop, J.M. Luis and B. Kirtman, *J. Chem. Phys.* **116** (2002), 9729.
[60] P. Norman and P. Cronstrand, *J. Ericsson, Chem. Phys.* **285** (2002), 207.
[61] R. Wortmann and D.M. Bishop, *J. Chem. Phys.* **108** (1998), 1001.
[62] J. Lipiński, *Int. J. Quant. Chem.* **34** (1988), 423.
[63] C.-K. Wang, P. Macak, Y. Luo and H. Ågren, *J. Chem. Phys.* **114** (2001), 9813.
[64] M.J. Kamlet and R.W. Taft, *J. Am. Chem. Soc.* **98** (1976), 377.
[65] A.A. Gorman, M.G. Hutchings and P.D. Wood, *J. Am. Chem. Soc.* **118** (1996), 8497.
[66] P. Suppan and J. Photochem, *Photobiol. A* **50** (1990), 293.
[67] R. Wortmann, P. Krämer, C. Glania, S. Lebus and N. Detzer, *Chem. Phys.* **173** (1993), 99.
[68] C.L. Thomsen, J. Thogersen and S.R. Keiding, *J. Phys. Chem. A* **102** (1998), 1062.
[69] A.M. Moran and A.M. Kelley, *J. Chem. Phys.* **115** (2001), 912.
[70] S.A. Kovalenko, R. Schanz, V.M. Farztdinov, H. Hennig and N.P. Ernsting, *Chem. Phys. Lett.* **323** (2000), 312.
[71] V.M. Farztdinov, R. Schanz, S.A. Kovalenko and N.P. Ernsting, *J. Phys. Chem. A* **104** (2000), 11486.
[72] Y. Luo, H. Ågren, S. Knuts and P. Jørgensen, *Chem. Phys. Lett.* **213** (1993), 356.
[73] P. Macak, Y. Luo, P. Norman and H. Ågren, *J. Chem. Phys* **113** (2000), 7055.
[74] P. Zuliani, M. Del Zoppo, C. Castiglioni, G. Zerbi, C. Andraud, T. Botin and A. Collet, *J. Phys. Chem.* **99** (1995), 16242.
[75] W. Bartkowiak and T. Misiaszek, *Chem. Phys.* **261** (2000), 353.
[76] M.J.S. Dewar, E.G. Zoebisch, E.F. Healy and J.J.P. Stewart, *J. Am. Chem. Soc.* **107** (1985), 3902.

Journal of Computational Methods in Sciences and Engineering 4 (2004) 559–568
IOS Press


# A sequential Monte Carlo/Quantum Mechanics study of the dipole polarizability of liquid benzene

Eudes E. Fileti and Sylvio Canuto*
*Instituto de Física, Universidade de São Paulo, CP 66318, 05315-970, São Paulo, SP, Brazil*

**Abstract.** Metropolis Monte Carlo classical simulation and quantum mechanical calculations are performed to obtain the dipole polarizability of liquid benzene. Super-molecular configurations are sampled from NVT Monte Carlo simulation of liquid benzene at room temperature and are used for subsequent quantum mechanical calculations. The auto-correlation function of the energy is used to analyze the statistical correlation between the configurations. Finite-field INDO and DFT calculations are performed in the statistically uncorrelated super-molecular clusters obtained in the simulation. The quantum mechanical results are shown to represent statistically converged values. The final results corroborate the recent contention that the dipole polarizability of liquid benzene suffers only a small change in the condensed phase.

Keywords: Monte Carlo simulation, polarizability, liquid benzene, density-functional theory, intermediate-neglect of differential overlap (INDO)

Mathematics Subject Classification: 31.10.+z, 33.15.-e, 71.10.-w, 71.15.-m

## 1. Introduction

The theoretical understanding of the electronic structure of gas phase, or isolated, molecules has progressed enormously in the last decades. Since the advent of the Schrödinger equation the progress in methods, techniques and algorithms allied to the computer hardware revolution has permitted theory to achieve a status complementary to laboratory experiments. However, most of the physical and chemical phenomena occur not for isolated molecules but interacting with the environment. This is the case, for instance, of most chemical experiments where solvents are everyday present. This leads to the necessity of treating solvents and much progress has been achieved using continuum models [1–5] where the solvent is described by its macroscopic constants such as the dielectric constant and index of refraction. In fact a liquid is more complex. The proper treatment of liquid systems has to consider its statistical nature [6,7]. Indeed, there are many possible geometrical arrangements of the molecules with with equivalent probability for non-zero temperature. Thus the liquid electronic properties are best described by a statistical distribution [8–12]. The structure of the liquid is represented by the radial distribution function and all properties are obtained from a statistical average [6,7].

Dipole polarizabilities are very important to understand the polarization of an electronic medium [13] and naturally relates to several molecular properties, including the dispersion contribution of the van der

*Corresponding author. Fax: +55 11 3091 6831; E-mail: canuto@if.usp.br.

Waals interaction [14]. Understanding the polarizability of a liquid system is thus very important. In this contribution we dedicate ourselves to the theoretical estimate of the dipole polarizability of liquid benzene. The choice of benzene is natural because all permanent odd moments are zero by symmetry. The permanent dipole moment and the first dipole hyper-polarizability, for instance, are zero. Thus the most important contribution to the polarization is the dipole polarizability. Of particular interest is the difference of the average dipole polarizability of the liquid compared to the gas phase. Whereas most continuum theories have obtained that the dipole polarizability is increased in the liquid, recent theoretical studies have challenged this and argued that in the liquid it should not show any appreciable change or be decreased instead [15–17]. The difficulties in obtaining properties of molecular liquid systems are related to the natural uncertainty associated to the complex and disordered liquid phase. In several previous studies it became clear that cluster or even micro-solvation models cannot describe in general the liquid properties [18,19]. The proper description of the liquid state needs a statistical procedure and assuring that the statistical properties are converged is necessary for reliable theoretical estimates. In this paper we use the sequential Monte Carlo quantum mechanics (S-MC/QM) methodology [20,21] where we first generate the structure of the liquid using Metropolis Monte Carlo. Next, statistically relevant configurations are separated and submitted to quantum mechanics calculations. Statistically converged results are obtained. Several QM calculations are necessary to obtain the ensemble average that is necessary to characterize the statistical nature of the liquid. This is an approach that is conceptually sound: the liquid system is explicitly considered and a proper statistical ensemble is used. The drawback is that such an approach is demanding computationally. In this present study the dipole polarizability of liquid benzene is obtained using density-functional theory (DFT) and the semi-empirical INDO models. However, the absolute accuracy is not the major point of interest here. Instead, we want to compare the dipole polarizability in the condensed liquid phase with that for the gas phase. Recent theoretical estimates point to a slight decrease of the polarizability [15–17] in the condensed medium.

## 2. Theoretical methodology

### 2.1. Monte Carlo simulation

Monte Carlo (MC) simulations are made using the Metropolis sampling technique and the periodic boundary conditions combined with the minimum image method in a cubic box [7]. The simulations are performed in the NVT ensemble. The total system consists of 343 benzene molecules at room temperature (298 K) and density of 0.8990 g/cm$^3$. The intermolecular interactions are described by the standard Lennard-Jones potential with 2 parameters for each atom $i$ ($\varepsilon_i$ and $\sigma_i$)

$$U_{ab} = \sum_i^a \sum_j^b 4\varepsilon_{ij} \left[ \left( \frac{\sigma_{ij}}{r_{ij}} \right)^{12} - \left( \frac{\sigma_{ij}}{r_{ij}} \right)^{6} \right]. \tag{1}$$

The pair-potential parameters are obtained using $\varepsilon_{ij} = (\varepsilon_i \varepsilon_j)^{1/2}$ and $\sigma_{ij} = (\sigma_i \sigma_j)^{1/2}$. We use the six-site OPLS [22] parameters and the experimental geometry. These are shown in Table 1. The intermolecular interactions are spherically truncated within a center of mass separation larger than the cutoff radius, $r_C = 18.4$ Å. Long-range corrections are calculated beyond this cutoff distance. In the simulation the molecules are kept with rigid geometries. The initial configurations are generated randomly, considering the position and orientation of each molecule. A new configuration is generated

Table 1
Geometry (in Å) and parameters of the Lennard-Jones potential of benzene

| Site | $x$ | $y$ | $z$ | $\varepsilon$ (kcal/mol) | $\sigma$ (Å) |
|---|---|---|---|---|---|
| C | 0.0000 | 1.4000 | 0.0000 | 0.110 | 3.750 |
| C | 1.2124 | 0.7000 | 0.0000 | 0.110 | 3.750 |
| C | 1.2124 | −0.7000 | 0.0000 | 0.110 | 3.750 |
| C | 0.0000 | −1.4000 | 0.0000 | 0.110 | 3.750 |
| C | −1.2124 | −0.7000 | 0.0000 | 0.110 | 3.750 |
| C | −1.2124 | 0.7000 | 0.0000 | 0.110 | 3.750 |
| H | 0.0000 | 2.4881 | 0.0000 | 0.000 | 0.000 |
| H | 2.1547 | 1.2440 | 0.0000 | 0.000 | 0.000 |
| H | 2.1547 | −1.2440 | 0.0000 | 0.000 | 0.000 |
| H | 0.0000 | −2.4881 | 0.0000 | 0.000 | 0.000 |
| H | −2.1547 | −1.2440 | 0.0000 | 0.000 | 0.000 |
| H | −2.1547 | 1.2440 | 0.0000 | 0.000 | 0.000 |

after randomly attempting to translate in all Cartesian directions and to rotate around a randomly chosen axis. The simulation consists of a thermalization phase of $1.7 \times 10^6$ MC steps, followed by an averaging stage of $17.15 \times 10^6$ MC steps. All Monte Carlo simulations and statistical correlation analysis are performed with the DICE [23] Monte Carlo statistical mechanics program. The great advantage of the sequential procedure is that all the important statistical information is available before running into the QM calculations [24–26]. This considerably reduces the number of super-molecular structures that will be submitted to the quantum mechanical calculations, because the configurations are selected according to their statistical correlation, obtained from the auto-correlation function of the energy [11,12,24,25]. Using a correlation step of $2.7 \times 10^5$, we separate a total of 62 configurations, with less than 5% of statistical correlation. These 62 structures are used in the quantum mechanical calculations. The solvation shells were defined from the analysis of the radial distribution function. The first solvation shell comprises a total of 14 benzene molecules. The second shell, a total of 63 benzene molecules. This is of course a very large system for most ab initio methodologies and some compromizing has to be used. We employ both the semi-empirical INDO and the first principles DFT method.

## *2.2. Quantum mechanics calculations*

The semi-empirical INDO calculations are made using Zerner's program in the original parameterization [27,28]. It has been shown that the INDO model gives reasonably good hyper-polarizabilities within the sum-over-state (SOS) methodology [29]. However, for the dipole polarizability of ground states all terms of the SOS are positive and leads to slow convergence of the sum procedure. The QM calculations of the dipole polarizabilities are then made using the finite-field approximation [30] in the INDO level. Five different electric fields are applied in each of the three cartesian orientations. The field values are $\pm$ 0.002, $\pm$ 0.001 and 0.0 a.u. The dipole polarizabilities reported are the results of a simple average over the QM INDO results for the 62 statistically uncorrelated super-molecular configurations. Five finite electric fields were used for the small (14 benzene molecules) and large (28 benzene molecules) super-molecular systems in the three cartesian components. A total of 1612 QM calculations are performed. To verify the major assumptions of this study we also employ the DFT gradient corrected Becke's three-parameter hybrid exchange-correlation functional combined with the Lee-Yang-Parr correlation term, the so-called B3LYP model [31,32]. The basis set used is the split-valence 6-31G. For the smaller case (14 benzene molecules) this leads to a total of 924 contracted gaussian-type functions. It should be stressed that the calculations for the super-molecular structures generated in the liquid are made using

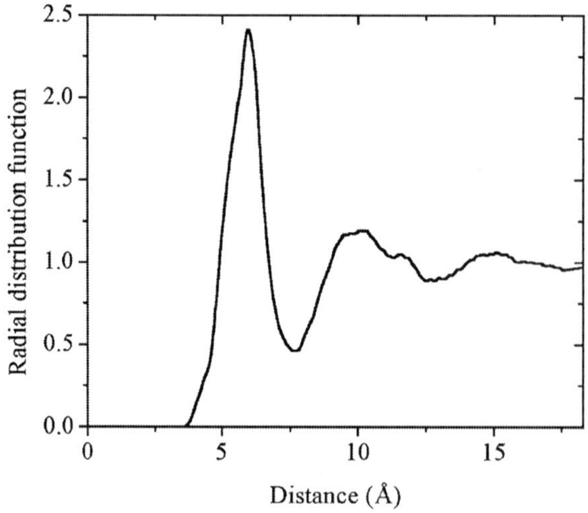


Fig. 1. Pair-wise radial distribution function between the center of mass of benzene.

Fig. 2. Illustration of two super-molecular configurations (left with 14 benzene molecules and right with 28 benzene molecules) extracted from the MC simulation.

the antisymmetrization of the entire system. For instance, in the case of the smaller system composed of 14 benzene molecules this corresponds to a total of 420 valence electron in the INDO model and 598 electrons in the B3LYP/6-31G case. The DFT calculations are made using Gaussian 98 [33].

## 3. Results and discussion

Figure 1 shows the calculated radial distribution function of the center-of-mass of benzene. A clear first solvation shell ending in 7.5 Å is seen. The second solvation shell ends at 12.9 Å. Spherical integration up to these distances give, respectively, 14 and 63 benzene molecules. Between these two minima there is a clear maximum at 10.0 Å and spherical integration of the radial distribution function up to this point gives a number of 28 benzene molecules in this solvation shell. Even at the semi-empirical INDO level we could not perform the finite-field calculations for the 63 benzene molecules. We have thus limited to

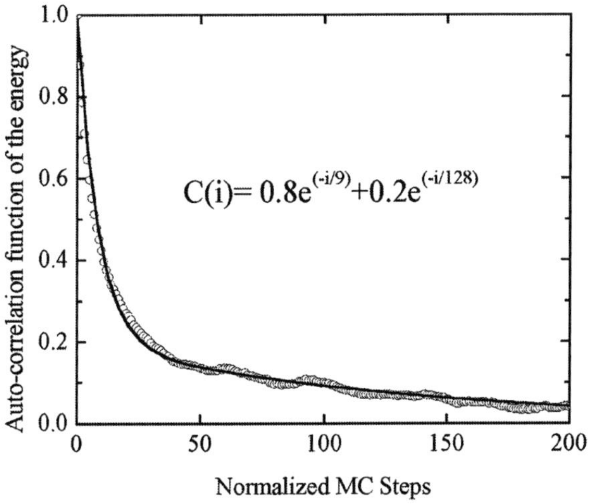


Fig. 3. Calculated auto-correlation function of the energy. MC step is normalized to the total number of molecules.

the first solvation shell (14 molecules) and the in-between shell with 28 molecules. Figure 2 illustrates one of the 62 super-molecular structures composed of 14 and 28 benzene molecules used in the QM calculations. Having defined the size of the super-molecular structures it is now important to establish the number of QM calculations that have to be performed to obtain the statistical convergence. The answer to this question is obtained from the auto-correlation function of the energy [11,12,24,25]. This function, over a chain of size $l$, is defined as

$$C(i) = \frac{\langle E_j E_{j+i}\rangle_{l-i} - \langle E_j\rangle_{l-i}\langle E_{j+i}\rangle_{l-i}}{\langle^2\rangle_l - \langle E\rangle_l^2} \tag{2}$$

where $i$, means the interval of MC configurations. For markovian processes this is an exponentially decaying function,

$$C(i) = \sum_j^n c_j e^{-i/tau_j} \tag{3}$$

The correlation step is

$$\tau \int_0^\infty C(i)di. \tag{4}$$

Figure 3 shows the calculated auto-correlation function of the energy of liquid benzene. From this and Eq. (4) it can be obtained that the correlation step in normalized MC steps is $\sim 40$. A normalized step corresponds here to 343 (the number of benzene molecules) MC steps. In calculating the averages we decided to select structures in an interval of 800 normalized steps corresponding to $2.744 \times 105$ MC steps. As the total number of configurations generated is $17.15 \times 10^6$ (or 50000 normalized steps), the averages are taken over 62 configurations, with a correlation of less than 5%. These 62 configurations will be subjected to quantum mechanical super-molecular calculations. These calculations are made at the INDO and B3LYP/6-31G levels using the finite field model. The wavefunction is antisymmetric

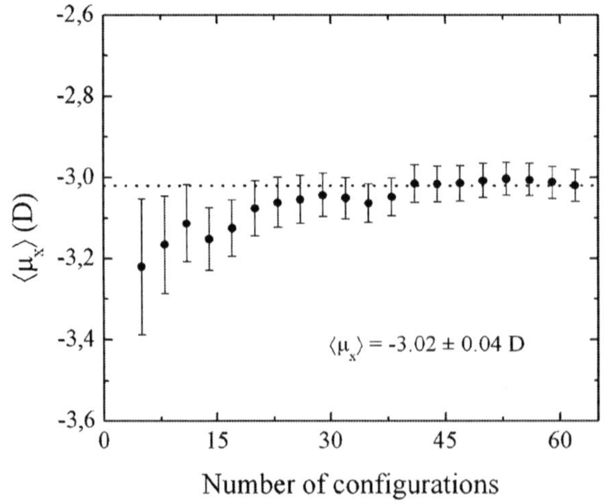


Fig. 4. Statistical convergence of the dipole polarization. Bar represents the statistical error. The example corresponds to the finite-field value of 0.002 a.u. in the small (14 benzene molecules) super-molecular configuration.

over the entire system and the finite field is applied in the entire super-molecular system. The dipole polarization is calculated and therefrom the dipole polarizability is obtained as the linear term of the polarization. Having the dipole polarizability for the entire system composed of N benzene molecules we obtain the condensed phase result of benzene dividing the result by the number N. This gives the polarizability per molecule. An important point is that there are no separate calculations for the solvent. We simply compare the results obtained for the super-molecules with the calculations for the isolated benzene system. The distinction between these two results is the influence of the condensation. The use of a properly antisymmetric wavefunction is important to include the exchange interaction between the different benzene molecules in the liquid and also to obtain the intermolecular interactions. The inclusion of the dispersion contribution to the van der Waals interaction is known to require correlation effects and it is likely to be excluded in the INDO calculations. As the DFT method is correlated it is ideally expected to include the dispersion part of the correlation effect. This is not assured however and the proper and explicit inclusion of van der Waals contribution is a central topic in DFT theoretical methodologies. We only emphasize that the use of the super-molecular approach with a proper antisymmetric wavefunction may include dispersion interaction at a lower level [34]. Indeed, for the benzene molecule the solvatochromic shift upon condensation, that is dictated by the dispersion interaction, is well described using the INDO theoretical methodology used here [11,25].

In the INDO calculations we use both the super-molecules with 14 and 28 benzene molecules. Figure 4 shows the convergence of the calculated dipole moment of the 14 benzene molecules in the external finite-field of 0.002 a.u. As can be seen the calculated dipole moment is statistically well converged after $\sim$ 40 QM calculations. The same procedure is repeated for all the other electric fields and also for the super-molecular configurations with 28 benzene molecules. Statistical convergence is assured in all cases. Using these results for the finite-fields, the dipole polarizability component is obtained as shown in Fig. 5. We report both invariants of the dipole polarizability, the $\alpha_{mean}$, and the anisotropy. The liquid value is the result of the statistical average $\langle\alpha_{mean}\rangle$ obtained using the calculated dipole polarizabilities of the super-molecules extracted from the MC simulation. The numerical results are finally summarized in Table 2. As expected the calculated dipole polarizability at the INDO semi-empirical level are not competitive with other numerical estimates. Parkinson and Zerner have considered the possibility of

Table 2
Calculated components of the dipole polarizability (in a.u.) of liquid benzene using finite-field approximation in the INDO and DFT models. $\langle \alpha_{mean} \rangle$ is the statistical average of the mean dipole polarizability and $\Delta\alpha$ is the statistical average polarizability anisotropy

| Model | $\Delta\alpha$ | $\langle \alpha_{mean} \rangle$ |
|---|---|---|
| INDO | | |
| 14 benzene | 2.1 | 42.7 |
| 28 benzene | 0.9 | 43.7 |
| B3LYP/6-31G | | |
| 14 benzene | 12.5 | 55.6 |
| Ref [17] | 36.5 | 63.7 |

Conversion factor: 1 Debye is 0.3935 a.u.
Calculated mean dipole polarizability for the gas phase is 43.0 a.u. (INDO) and 53.4 a.u. (B3LYP/6-31G).
Experimental dipole polarizability for the gas phase value is 69.7 a.u. [36].

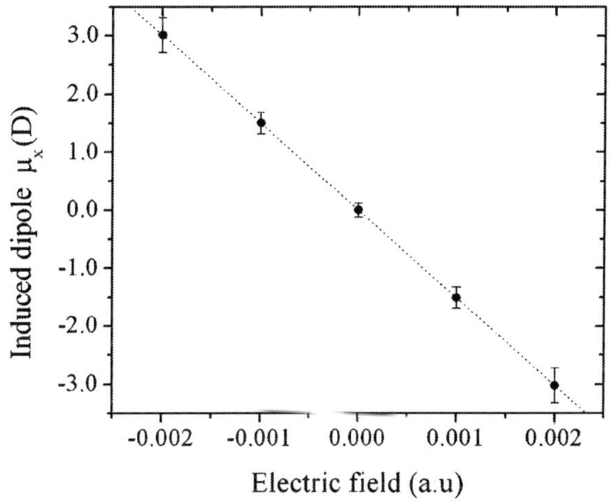


Fig. 5. Polarization as a function of the applied field for the small (14 benzene molecules) configuration. Each point shown is the result of a converged statistical average of 62 quantum mechanics calculations (See Fig. 4).

extending these calculations to the random-phase approximation (RPA) level [35]. Our theoretical result for the gas phase is 43.0 a.u , which is equivalent to the RPA result but still too small compared to the experimental result [36] of 69.7 a.u. At this point the absence of more diffuse function is responsible for this decreased value. In the condensed phase this lack is less important. However, in comparison with the condensed phase result we can see the trend that the dipole polarizability assumes essentially the same value. Two aspects are worth noting. First, the emphasis in this present study lies in a realistic representation of the liquid. Hence it is worth noticing that the dipole polarizability of the liquid benzene obtained here is nearly isotropic. For the larger super-molecular system of 28 benzene molecules the anisotropic component of the polarizability is only 0.9 a.u. This is indicative that a reasonable representation of the bulk polarizability is obtained here. Second, it should be pointed out that the polarizability is represented by a statistical distribution. Figure 6 illustrates the distribution of

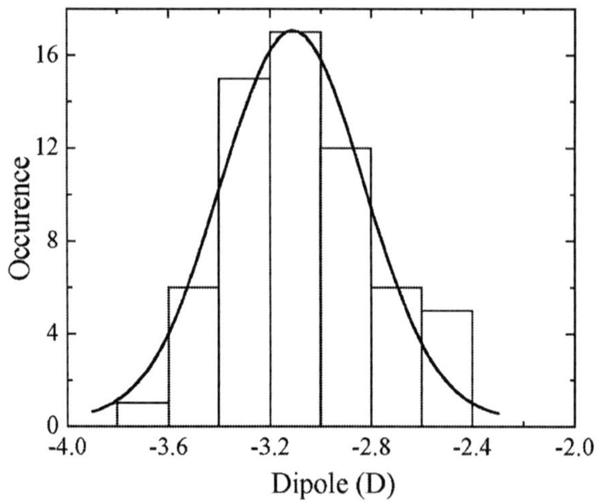


Fig. 6. Statistical distribution of calculated dipole moment for the field of 0.002 a.u in the x-direction.

the polarization in the x-direction for the field of 0.002 a.u. Although the average value is, in this case, $\langle \mu_x \rangle = -3.08$ D, the distribution of polarization values is reasonably broad.

The B3LYP/6-31G calculations for the condensed phase are performed only for the smaller super-molecules composed of 14 benzene molecules. The results for the mean dipole polarizability are shown in Fig. 7 and seems to converge to the average of $55.6 \pm 0.5$ a.u. As a comparison the gas phase result obtained for the isolated benzene molecule is 53.4 a.u. It is well-known that diffuse functions play an important role in the calculations of dipole polarizabilities. In this present study we are limited to the size of basis function and could only use a relatively small basis without a proper description of the diffuse part of the basis set. Indeed in this study the basis set size for the small cluster of 14 benzene molecules includes nearly one-thousand basis function and could not be extended any further. Note that the result for the dipole polarizability of a single super-molecule requires, in fact, several single-point calculations for the different finite field values. Because the super-molecules including several benzene molecules are extended compared to the single molecule case, it is expected that the lack of diffuse function will be more harmful for the gas phase than the condensed phase. Hence, the difference of 2.2 $\pm$ 0.5 a.u between the condensed and gas phases results is likely to be slightly overestimated. Again, the first principle DFT calculations, similar to the semi-empirical INDO results, give an average value for the mean polarizability that is essentially the same result as in the gas phase.

As it was first discussed by Jensen et al. [17] using time-dependent DFT results the dipole polarizability of condensed phase benzene is slightly decreased compared to the gas phase result by 8%. Morita and Kato [15] have earlier considered the medium (water) perturbation on the dipole polarizabilities of small systems like Ne, Ar, CH4 and anions like $Cl^-$. Similarly, they conclude that for the case of neutral systems in water there is a decrease in the dipole polarizability between 13 and 18%. For the anion, $Cl^-$, in water the decrease is as much as 37%. van Duijnen et al. [37] contends that the dipole polarizability of formaldehyde in water shows no appreciable change and that if polarizability is to change due to medium effect it should be decreased. In line with this result we find that the dipole polarizability of liquid benzene shows no appreciable change and, in fact, using statistically converged result we find the results to be equivalent.

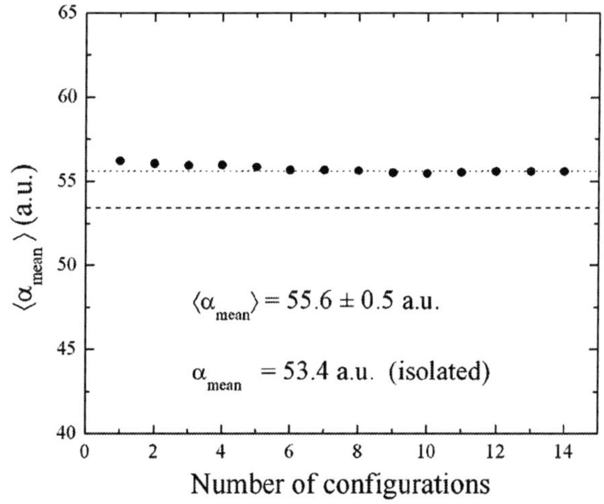


Fig. 7. Calculated results for the mean dipole polarizability per benzene molecule for the super-molecular configurations using the B3LYP/6-31G theoretical model. Dashed line represents the gas phase value.

## 4. Conclusion

Monte Carlo classical simulation is performed to generate the structure of liquid benzene at room temperature. Configurations are sampled using the auto-correlation function of the energy and the structures used in the subsequent quantum mechanical calculations are less than 5% statistically correlated. INDO finite-field calculations are made in the super-molecular structures including all benzene molecules within 10 Å. For each finite-field value the electronic polarization is verified to be statistically converged. As a complement, the quantum mechanical results are performed also at the density-functional B3LYP/6-31G level. The final results indicate that the dipole polarizability of the liquid is equivalent to the gas phase result. This result corroborates the recent contention that the dipole polarizability of liquid benzene suffers only a small change in condensed phase.

## Acknowledgments

This work has been partially supported by CNPq and FAPESP (Brazil). Discussions with Dr. K. Coutinho and Dr. R. Rivelino are greatfully acknowledged.

## References

[1] L. Onsager, *J. Am. Chem. Soc.* **58** (1936), 1486.
[2] O. Tapia and O. Goscinski, *Molec. Phys.* **29** (1975), 1653.
[3] J.L. Rivail and D. Rinaldi, *Chem. Phys.* **18** (1976), 233.
[4] J. Tomasi and M. Persico, *Chem. Rev.* **94** (1994), 2027.
[5] M.M. Karelson and M.C. Zerner, *J. Phys. Chem.* **96** (1992), 6949.
[6] D.M. Heyes, *The Liquid State. Applications of Molecular Simulations*, John Wiley, New York, 1998.
[7] M.P. Allen and D.J. Tildesley, *Computer Simulation of Liquids*, Clarendon Press, Oxford, 1987.
[8] J.T. Blair, K. Krogh-Jespersen and R.M. Levy, *J. Am. Chem. Soc.* **111** (1989), 6948.
[9] J. Gao, *J. Am. Chem. Soc.* **116** (1994), 9324.
[10] J. Zeng, N.S. Hush and J.R. Reimers, *J. Chem. Phys.* **99** 1496, (1993).

[11] K. Coutinho, S. Canuto and M.C. Zerner, *Int. J. Quantum Chem.* **65** (1997), 885.
[12] K. Coutinho and S. Canuto, *Adv. Quantum Chem.* **28** (1997), 89.
[13] A.D. Buckingham, *Adv. Chem. Phys.* **12** (1967), 107; D.M. Bishop, *Adv. Quantum Chem.* **25** (1994), 1.
[14] A.J. Stone, *The Theory of Intermolecular Forces*, Clarendon Press, Oxford, 1996.
[15] A. Morita and S. Kato, *J. Chem. Phys.* **110** (1999), 11987.
[16] P. Th. van Duijnen, A.H. De Vries, M. Swart and F. Grozema, *J. Chem. Phys.* **117** (2002), 8442.
[17] L. Jensen, M. Swart, P. Th. van Duijnen and J.G. Snijders, *J. Chem. Phys.* **117** (2002), 3316.
[18] K. Coutinho, R.C. Guedes, B.J. Costa Cabral and S. Canuto, *Chem. Phys. Lett.* **369** (2003), 345.
[19] K. Coutinho and S. Canuto, *J. Chem. Phys.* **113** (2000), 9132.
[20] S. Canuto, K. Coutinho and D. Trzresniak, *Adv. Quantum Chem.* **41** (2002), 161.
[21] T. Malaspina, K. Coutinho and S. Canuto, *J. Chem. Phys.* **117** (2002), 1692.
[22] W.L. Jorgensen, J.D. Madura and C.J. Swenson, *J. Amer. Chem. Soc.* **106** (1984), 6638.
[23] K. Coutinho and S. Canuto, *DICE: A Monte Carlo program for molecular liquid simulation*, University of São Paulo, 1997.
[24] K. Coutinho, M.J. Oliveira and S. Canuto, *Int. J. Quantum Chem.* **66** (1998), 249.
[25] K. Coutinho, S. Canuto and M.C. Zerner, *J. Chem. Phys.* **112** (2000), 9874.
[26] S. Canuto and K. Coutinho, *Int. J. Quantum Chem.* **77** (2000), 192.
[27] J. Ridley and M.C. Zerner, *Theor. Chim. Acta* **32** (1973), 111.
[28] M.C. Zerner, *INDO-UF: A semi-empirical program package of the University of Florida*, Gainesville, FL 32611.
[29] S. Di Bella, T.J. Marks and M.A. Ratner, *J. Am. Chem. Soc.* **116** (1994), 4440.
[30] H.D. Cohen and C.C.J. Roothaan, *J. Chem. Phys.* **43** (1965), 34; I. Cernusak, G.H.F. Diercksen and A.J. Sadlej, *Phys. Rev. A* **33** (1986), 814; G. Maroulis, *J. Phys. B: At. Mol. Opt. Phys.* **34** (2001), 3727.
[31] A.D. Becke, *Phys. Rev. A* **38** (1988), 3098.
[32] C. Lee, W. Yang and R.G. Parr, *Phys. Rev. B* **37** (1988), 785.
[33] M.J. Frisch et al., GAUSSIAN 98, Revision A7, Pittisburgh, PA: Gaussian Inc, 1998.
[34] S. Canuto, K. Coutinho and M.C. Zerner, *J. Chem. Phys.* **112** (2000), 7293.
[35] W.A. Parkinson and M.C. Zerner, *Chem. Phys. Lett.* **139** (1987), 563.
[36] D.R. Lide, *Handbook of Chemistry and Physics*, 73rd edition, CRC-Press, Boca Raton, 1992–1993.
[37] P. van Duijnen, M. Swart and F. Grozema, in: *ACS Symposium Series*, J. Gao and M.A. Thompson, (Vol. 712), ACS, Washington, D.C., 1998, p. 220.

Journal of Computational Methods in Sciences and Engineering 4 (2004) 569–588
IOS Press



# Hyperpolarizability of novel *carbo*-meric push-pull chromophores

Christine Lepetit, Pascal G. Lacroix, Viviane Peyrou, Catherine Saccavini and Remi Chauvin*
*Laboratoire de Chimie de Coordination, UPR 8241 CNRS, 205 Route de Narbonne, F-31077 Toulouse cedex 04, France*

**Abstract.** A systematic *carbo*-meric comparison of geometrical and NLO properties of push-pull aromatic NLO chromophores is undertaken at the B3PW91/6-31G* level and the INDO-SOS level respectively. In a first approach, merocyanine-like chromophores all containing a $p$-oxyphenyl donor and a N-methyl-4-pyridyl acceptor with various conjugated hydrocarbon bridges are considered. *Carbo*-merization of ethylene, 2,5-thiophenylene or 1-oxo-2,5-cyclopentadienylene bridges results in a *ca* ten times exaltation of static hyperpolarizability ($\beta_0$). Surprisingly, *carbo*-merization of the $p$-phenylene unit of $^-$O-$C_6H_4$–$C_6H_4$–$C_5H_4N^+$–Me results in a three-fold decrease of $\beta_0$. The negative *carbo*-meric effect is even more pronounced for the $^-$O-$C_6H_4$–$C_6H_4$–C≡C–$C_5H_4N^+$–Me ethynylogue. This effect is however relative to the exceptional value of the latter parent molecule ($\beta_0 = 825 \times 10^{-30}$ cm$^5$ esu$^{-1}$). The $^-$O-$C_6H_4$–C≡C-$C_6H_4$–$C_5$ $H_4N^+$–Me isomer exhibits a similar hyperpolarizability ($\beta_0 = 774 \times 10^{-30}$ cm$^5$ esu$^{-1}$). Inserting an additional $C_2$ unit results in the double ethynylogue structure $^-$O-$C_6H_4$–C≡C–$C_6H_4$–C≡C–$C_5H_4N^+$–Me, for which $\beta_0$ = 1982 $\times 10^{-30}$ cm$^5$ esu$^{-1}$. The "phenylethynylene effect" becomes dramatic in the $^-$O-$C_6H_4$–C≡C–($C_6H_4$–C≡C)$_2$–$C_5H_4N^+$–Me homologue which exhibits an unprecedented calculated value among small organic chromophores, namely: $\beta_0 = 33856 \times 10^{-30}$ cm$^5$ esu$^{-1}$. A generalized absolute bond length alternation parameter (ABLA) is propounded and calculated for a homogeneous family of merocyanine-like chromophores. An astonishing regular variation of $\beta_0$ *vs* ABLA is empirically observed: drawing nearer to a critical value ABLA° ≈ 0.040 Å results in a steep increase of $\beta_0$. A Lorentzian-based fit is empirically found ($R > 0.999$) and heuristically derived from simple approximations in a two-level model. The curve interpolates points corresponding to all kinds of chromophores bearing $p$-$C_6H_4O$ and 4-$C_5H_4N$ termini, including the non-zwitterionic isomer MeO-$C_6H_4$–$C_{18}H_4$–$C_5H_4N$, the experimental synthesis of which is addressed. As in the case of $p$-nitroaniline, a positive *carbo*-meric effect on $\beta_0$ is restored upon *carbo*-merization of the $p$-phenylene bridge of the non-zwitterionic $Me_2N$-$C_6H_4$–$C_6H_4$–$C_6H_4$–$NO_2$ congener of $^-$O-$C_6H_4$–$C_6H_4$–$C_5H_4N^+$–Me. It is also shown that $\beta_0$ exaltation is larger when *carbo*-merization is applied to the terminal benzene ring of 4-nitro-biphenyl derivatives, thus confirming the excellent intrinsic donor properties of the *carbo*-benzene ring.

Keywords: Non-linear Optics, *carbo*-mers, bond-length alternation, chromophores, aromaticity, hyperpolarizability, ZINDO/DFT calculations

Mathematics Subject Classification: 92E10, 78A60

## 1. Introduction

The design of molecular materials with second-order non-linear optical properties [1], such as bulk hyperpolarizability ($\chi^2$), is currently based on a two-step process. The first step consists in the search for molecular units of high quadratic hyperpolarizability ($\beta$), while the second step consists in embedding selected chromophores in a non-centrosymmetric medium (crystal with chiral space groups, poled

*Corresponding author. Tel.: +33 05 61 33 31 13; Fax: +33 05 61 55 30 03; E-mail: chauvin@lcc-toulouse.fr.

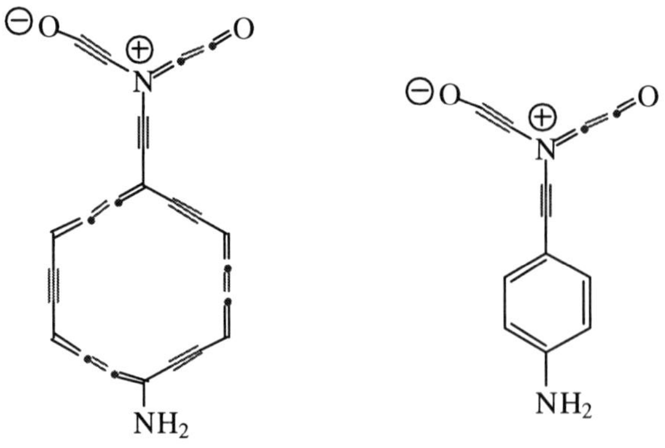


Scheme 1. Representative efficient *carbo*-meric chromophores from $p$-nitro-aniline [2].

polymers). Various approaches for the first step can be classified according to the type of molecular non-centrosymmetry. The most traditional approach is the search for dipolar (push-pull) chromophores with a residual symmetry ($\boldsymbol{C}_s, \boldsymbol{C}_{2v}, \ldots$). In this approach, the effects of increasing the relative donor/acceptor strength and lengthening the $\pi$-conjugated bridge, have been extensively investigated. On the basis of the approximate two-level expression of the static hyperpolarizability Eq. (1a), these effects consist in increasing the numerator and lowering the numerator, respectively.

$$\beta_0 = C, \frac{f \cdot \Delta\mu,}{(\Delta E)^3}, \tag{1a}$$

where $f$ is the oscillator strength, $\Delta\mu$ is the change in dipole moment occuring upon charge-tansfer transition, $\Delta E$ is the transition energy, and C is a constant depending on the unit system.

A recently disclosed alternative strategy which remains *a priori* consistent with the approximate Eq. (1a), consists in starting from known efficient donor-acceptor chromophores and "inflating" them by *carbo*-merization [2], namely by replacing C–C and C=C bonds for their respective C–C≡C–C and C=C=C=C *carbo*-meric units [3]. It is worth noting here that in a related, but definitely different approach, many studies focussed on non-linear optical properties of carbon-rich chromophores for themselves [4]. The *carbo*-mer expansion can affect the whole parent structure of the chromophore or be localized to few critical domains out of five: the donor, the acceptor, the bridge, the donor-bridge bond, the acceptor-bridge bond. This strategy has been theoretically illustrated and analyzed for $p$-nitroaniline. It was shown that a eighty-fold exaltation of the static quadratic hyperpolarizability ($\beta_0$) could be achieved by *carbo*-merization of $p$-nitro-aniline, and that this exaltation is preserved when the *carbo*-merization is restricted to three essential domains: the acceptor ($NO_2 \Rightarrow NC_4O_2$), the acceptor-bridge bond ($O_2N$–phenylene $\Rightarrow$ $O_2N$–C≡C–phenylene) and the bridge ($p$-phenylene $\Rightarrow$ *carbo*-$p$-phenylene) [2]. *Carbo*-merization of the $p$-phenylene bridge is highly essential as evidenced by comparing the hyperpolarizability coefficients of **1a** and **1b**: $\beta_0$(**1b**) = 35.6 $\times$ $10^{-30}$ cm$^5$ esu$^{-1}$, and $\beta_0$(**1a**) = 281.9 $\times$ $10^{-30}$ cm$^5$ esu$^{-1}$ (Scheme 1). In this study, it was also shown that the *carbo*-benzene ring possesses unprecedented intrinsic donating properties among pure hydrocarbon groups.

A further exploration of the *carbo*-merization approach is here related under the following headings.

2. Computational details

3. Results

3.1. *Carbo*-meric effects of conjugated hydrocarbon bridges on quadratic hyperpolarizabililty

## 2. Computational details

Geometries of the chromophores were extracted from the Cambridge Data Base (CDB) when mentioned or fully optimized, in the singlet state, at the at the B3PW91/6-31G* level using Gaussian98 [5]. Vibrational analysis was performed at the same level in order to check the obtention of a minimum on the potential energy surface. Some representative calculated structures are depicted in Fig. 1(a–c).

The all-valence INDO (intermediate neglect of differential overlap) method [6], in connection with the sum-over-state (SOS) formalism [7] was employed for the calculation of the molecular hyperpolarizabilities. Details of the computationally efficient INDO-SOS-based method for describing second-order molecular optical nonlinearities have been reported elsewhere [8]. Calculations were performed using the INDO/1 Hamiltonian incorporated in the commercially available MSI software package ZINDO [9]. The monoexcited configuration interaction (MECI) approximation was employed to describe the excited states. The 500 lowest energy one-electron transitions between the 23 highest occupied molecular orbitals and the 23 lowest unoccupied ones were chosen to undergo CI mixing. It was checked to be sufficient for reaching convergence of $\beta$.

## 3. Results

### *3.1. Carbo-meric effects of conjugated hydrocarbon bridges on quadratic hyperpolarizabilty*

#### *3.1.1. Selection of phenolate/N-methylpyridinium as a donor/acceptor couple*

As reminded in the introduction, *carbo*-merization of both the phenylene unit and the nitro group are essential for a spectacular exaltation of the static hyperpolarizability of *p*-nitroaniline [2]. Nevertheless, it was shown that whereas the experimental synthesis of the *carbo*-nitro group (-$NC_4O_2$) is highly challenging, its stability under standard conditions is questionable. Indeed, high fluxionality of the putative $HNC_4O_2$ molecule could be inferred from various calculation levels (DFT, QCISD, CASSCF) [2]. In contrast, based on recent synthetic results, *carbo*-merization of the benzene bridge is experimentally feasible (*vide infra*). Since all known *carbo*-benzene derivatives are substituted by aromatic groups ($C_6H_5$, *p-tert*-Bu-$C_6H_4$, ...), both the $NH_2$ and $NO_2$ groups of *p*-nitroaniline are here replaced for aromatic donor and acceptor groups respectively. Owing to their well established NLO efficiency, N-methylpyridinium and phenolate are selected as respective trial groups.

The *carbo*-merization of other conjugated bridges than *p*-phenylene is also considered (Scheme 2).

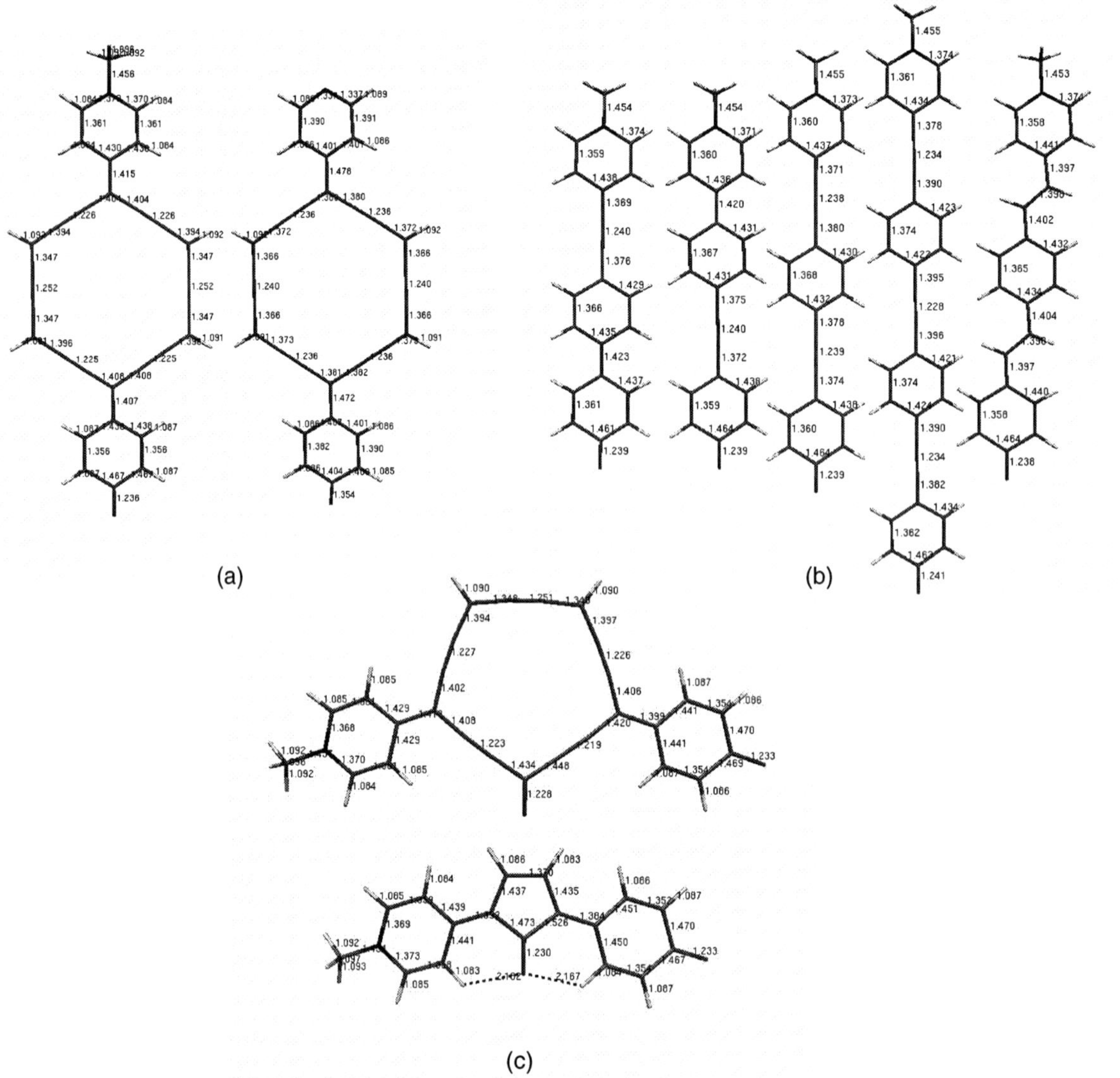


Fig. 1. Fully optimized structures at the B3PW91/6-31G* level of: (a) isomeric methyl derivatives of *p*-(4-hydroxyphenyl)-(4-pyridyl)*carbo*-benzene **2a** and **3a**. Interatomic distances are in Å; (b) various phenolate/N-methylpyridinium chromophores containing arylethynyl or arylethenyl units in their bridges. From left to right: **4b**, **4c**, **4d**, **4e**, **4f**; (c) of 2-(4-oxyphenyl)-5-(N-methyl-4-pyridyl)cyclopentadienone **7a** and its bridge-*carbo*-mer **7b**. Interatomic distances are in Å.

### 3.1.2. *p-Phenylene bridge*

Gas-phase geometry and hyperpolarizability of **2a** and of its parent molecule **2b** (Scheme 3) were computed and compared with those of their respective non-zwitterionic isomers **3a** and **3b** (Table 1). In the case of the parent *p*-nitroaniline, it was shown that the bridge – acceptor bond was an essential *carbo*-merization domain for hyperpolarizability exaltation [2]. Local *carbo*-merization (ethynylogation) of the methylpyridinium – phenylene (resp. *carbo*-phenylene) bond of **2b** (resp. **2a**) was therefore considered in **4b** (resp. **4a**) (Scheme 3).

Table 1
Computational data on electronic properties of "strong" push-pull chromophores (see text)

| Compound | $\beta$ | $\beta_{2level}$ | Dominant transition | $\lambda$ (nm) | $f$ | $\Delta\mu$ (D) | Composition |
|---|---|---|---|---|---|---|---|
| **1a** | 281.9 | 560.4 | $1 \rightarrow 4$ | 685 | 0.94 | 19.5 | 0.943 HOMO → LUMO |
| **1b** | 35.6 | 74.5 | $1 \rightarrow 6$ | 454 | 0.6 | 15.2 | 0.873 HOMO-1 → LUMO |
| **2a** | 131.2 | 387.6* | $1 \rightarrow 2$ | 587 | 3.29 | 7.9 | 0.961 HOMO → LUMO |
| **2b** | 304.3 | 362.5 | $1 \rightarrow 2$ | 589 | 2.61 | 8.1 | 0.961 HOMO → LUMO |
| **3a** | 24.4 | 77.8* | $1 \rightarrow 7$ | 375 | 4.06 | 3.5 | 0.764 HOMO-1 → LUMO+1 |
| **3b** | 4.0 | 13.0* | $1 \rightarrow 4$ | 279 | 1.29 | 5.17 | 0.939 HOMO → LUMO |
| **4a** | 61.4 | 401.6* | $1 \rightarrow 2$ | 646 | 4.02 | 5.5 | 0.957 HOMO → LUMO |
| **4b** | 824.7 | 1026.0 | $1 \rightarrow 2$ | 661 | 3.36 | 18.6 | 0.956 HOMO → LUMO |
| **4c** | 767.9 | 923.9 | $1 \rightarrow 2$ | 658 | 3.40 | 11.5 | 0.958 HOMO → LUMO |
| **4d** | 1981.8 | 2526.3 | $1 \rightarrow 2$ | 744 | 4.04 | 18.5 | 0.957 HOMO → LUMO |
| **4e** | 33855.6 | 38767.0 | $1 \rightarrow 2$ | 1226 | 4.15 | 60.7 | 0.988 HOMO → LUMO |
| **4f** | 370.8 | 221.6 | $1 \rightarrow 2$ | 688 | 3.62 | 2.3 | 0.932 HOMO → LUMO |
| **5a** | 65.6 | 129.7 | $1 \rightarrow 2$ | 512 | 1.86 | 6.5 | 0.962 HOMO → LUMO |
| **5b** | 11.8 | 27.8* | $1 \rightarrow 2$ | 481 | 2.03 | 1.81 | 0.966 HOMO → LUMO |
| **7a** | 99.6 | 267.1* | $1 \rightarrow 2$ | 587 | 2.01 | 8.57 | 0.942 HOMO → LUMO |
| **7b** | 9.4 | 45.2* | $1 \rightarrow 2$ | 565 | 0.43 | 13.7 | 0.934 HOMO → LUMO |
| **8a** | 73.7 | 334.0* | $1 \rightarrow 2$ | 592 | 1.57 | 11.5 | 0.934 HOMO → LUMO |
| **8b** | 7.0 | 95.7* | $1 \rightarrow 2$ | 546 | 1.93 | 4.1 | 0.959 HOMO → LUMO |

*: $\beta_{2level}/\beta > 3$. $\beta_{2level}$ and $\beta_{3level}$ are large and opposite resulting in a small $\beta_{total}$. This suggests that the corresponding compound cannot be described by the two-level model.

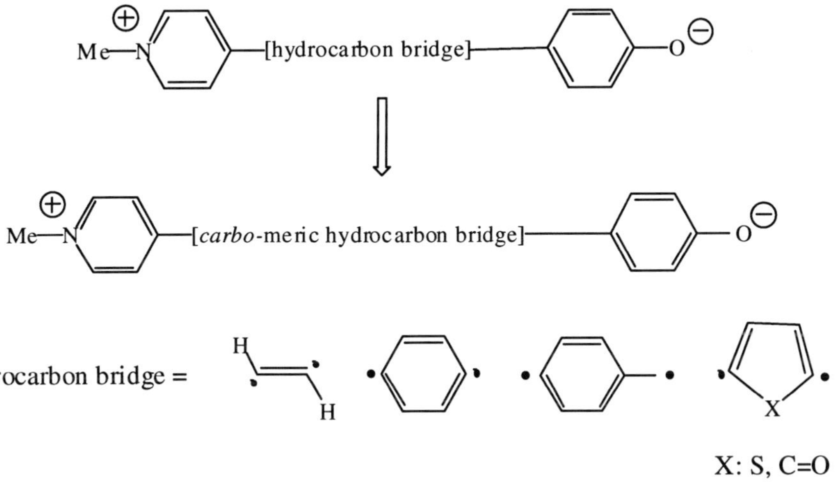


Scheme 2. *Carbo*-meric hydrocarbon NLO bridges considered in this work.

**2b** and **2a** possess quite high $\beta_0$ values, but rather surprisingly, the value of the parent molecule **2b** is twice higher ($\beta_0$ (**2b**) $= 304.3 \times 10^{-30}$ cm$^5$ esu$^{-1}$) than the value of its bridge *carbo*-mer **2a** ($\beta_0$(**2a**) $= 131.2 \times 10^{-30}$ cm$^5$ esu$^{-1}$). The calculated hyperpolarizability of **2b** for an external field wavelength of 1.17 $\mu$m ($2145 \times 10^{-30}$ cm$^5$ esu$^{-1}$) has the same order of magnitude as the experimental value at the external field wave length of 1.064 $\mu$m reported by Feng et al. ($1664 \times 10^{-30}$ cm$^5$ esu$^{-1}$) [10].

As expected, the hyperpolarizability of both the non-zwitterionic isomers **3b** and **3a** is quite low, but the value for the bridge *carbo*-mer ($\beta_0$ (**3a**) $= 23.8 \times 10^{-30}$ cm$^5$ esu$^{-1}$) is five times higher than the value for the parent molecule ($\beta_0$(**3b**) $= 4.0 \times 10^{-30}$ cm$^5$ esu$^{-1}$).

The opposite *carbo*-meric effects observed for **2b** => **2a** and **3b** => **3a**, are related with a simple polarity feature: whereas the ground state dipole moment of **2a** points from the nitrogen atom towards the

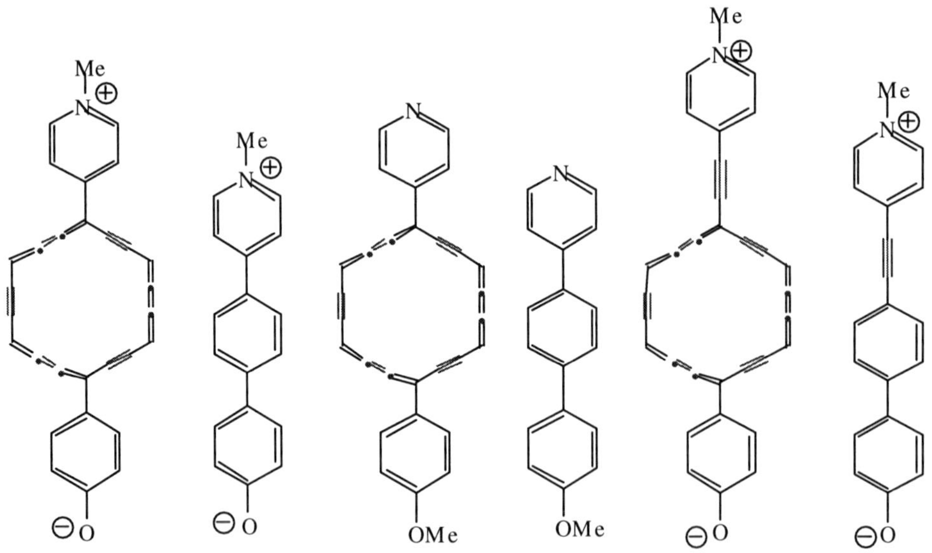


**2a** **2b** **3a** **3b** **4a** **4b**

Scheme 3. *Carbo*-meric NLO chromophores and their parent structures.

oxygen atom, the opposite holds in its non-zwitterionic isomer **3a**. A positive bridge *carbo*-merization effect might thus apply to the formally apolar push-pull chromophores ($p$-nitro-aniline, **1b**, **3b**) but not to the zwitterionic ones (**2b**). This trend is even more striking for the phenylethynylene-bridged molecule **4b**: $\beta_0$(**4b**) $= 824.7 \times 10^{-30}$ cm$^5$ esu$^{-1}$ *vs* $\beta_0$(**4a**) $= 61.4 \times 10^{-30}$ cm$^5$ esu$^{-1}$.

The surprisingly high $\beta_0$ value of the simple diarylacetylene chromophore **4b** deserves a peculiar attention. **4c** is an isomer of **4b**, where the triple bond is now inserted into the donor – bridge bond of **2b**. The static hyperpolarizability of **4c** is quite high as well, and comparable with that of **4b**: $\beta_0$(**4c**) $= 733.7 \times 10^{-30}$ cm$^5$ esu$^{-1}$ (Table 1). These unexpected results prompted us to study the higher phenylethynylene homologs **4d** and **4e** (Scheme 4, Table 1). The effect of the $C_2$ insertion into both the donor – bridge and the acceptor – bridge bonds of **2b** turns out to be additive in nature, resulting in a drastic hyperpolarizability exaltation: $\beta_0$(**4d**) $= 1982.3 \times 10^{-30}$ cm$^5$ esu$^{-1}$. The exaltation becomes dramatic for **4e**: $\beta_0$(**4e**) $= 33855.6 \times 10^{-30}$ cm$^5$ esu$^{-1}$! To the best of our knowledge, such an outstanding calculated hyperpolarizability value for such a low molecular weight (MW(**4e**) = 409.5) is unprecedented. Moreover, to the best of our knowledge, the synthesis of such simple compounds (**4b-4e**) has even not been reported.

Diarylacetylene chromophores were previously claimed to be about 40% less efficient than their *trans*-stilbene counterparts in terms of quadratic hyperpolarizability [11]. The opposite trend is here exemplified: the bis(arylethynyl)benzene chromophore **4d** is five times more efficient than its bis(arylethenyl)benzene congener **4f** of similar molecular weight (MW(**4d**) $= 309.4 \approx$ MW(**4f**) $=$ 313.4). This result is to be compared with a result reported by Lecours et al. showing that the hyperpolarizabilitiy of push-pull bis(arylethynyl)porphyrins 20 times greater than that of bis(arylethenyl) congeners [12].

### *3.1.3. Ethylene bridge*

Push-pull stilbene-based chromophores are among the most efficient ones. Merocyanine **5b** (Scheme 5) in its gas-phase optimized geometry exhibits a moderate $\beta_0$ value ($\beta_0$(**5b**) $= 11.8 \times 10^{-30}$ cm$^5$

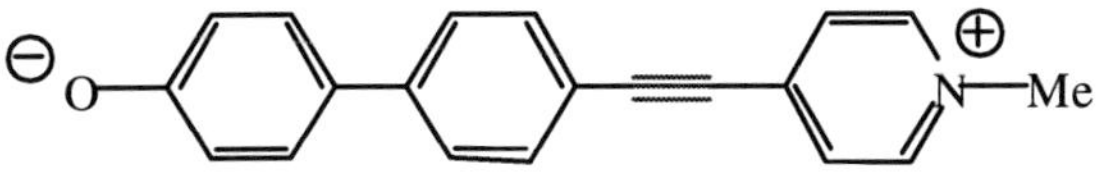
**4b**

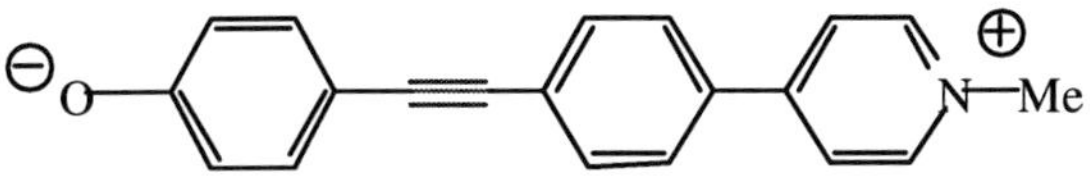
**4c**

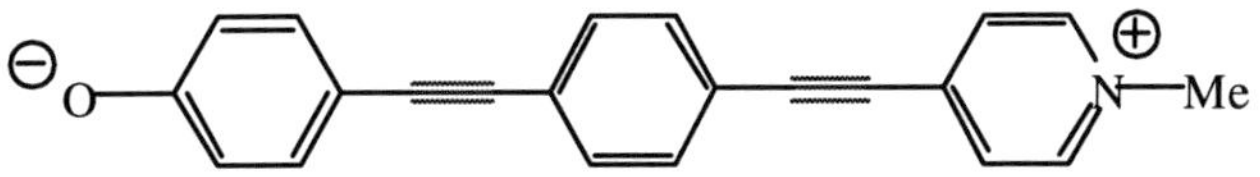
**4d**

**4e**

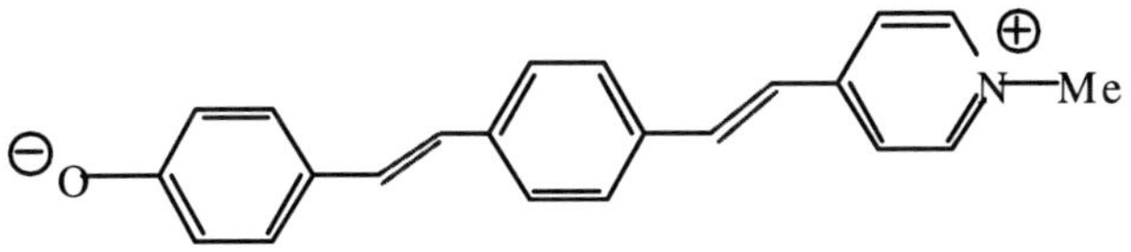
**4f**

Scheme 4. Representative "zwitterionic" highly efficient NLO chromophores.

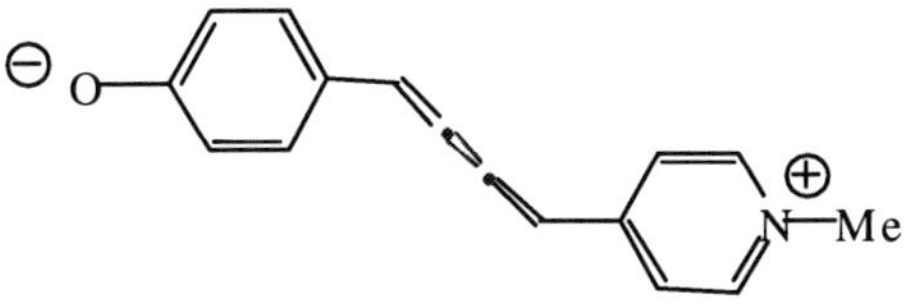
**5a**

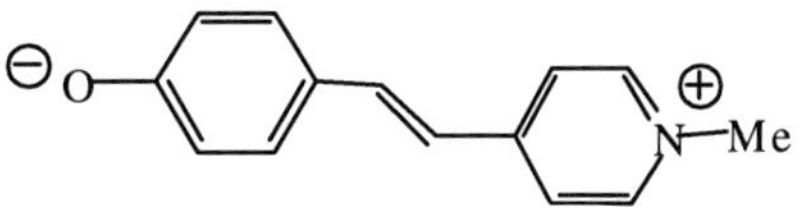
**5b**

Scheme 5. *Trans*-stilbenic chromophore and its bridge *carbo*-mer.

esu$^{-1}$). *Carbo*-merization of the ethylenic bridge of **5b** leads to the *trans*-butatriene bridge of **5a**, and results in a six-fold exaltation of the hyperpolarizability ($\beta_0$(**5a**) $= 65.6 \times 10^{-30}$ cm$^5$ esu$^{-1}$). It is worth noting here that in its crystal state geometry, the parent compound (**5b'**) experiences a dramatic twenty-fold enhancement of its putative gas-phase hyperpolarizability $\beta_0$(**5b'**) $= 210.0 \times 10^{-30}$ cm$^5$ esu$^{-1}$. As already pointed out by Gorman et al. [13], charge separation is strongly disfavored in the gas phase as there is no solvent or crystal field to stabilize it. Thus, ignoring the influence of the medium polarity on the chemical structure of merocyanines will tend to underestimate their calculated gas phase hyperpolarizability.

### *3.1.4. Five-membered "aromatic" ring bridges*

Just as benzene and anionic cyclopentadienyl rings are aromatic, *carbo*-benzene and cationic *carbo*-cyclopentadienyl rings were shown to be definitely aromatic in either structural, magnetic or energetic sense [14]. Assuming that the (anti-)aromatic character of the bridge plays a crucial role for the NLO efficiency of push-pull chromophores, the cyclopentadienyl bridge can be naturally considered. Nevertheless, the ionic nature of this bridge makes difficult a straightforward comparison with compounds containing neutral benzene or *carbo*-benzene rings in their bridges. Neutral cyclopentadienyl and *carbo*-cyclopentadienyl derivatives can exhibit zwitterionic, but still neutral, resonance forms where the rings are locally equivalent to the parent ionic aromatic rings. The cyclopentadienone ring **6b** is considered

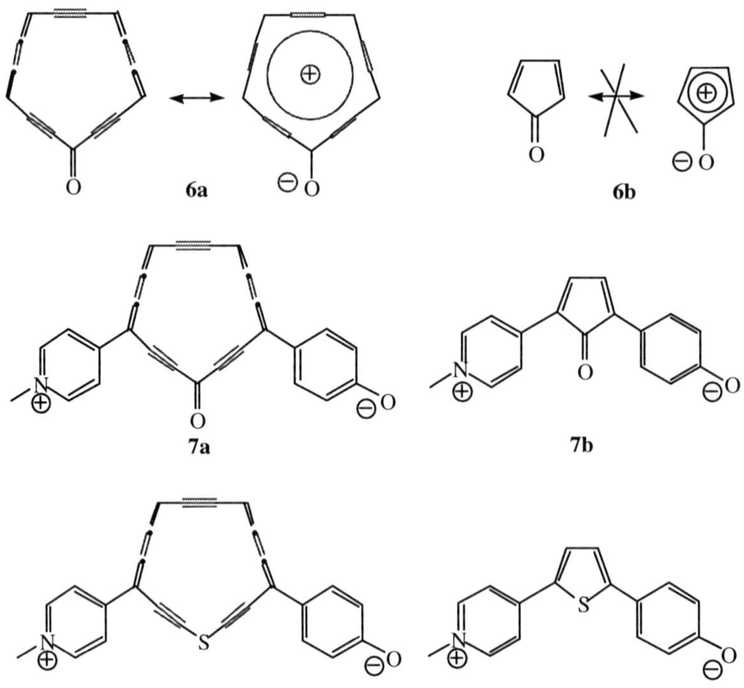


Scheme 6. Chromophores derived from cyclopentadienone (see Fig. 1(c)) and thiophene, and their bridge-*carbo*-mers.

here (Scheme 6). The natural polar resonance form of the parent compound **6b** posseses 4 $\pi_z$ electrons in the ring and is anti-aromatic according to the Hückel rule. By contrast, the natural polar resonance form of the ring *carbo*-mer **6a** possesses 14 $\pi_z$ electrons in the ring and is aromatic according to the Hückel rule. The phenolate and methylpyridinium termini are attached at the positions 2 (resp. 4) and 5 (resp. 13) of cyclopentadienone **6b** (resp. *carbo*-cyclopentadienone **6a**), thus defining the chromophore **7b** (resp. **7a**). The hyperpolarizability values of **7b** and **7a** should partly reflect the effect of anti-aromaticity and aromaticity, respectively.

As thiophene is recognized as a relevant (poorly aromatic) monomeric unit in organic conductors, this five-membered heterocycle deserves to be compared with cyclopentadienone. Moreover, the thiophene ring has been recognized to act as better bridge than benzene in known NLO chromophores [15]. The thiophene and *carbo*-thiophene derivatives **8a-b** are thus also considered here.

Since cyclopentadienone **6b** and its ring *carbo*-mer **6a** are themselves dipolar, their quadratic hyperpolarizability coefficient has been first computed. The low values found ($\beta_0$(**6b**) $= 0.61 \times 10^{-30}$ cm$^5$ esu$^{-1}$, $\beta_0$(**6a**) $= 9.8 \times 10^{-30}$ cm$^5$ esu$^{-1}$) show that their ring could indeed act as spectator bridges between positions 2 and 5 in **7b** or between positions 4 and 13 in **7a**, respectively.

The *carbo*-meric structures **7a** and **8a** exhibit similar but moderate hyperpolarizabilities (99.6 and 73.7 $\times 10^{-30}$ cm$^5$ esu$^{-1}$, respectively). A ten-fold positive *carbo*-meric effect is however observed in both cases ($\beta_0$(**7b**) $= 9.4 \times 10^{-30}$ cm$^5$ esu$^{-1}$, $\beta_0$(**8b**) $= 7.0 \times 10^{-30}$ cm$^5$ esu$^{-1}$). This positive effect in the cyclopentadienyl and thiophene series is opposite to that occuring in the benzene series (see Section 2.1.2). However, while aromatic character is preserved upon *carbo*-merization of benzene, it is reversed upon *carbo*-merization of cyclopentadienone (see above). Since ground-state aromaticity is not a driving force for NLO efficiency, it seems that the aromatic character of the *carbo*-cyclopentadienyl unit is suppressed upon 4,13-substitution by the phenolate and pyridinium groups. Thus, the fifteen-membered

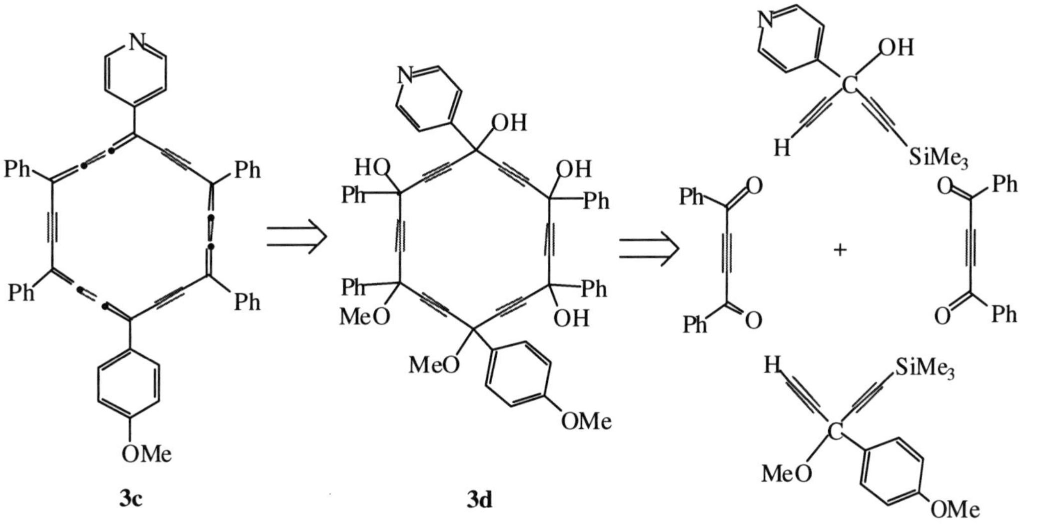


Scheme 7. Principle of the experimental synthesis of the *carbo*-benzene chromophore **3c**.

rings of **7a** and **8a** would not be significantly cyclically delocalized, and just act as rigid pseudo-linear $C_{10}$ conjugated bridges. The $\beta_0$ exaltation upon *carbo*-merization of **7b** and **8b** would therefore merely correspond to a classical lengthening effect.

### *3.2. Synthetic feasibility*

We recently undertook the challenging experimental synthesis of the *carbo*-meric chromophore unit occuring in **3a** and **2a**. To date, only few *carbo*-benzene derivatives have been reported. All of them contain a quasi $D_{6h}$-symmetric $C_{18}$ ring substituted by aryl groups such as Ph, *p-tert*-Bu-$C_6H_4$ [16], and the unsubstituted *carbo*-benzene $C_{18}H_6$ still remains as a putative species [17]. In order to minimize the number of synthetic steps, experimental efforts hitherto focussed on highly symmetrical derivatives ($D_{6h}$, $D_{3h}$). Owing to the difficulty of introducing hydrogen atoms at the periphery of the $C_{18}$ ring, the tetraphenyl derivative **3c** (resp. **2c**) of **3a** (resp. **2a**) was devised as a reasonable target (Scheme 7). The envisioned strategy is based on the desymmetrization of an optimized approach which allowed us to prepare hexaphenyl-*carbo*-benzene $C_{18}Ph_6$ in 9 steps, and the first $C_{2v}$-symmetric tetraphenyl-*carbo*-benzene $C_{18}Ph_4H_2$ in 14 steps [18].

The first steps here consist in sequential alkynyl-propargyl coupling reactions [19] leading to *p*-anisyl- and 4-pyridyl- dialkynylcarbinol intermediates (Scheme 7). Due to the presence of a somewhat electrophilic pyridine substituent, some difficulties were encountered at the cyclizing step leading to the highly functional hexaoxy-[6]pericyclyne **3d** [20]. Nevertheless, the use of non-nucleophilic bases for the double deprotonation of the pentayne precursor allowed for the isolation of **3d** which was characterized by NMR spectroscopy and mass spectrometry. Detailled procedures will be shortly reported [21].

Very recently, reductive aromatization of the hexaoxy[6]pericyclyne **3d** with $SnCl_2$/HCl afforded the first milligrams of a highly colored compound, whose spectroscopic analyses are consistent with the desired structure. Although no X-ray crystal structure could definitely confirm the structure to date, the $^1$H NMR spectrum displays deshielded signals at 9.2 and 9.4 ppm which can be assigned to *ortho*-protons of the aromatic substituents (phenyls, *p*-anisyl or 4-pyridyl). In the solid state, depending on the crystalization solvents, the compound appears as either small plates or thin hair-like black-violet

Table 2
Parameters used for the calculation of the Absolute Bond Length Alternation through Eq. (2a) (see text)

| A – B | o/o+1 | $R_{opt}$ | $\alpha^o$ |
|---|---|---|---|
| $Csp^2$⋯$Csp^2$ | 1/2 | 1.387 | 1.0000 |
| $Csp^2$⋯$Nsp^2$ | 1/2 | 1.334 | 1.0397 |
| $Csp^2$⋯O | 1/2 | 1.265 | 1.0964 |
| $Csp$⋯$Csp^2$ | 1/2 | 1.345[(a)] | 1.0312 |
| $Csp$⋯$Csp$ | 2/3 | 1.240[(b)] | 1.1186 |

[(a)] Determined from the Harmonic Oscillator Model from standard bond lengths with assumed integral bond order: d(HCC–CH:$CH_2$) = 1.434 Å, d($H_2$C:C=CH·$CH_3$) = 1.301 Å, [(b)] Determined from the Harmonic Oscillator Model from standard bond lengths with assumed integral bond order: d($H_2$C:C=C:$CH_2$) = 1.28 Å, and d($H_3$C·C≡C·$CH_3$) = 1.214 Å.

crystals. Although poorly soluble, they afford intense red solutions in various solvents ($CHCl_3$, $CH_2Cl_2$, tetrahydrofurane, . . .). The UV-visible spectrum of a chloroform solution of **3c** displays an intense band at $\lambda_{max}$ = 476 nm. Conversion of **3c** to its zwitterionic isomer **2c** would allow for an experimental EFISH measure of the hyperpolarizabilty of the latter and for comparison with the calculated gas-phase value (see above).

These results just illustrate that the molecules studied are not merely theoretical in nature.

## 4. Discussion

### 4.1. Structural analysis. Generalized BLA coefficient

The purpose is here to bring out a possible correlation between the static hyperpolarizability coefficient $\beta_0$ as an *electronic* property and some relevant function of bond distances as a *structural* property.

Using a two-level model, the quadratic hyperpolarizability ($\beta$) of most donor-acceptor organic chromophores can be related to a single transition between a ground state and a single excited state having a charge transfer character. The ground state can be described in term of a linear combination of two resonance structures, one of which is formally non-polar, and the other is formally zwitterionic. As the donor and acceptor strength increases, so will the contribution of the zwitterionic resonance form to the ground state. It was shown that there is an optimal combination of donor/acceptor strengths for a given bridge that maximizes the NLO response [22]. $\beta$ will therefore be correlated with the degree of bond length alternation ($d^{single}$ – $d^{double}$), in the polyene-like bridge. Less polar chromophores containing potentially aromatic rings are described by a lower contribution of the zwitterionic form to the ground state, due to the energetic price associated with the loss of aromaticity. These molecules will therefore tend to be less bond-alternated, for a given set of donor and acceptor pair, than in a polyene of comparable length. As a major consequence, their hyperpolarizability will be reduced. Marder suggested that, contrary to push-pull polyenes for which there is an optimized bond length alternation of about 0.044 Å corresponding to a maximum $\beta$ value [13,23], aromatic molecules with sufficiently strong donors and acceptors have not been synthesized to maximize the NLO response [24].

The Bond Length Alternation parameter is currently defined for linear polymethyne bridges, namely containing $Csp^2$⋯$Csp^2$ bonds only. A more general definition is required for any $\pi$-systems, whatever is their topology (linear or branched) and the hybridization state ($sp^2$, $sp$) or the elemental nature (C, N, O, ...) of the constituting atoms. In particular, the generalized definition should apply to aromatic rings, the intrinsic structural aromaticity of which could be analyzed in terms of bond length alternation [25]. First, the "bond localization parameter" of a bond is defined as the ratio of its actual length $d_i$ to the optimal length $R_{opt}(i)$ of a virtual bond of the same type (identical natures and hybridization states of the constituting atoms) in a perfectly resonating structure. The average Absolute Bond Length Alternation (ABLA) is the mean absolute difference between the localization parameter of a bond and those of all the adjacent bonds, normalized to the optimal methyne-methyne bond length.

$$\text{ABLA} = \frac{1}{n}\sum_{i}\sum_{k_i}|\alpha^{\sigma(i)}d_i - \alpha^{\sigma(k_i)}\cdot d_{k_i})| \tag{2a}$$

where

- $i$ runs over all the bonds sustaining the $\pi$ system, and $k_i$ over the bonds adjacent to the ith bond ($1 \leqslant k \leqslant 1, 2, 3$ or 4). $n$ is the number of conjugated bonds.
- $o(i)$ is the minimum bond order of the ith bond ($o(i) = 1, 2$).
- $R_{opt}(i)$ is the Krygowskii's optimal bond length for the ith bond. $R_{opt}(i)$ corresponds to the half-integral bond order $o(i)+1/2$ occuring in a perfectly resonating structure. $R_{opt}$ values are taken from reference when available [26]. For the present purpose, $R_{opt}(Csp^2 \cdots Csp)$ and $R_{opt}(Csp \cdots Csp)$ are computed within the framework of the harmonic oscillator model from reference bond lengths (Table 2).
- $\alpha^{\sigma(i)} = \frac{R_{opt}(Csp^2...Csp^2)}{R_{opt}(i)}$ (= cste = 1 for polymethyne bridges, for which $o(i) = 1$)

In the case of a simple linear polymethyne chain with alternating short and long bond distances, the partial ABLA parameter (restricted to the bridge) is reduced to the absolute value of the classical BLA parameter Eq. (2b):

$$BLA = \frac{1}{n}\Sigma(d_i^{single} - d_{i+1}^{double}) \tag{2b}$$

The parameter values are listed in Table 2.

The ABLA values were computed for the homogeneous set of push-pull chromophores with pyridinium and phenolate termini. A surprising regular variation of $\beta_0$ *vs* ABLA is obtained: it is smooth at low and high ABLA, but becomes very steep as ABLA draws nearer to a critical value, ABLA$^\circ \approx 0.04$ Å. It is known that for a linear topology, not only $\mu\beta_0$ [27], but also $\Delta\mu$ [28], vary as the bridge length, N (number of bonds), between the donor and the acceptor groups. Smooth variations of $\beta_0$and $\mu\beta_0$ vs $|\Delta\mu|$ over restricted sets of chromophores were also reported [29]. As previously shown for phenylacetylene chromophores [30], $\beta_0$ could thus vary as $N^m$. Assuming proportionality, i.e. $m = 1$, the plot $\beta_0$/N *vs* ABLA here affords a still smoother variation. After several unsatisfactory trial analytical fits (polynomial, harmonic, gaussian), a squared lorentzian fit Eq. (3) was incidentally found to afford R = 0.9996 for the critical value ABLA$^\circ$ = 0.039528 Å Fig. 2. This value is close to the value of 0.04 Å found by Marder for phenolate/N-methylpyridinium chromophores containing linear polymethyne bridges [13].

$$\beta_0 = 1.4015.10^{-6}.N\ \text{ABLA}/\{(\text{ABLA} - 0.039528)^2 + 0.002226\}^2 \tag{3}$$

$$R = 0.9996$$

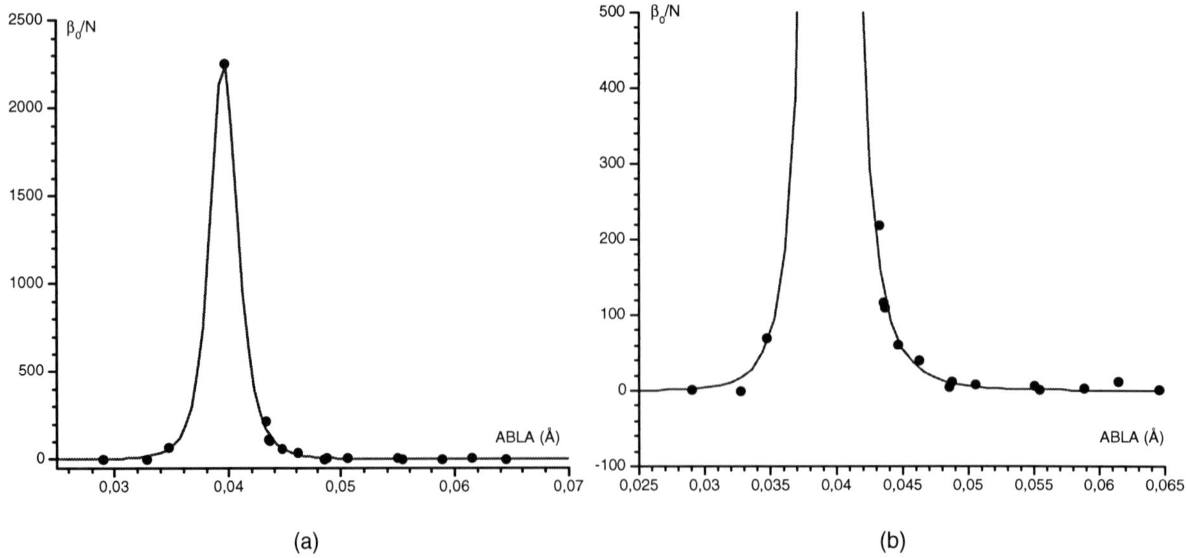


Fig. 2. Plots of $\beta_0$/N vs ABLA for merocyanine-like chromophores listed in Table III. For both scales **a** and **b**, the curve fit corresponds to (Eq. 3).

It must be cautiously emphasized that, according to both scales a) and b) of Fig. 2, the lorentzian correlation definitely fits much more than a single point above a scatter of points around a quasi uniform base line. The form of Eq. (3) can be heuristically derived within the framework of the two-level model. Assuming a Mulliken-Rieke expression of the oscillator strength ($f = \Delta E\, \mu_{gn}^2$), Eq. (1a) can be rewritten as Eq. 1(b):

$$\beta_0 = \frac{\mu_{gn}^2 \cdot \Delta\mu,}{(\Delta E)^2} \tag{1b}$$

where $\mu_{gn}$ is the dipole transition moment [31].

Applying Valence Bond theory of resonance to polyenic chromophores [32], Barzoukas et al. proposed a generalization of the BLA coefficient [33]. The process is here resumed for the simplest chromophore (Scheme 8). The ground state and excited state wave functions are expressed as linear combination of the wave functions of the orthogonal zwitterionic ($Z$) and non-zwitterionic ($N$) limiting resonance forms as indicated in Scheme 8.

Setting MIX $= -\cos\theta$, the authors showed that the terms of Eq. 1(b) vary as: $\mu_{gn}^2 = \mu_{CS}^2$ (1-MIX$^2$), $\Delta\mu = \mu_{CS}$ MIX and $(\Delta E)^2 = 4\, t^2/(1\text{-MIX}^2)$, where $\mu_{CS} = \mu_Z - \mu_N$ and $t = \langle\Psi_Z|H|\Psi_N\rangle > 0$. The final expression reads:

$$\beta_0 = 3\mu_{CS}^3/(32t^2)\ \text{MIX}\ (1 - \text{MIX}^2)^2$$

Since $\mu_{CS}$ and $t$ refer to pure limiting resonance forms, they mainly depend on the topology of the $\pi$ system bridging fixed donor and acceptor groups. On the basis of ref. [27–30] (see discussion above), we may assume here that they vary as N, the number of conjugated bonds bridging the donor ($PhO^-$) and the acceptor ($C_5H_4N^+Me$). Thus:

$$\beta_0/N = \text{K MIX}\ (1 - \text{MIX}^2)^2$$

Table 3
Calculated electronic and geometrical data for merocyanine-like chromophores correlated in Fig. 2 Eq. (3)

| | $\beta_0$ ($\times 10^{-30}$ cm$^5$ esu$^{-1}$) | $\mu^{(b)}$ (D) | ABLA (Å) | N |
|---|---|---|---|---|
| **2b** | 304.3 | 19.79 | 0.0446 | 5 |
| **4b** | 824.7 | 22.71 | 0.0435 | 7 |
| **3b** | 4.0 | 4.26 | 0.0327 | 5 |
| **4c** | 773.8 | 23.59 | 0.0436 | 7 |
| **4d** | 1982.3 | 26.53 | 0.0432 | 9 |
| **4f** | 370.9 | 23.89 | 0.0462 | 9 |
| **4e** | 33856.0 | 37.29 | 0.0396 | 15 |
| **5b** | 11.8 | 15.40 | 0.0588 | 3 |
| **5b'** (RX) | 210.9 | 19.00 | 0.0347 | 3 |
| **7b** | 9.4 | 14.62 | 0.0645 | 5$^{(a)}$ |
| **8b** | 7.0 | 17.68 | 0.0554 | 5$^{(a)}$ |
| **2a** | 131.2 | 23.89 | 0.0487 | 11 |
| **3a** | 24.2 | 6.53 | 0.0290 | 11 |
| **4a** | 61.4 | 27.70 | 0.0486 | 13 |
| **5a** | 65.6 | 16.27 | 0.0614 | 5 |
| **7a** | 99.6 | 19.83 | 0.0505 | 11$^{(a)}$ |
| **8a** | 73.7 | 18.28 | 0.0550 | 11$^{(a)}$ |

$^{(a)}$The $\pi$ system of the five-membered ring bridges was restricted to the five-bond path (resp. eleven-bond path) involved in the natural resonance relating the zwitterionic form with the apolar form of **7b–8b** (resp. **7a–8a**). $^{(b)}$Dipole moment of the ground state.

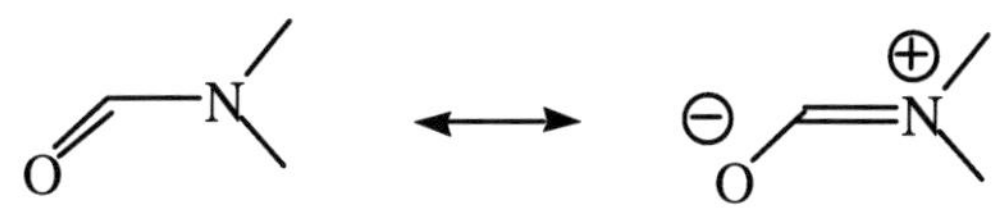

| | $\Psi_N$ | $\Psi_Z$ |
|---|---|---|
| ground state ($\Psi_g$): | $\cos(\theta/2)$ | $\sin(\theta/2)$ |
| excited state ($\Psi_e$) : | $-\sin(\theta/2)$ | $\cos(\theta/2)$ |

Scheme 8. Combination of non-zwitterionic ($N$) and zwitterionic ($Z$) limiting resonance form wave functions in the wavefunctions of the ground state and the excited state of a model cyanine.

For a given chromophore, the authors then assumed the proportionality MIX $= 0.11$ BLA, which results in a polynomial function $\beta_0$/N *vs* BLA. In the present case, however, no satisfactory polynomial fit of $\beta_0$/N *vs* ABLA can be found. The linear definition

$$\mathrm{MIX} = k\ \mathrm{ABLA},$$

is therefore not relevant for our rather broad family of different molecules. Instead, ABLA values are related to positive MIX values as solutions of a normalized polynomial equation:

$$\mathrm{MIX}\,(1 - \mathrm{MIX}^2)^2 = k^2 \frac{\mathrm{ABLA}}{[c^2 + (\mathrm{ABLA} - \mathrm{ABLA}^{\circ})^2]^2}$$

where $k$ and $c$ would depend on the donor and acceptor groups considered.

Table 4
Electronic properties of push-pull chromophores with nitrophenyl-based accepting groups (see text)

| Compound | $\beta$ | $\beta_{2level}$ | Dominant transition | $\lambda$ (nm) | f | $\Delta\mu$ | Composition |
|---|---|---|---|---|---|---|---|
| **9a** | 169.7 | 430.4* | $1 \rightarrow 4$ | 442 | 0.92 | 23.9 | 0.763 HOMO → LUMO+2 |
| **9b** | 27.3 | 101.9* | $1 \rightarrow 4$ | 349 | 1.04 | 19.3 | 0.692 HOMO → LUMO |
| **10** | 10.2 | 27.1 | $1 \rightarrow 4$ | 322 | 0.64 | 13.3 | 0.906 HOMO → LUMO |
| **11** | 35.5 | 133.1* | $1 \rightarrow 7$ | 384 | 1.47 | 11.1 | 0.665 HOMO → LUMO+2 |
| **12** | 81.5 | 313.0* | $1 \rightarrow 5$ | 426 | 1.50 | 21.5 | 0.621 HOMO → LUMO+2 |
| **13** | 192.2 | 266.0 | $1 \rightarrow 4$ | 687 | 1.7 | 5.0 | 0.928 HOMO → LUMO |
| **14** | 3.2 | 5.5 | $1 \rightarrow 5$ | 289 | 0.31 | 9.4 | 0.978 HOMO-1 → LUMO |
| **15** | 33.5 | 107.3* | $1 \rightarrow 7$ | 404 | 1.44 | 10.8 | 0.662 HOMO-1 → LUMO+2 |

*: $\beta_{2level}/\beta > 3$. $\beta_{2level}$ and $\beta_{3level}$ are large and opposite resulting in a small $\beta_{total}$. This suggests that the corresponding compounds cannot be described by the two-level model.

Physically, the ABLA numerator features the fact that at low ABLA and MIX values (→0), the leading term in Eq. (1b) is $\Delta\mu$ (namely MIX itself). The biquadratic denominator features the fact that at higher ABLA and MIX values, the leading term is $(\mu_{gn}/\Delta E)^2$ (namely $(1–\text{MIX}^2)^2$). Solving the above equation affords the (positive) MIX value, namely the mixing between the two limiting resonance forms $(N)$ and $(Z)$, from the geometrical ABLA parameter.

### 4.2. The p-carbo-phenyl group as a donor substituent for NLO chromophores

In Section 3.1, it was suggested that *carbo*-merization of a $p$-phenylene bridge results in either a lowering or a moderate exaltation of the quadratic hyperpolarizability depending on the formal polarity of the chromophore. New formally apolar chromophores with a nitro group as the accepting group are considered here (Schemes 9, 10). The static hyperpolarizability of molecules **9–15** in their fully optimized geometry at the B3PW91/6-31G* level are given in Table 4.

The 4-amino-4”-nitro terphenylene non-zwitterionic chromophore **9b** has been known for some time [34]. Its low quadratic hyperpolarizability coefficient measured in the solid state ($\beta_{1.91\mu m} = 11.0 \times 10^{-30}$ cm$^5$ esu$^{-1}$) is qualitatively confirmed by our calculation in the gas phase ($\beta_0$(**9b**) = 27.3 × $10^{-30}$ cm$^5$ esu$^{-1}$). It is ten times less efficient than its zwitterionic congener **2b**. Upon *carbo*-merization of the $p$-phenylene bridge of **9b**, a six-fold hyperpolarizability exaltation is predicted: $\beta_0$(**9a**) = 169.7 × $10^{-30}$ cm$^5$ esu$^{-1}$. This confirms that a positive effect of bridge *carbo*-merization is generally valid for formally apolar push-pull chromophores ($p$-nitroaniline, **1b**, **3b**). Nitrobenzene **14**, a formally apolar “chromophore” with an extremely weak donor-acceptor character ($\beta_0$(**14**) = 3.2 × $10^{-30}$ cm$^5$ esu$^{-1}$) gives additional support to this trend: the *nitro-carbo*-benzene **15** is ten times more efficient than **14**: $\beta_0$(**15**) = 33.5 × $10^{-30}$ cm$^5$ esu$^{-1}$.

In this case, however, the phenyl and *carbo*-phenyl groups should be regarded not as bridges, but as donors. Keeping in mind that investigations on the parent $p$-nitroaniline chromophore [2] showed that the unsubstituted *carbo*-phenyl group is extremely efficient as an intrinsic donor, its donor capabilities are examined below in a systematic fashion.

Let us consider $p$-nitrobiphenyl **10**. This structure contains both a phenyl donor and a $p$-phenylene bridge. *Carbo*-merization of either unit results in the same increase of molecular mass (MW(**11**) = MW(**12**) = 343). Nonetheless, as shown in Scheme 10, *carbo*-merization of the phenyl donor end of **10** results in a stronger hyperpolarizability exaltation ($\beta_0$(**12**)/$\beta_0$(**10**) = 8.0) than does *carbo*-merization of the $p$-phenylene bridge ($\beta_0$(**11**)/$\beta_0$(**10**) = 3.5). Moreover, it follows that *carbo*-merization of the remaining phenyl donor end of **11** results in a stronger hyperpolarizability exaltation ($\beta_0$(**13**)/$\beta_0$(**11**) = 5.4) than does *carbo*-merization of the remaining $p$-phenylene bridge of **12** ($\beta_0$(**13**)/$\beta_0$(**12**) = 2.4).

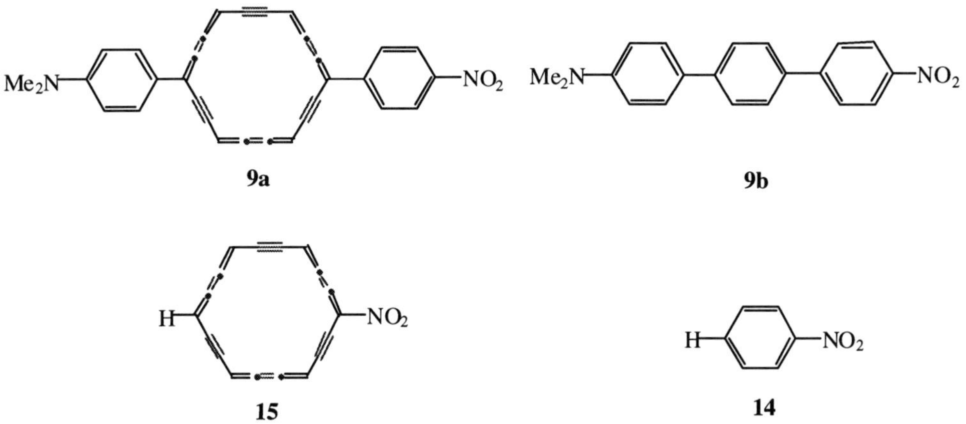


Scheme 9. Formally apolar nitrobenzene-derived aromatic chromophores.

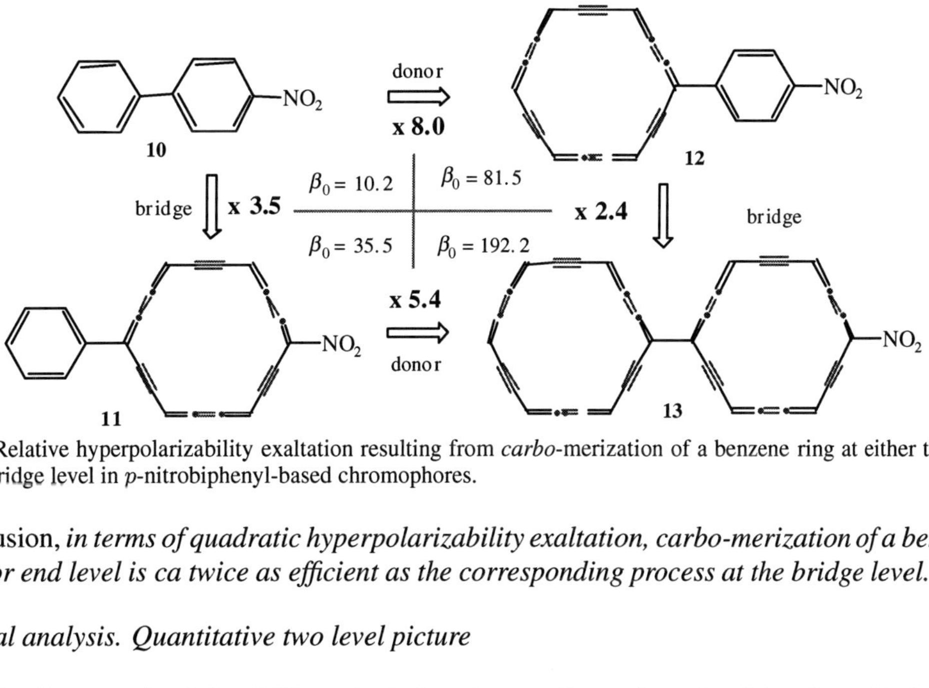


Scheme 10. Relative hyperpolarizability exaltation resulting from *carbo*-merization of a benzene ring at either the donor end level or the bridge level in *p*-nitrobiphenyl-based chromophores.

In conclusion, *in terms of quadratic hyperpolarizability exaltation, carbo-merization of a benzene ring at the donor end level is ca twice as efficient as the corresponding process at the bridge level.*

### *4.3. Orbital analysis. Quantitative two level picture*

Within the framework of the SOS perturbation theory, $\beta$ can be related to all excited states of a molecule, and can be partitioned into two contributions, so-called $\beta_{2level,}$ and $\beta_{3level,}$ according to the schematic relation Eq. (4).

$$\beta = -\frac{\pi^2}{2h^2}\sum_{n\neq g}\{function\ of\ r_{gn}\} + \sum_{n\neq g}\sum_{\substack{n'\neq g\\ n'\neq n}}\{function\ of\ r_{gn}\times r_{nn'}\times r_{n'g}\}$$

in which $r_{gn} = \langle\Psi_g|\mathbf{r}|\Psi_n\rangle$ is the orbital moment associated to a transition between electronic states $n$ and $g$. The first set of braces in Eq. (4) is the two-level contribution ($\beta_{2level}$)) and the second set is the three-level contribution ($\beta_{3level}$). Each two-level component of $\beta_{2level}$ involves two states, the ground state ($g$) plus one excited state ($n$), while $\beta_{3level}$ contains components in which the ground state and

Table 5
$\beta_{2level}$ transition multiplicity for molecules featuring a high contribution of the $\beta_{3level}$ component in their $\beta_0$ value (see text)

| Compound | $\beta_{2level}/\beta_{3level}$ | Weights (%) of the 5 Most Important Transitions | | | | |
|---|---|---|---|---|---|---|
| **2a** | 1.5 | 86.5 | 3 | 2 | 1 | 1 |
| **3a** | 1.5 | 62 | 7 | 3 | $< 1$ | $< 1$ |
| **3b** | 1.5 | 57 | 7 | 2 | 1 | 1 |
| **4a** | 1.2 | 85 | 3 | 1 | 1 | $< 1$ |
| **5b** | 0.8 | 75 | 2 | 2 | 2 | 1 |
| **7a** | 1.5 | 80 | 1 | 1 | $< 1$ | $< 1$ |
| **8a** | 1.3 | 78 | 9 | 6 | $< 1$ | $< 1$ |
| **7b** | 1.3 | 61 | 25 | 3 | 1 | $< 1$ |
| **8b** | 1.1 | 87 | 4 | 1 | 1 | $< 1$ |
| **9a** | 1.7 | 29 | 21 | 18 | 13 | 9 |
| **9b** | 1.4 | 59 | 17 | 5 | 3 | 3 |
| **11** | 1.4 | 52 | 31 | 5 | 4 | 3 |
| **12** | 1.3 | 49 | 22 | 13 | 10 | 2 |
| **15** | 1.5 | 59 | 33 | 6 | 3 | 1 |

two excited states ($n$ and $n'$) are involved. In most chromophores, $\beta_{2level}$ is related to a single (usually HOMO→LUMO based) $g \rightarrow n$ transition, having a strong charge transfer character ($r_{gn} \gg r_{gn'}$). Therefore, all the $r_{gn}$ x $r_{nn'}$ x $r_{n'g}$ terms are weak, and the three-level contribution is smaller than the two-level counterpart. Indeed, it is found that three-level contributions usually sum to roughly 50% as much as two-level contributions, and are of opposite sign. Based on this observation, a widely used assumption is that understanding $\beta_{2level}$ provides qualitative understanding of $\beta$. This model leads to an analysis of the NLO response ultimately related to the HOMO→LUMO transition, for most "push-pull" organic chromophores.

Computational data for the $\beta$ calculations are listed in Tables 1 and 4. At first sight, a striking feature is the large $\beta_{3level}$ value associated with several structures, for which the analysis of the NLO response in term of a single transition must therefore be qualified. From a general viewpoint, the reason for large $\beta_{3level}$ values is not systematically understood [35]. The $\beta_{2level}$ transition multiplicity of 14 compounds with unusual $\beta_{3level}$ contributions ($\beta_{2level}/\beta_{3level} \ll 2$) is detailed in Table 5.

The greatest multiplicity (4 relevant charge transfer transitions) occurs for chromophores **9a** and **12**. They both possess the same *carbo*-benzene-4-nitrophenyl skeleton, which seems to be responsible for this unusual feature. Molecules **2a**, **4a**, **5b**, **7a**, **8a**, and **8b**, are more traditional, in that sense that they possess only one transition with charge transfer character, as expected for any "push pull" chromophore. The presence of a large $\beta_{3level}$ is therefore completely unexpected. In Eq. (4), however, the excited state to excited state ($n \rightarrow n'$) contribution to the $\beta_{3level}$ term, which is difficult to rationalize, is likely involved. Unfortunately, direct information on these transitions is not currently available, and therefore, a chemically oriented description of this three-level expression is not immediately evident. Many computational investigations conducted in our laboratory have indicated that significant $\beta_{3level}$ contributions are occasionally observed for molecules, in which the strong donor-acceptor character is missing. Surprisingly, this is not the case here. However, the key parameter for the estimation of the NLO response remains $\beta$ and not $\beta_{2level}$ or $\beta_{3level}$.

For most push-pull molecules with pseudo-$C_{2v}$ symmetry, the charge transfers arising from the electronic transitions are directed along the molecular symmetry axis. $\beta$ therefore exhibits a strong 1-dimensional character resulting in a single nonzero tensor component $\beta_{zzz}$ along the dipolar $z$ axis. Nevertheless, the general expression of the vector component of $\beta$ along the dipolar $z$ axis reads:

$$\beta_z = \beta_{zzz} + 1/3[\beta_{zxx} + \beta_{xzx} + \beta_{xxz} + \beta_{zyy} + \beta_{yzy} + \beta_{yyz}]$$

which in turn, can be splitted as a $\beta_{z,2level}$ term plus a $\beta_{z,3level}$ term. Assuming a xz molecular plane, the perpendicular charge transfers (along the $y$ axis) are extremely small, so $\beta_{zyy}$, $\beta_{yzy}$ and $\beta_{yyz}$ are always negligible. For chromophores exhibiting an elongated conjugation pathway in the $x$ direction like those being under investigation in the present study, however, the off-diagonal term 1/3$[\beta_{zxx} + \beta_{xzx} + \beta_{xxz}]$ could significantly contribute to the total hyperpolarizability [36]. Nevertheless, it has to be reminded that even in the case of dipolar multidimensional chromophores, the optical nonlinearity has generally been described in term of a two-state model [37].

## 5. Conclusion

From the above exploration of *carbo*-meric effects on quadratic hyperpolarizabilty, one may bring out the five following results and prospects.

(i) The *carbo*-phenyl group ($C_{18}H_5$) definitely acts as an efficient hydrocarbon donor, at least with respect to the nitro ($NO_2$) acceptor group.

(ii) *Carbo*-merization of a bridging $p$-phenylene unit in push-pull chromophores may result in opposite effects on the $\beta_0$ value: an exaltation for formally apolar chromophores ($R_2N$, $p$-$C_6H_4OMe$, $C_6H_6$, $C_{18}H_6$/$NO_2$,4-$C_5H_5N$), and a lowering for merocyanine-like zwitterionic chromophores ($p$-$C_6H_4O^-$/4-$C_5H_5N^+$-Me).

(iii) The *carbo*-meric lowering effect on $\beta_0$ observed for merocyanine-like chromophores is specific to the gas-phase conditions of the calculations. A polar medium (crystal or solution state) is anticipated to result in a different behaviour or even revert the lowering effect, by acting at either the geometrical level or, more likely [38], at the electronic/orbital level. The experimental synthesis of $^-$O-$C_6H_5$-$C_{18}Ph_4$–$C_5H_4N^+$–Me thus remains a challenging task, which is in progress.

(iv) To date, the unusual large $\beta_{3level}$ component observed for many chromophores containing a *carbo*-aromatic unit does not receive a simple explanation. Beyond the need for a convincing analysis, one may more generally wonder what molecular property could specifically arise from a large $\beta_{3level}$. To the best of our knowledge, this remains a challenging issue for theorists.

(v) The correlation between the electronic hyperpolarizability property ($\beta_0$) and the geometrical bond length alternation property (ABLA) reveals a dramatic, but continuous, increase of $\beta_0$ when approaching the critical merocyanine ABLA° value (0.04 Å). A chemical way to reach this ABLA° value might be the insertion of phenylacetylene units in the bridge. The surprisingly simple molecule $^-$O-$C_6H_5$-(-C≡C–$C_6H_4$)$_2$–C≡C–$C_5H_4N^+$–Me, (MW(**4e**) = 409.5, ABLA(**4e**) = 0.03957 Å, $\beta_0$(**4e**) = 33856 × $10^{-30}$ $cm^5$ $esu^{-1}$), is likely very close to the optimal structure. It naturally becomes a target for future theoretical (transparency, sensitivity to solvent effects, ...) and experimental (synthesis...) studies. More generally, the relevance of the ABLA parameter deserves to be tested in other series (non-merocyanine-like) of push-pull chromophores.

### Aknowledgements

The authors would like to thank CALMIP (calcul intensif en Midi-Pyrénées, Toulouse – France) and IDRIS (Institut du Développement et des Ressources en Informatique Scientifique – Orsay – France) for computing facilities and Philippe Arnaud for computational assistance. R.C. also thanks the Ministère de l'Enseignement, de la Recherche et de la Technologie for an ACI funding.

## References


[1] D.J. Williams, Organic Polymeric and Non-Polymeric Materials with Large Optical Nonlinearities, *Angew. Chem. Int. Ed. Engl.* **23** (1984), 690–703.

[2] J.-M. Ducéré, C. Lepetit, P.G. Lacroix, J.-L. Heully and R. Chauvin, Quadratic Hyperpolarizability of Carbomeric Structures, *Chemistry of Materials* **14** (2002), 3332–3338.

[3] R. Chauvin, Carbomers, I. A General Concept of Expanded Molecules, *Tetrahedron Letters* **36** (1995), 397–400.

[4] For more or less recent examples, see: (a) E.M. Graham, V.M. Miskowski, J.W. Perry, D.R. Coulter, A.E. Stiegman, W.P. Schaefer and R.E. Marsh, Unusual Structural Distortions Induced by Charge-Transfer Interactions through Conjugated Molecules: Crystal Structures of $NH_2C_6H_4(C{\equiv}C)_nC_6H_4NO_2$ ($n = 0 - 3$), *Journal of the American Chemical Society* **111** (1989), 8771–8779; (b) A.E. Stiegman, E. Graham, K.J. Perry, L.R. Khundkar, L.-T. Cheng and J.W. Perry, The Electronic Structure and Second-Order Nonlinear Optical Properties of Donor-Acceptor Acetylenes: A Detailed Investigation of Structure-Property Relationships, *Journal of the American Chemical Society* **113** (1991), 7658–7666; (c) C. Dehu, F. Meyers and J.L. Brédas, Donor-Acceptor Diphenylacetylenes: Geometric Structure, Electronic Structure, and Second-Order Nonlinear Optical Properties, *Journal of the American Chemical Society* **115** (1993), 6198–6206; (d) B. Ma, H.M. Sulzbach, Y. Xie and H.F. Schaefer III., $\pi$ Electron Delocalization and Compression in Acyclic Acetylenic Precursors to Multidimensional Carbon Networks: Comparison with Experiment for the Recently Synthesized Tris (trimethylsilyl)-Substituted Tetraethynylmethane. Structures, Thermochemistry, Infrared Spectra, Polarizabilities, and Hyperpolarizabilities, *Journal of the American Chemical Society* **116** (1994), 3529–3538; (e) E.M. Maya, E.M. Garcia-Frutos, P. Vazquez, T. Torres, G. Martin, G. Rojo, F. Agullo-Lopez, R.H. Gonzalez-Jonte, V.R. Ferro, J.M. Garcia de la Vega, I. Ledoux and J. Zyss, Novel Push-Pull Phthalocyanines as Targets for Second-Order Nonlinear Applications, *Journal of Chemical Physics A* **107** (2003), 2110–2117.

[5] M.J. Frisch, G.W. Trucks, H.B. Schlegel, G.E Scuseria, M.A. Robb, J.R. Cheeseman, V.G. Zakrzewski, J.A. Montgomery Jr., R.E. Stratmann, J.C. Burant, S. Dapprich, J.M. Millam, A.D. Daniels, K.N. Kudin, M.C. Strain, O. Farkas, J. Tomasi, V. Barone, M. Cossi, R. Cammi, B. Mennucci, C. Pomelli, C. Adamo, S. Clifford, J. Ochterski, G.A. Petersson, P.Y. Ayala, Q. Cui, K. Morokuma, D.K. Malick, A.D. Rabuck, K. Raghavachari, J.B. Foresman, J. Cioslowski, J.V. Ortiz, A.G. Baboul, B.B. Stefanov, G. Liu, A. Liashenko, P. Piskorz, I. Komaromi, R. Gomperts, R.L. Martin, D.J. Fox, T. Keith, M.A. Al-Laham, C.Y. Peng, A. Nanayakkara, C. Gonzalez, M. Challacombe, P.M. W. Gill, B. Johnson, W. Chen, M.W. Wong, J.L. Andres, C. Gonzalez, M. Head-Gordon, E.S. Replogle and J.A. Pople, Gaussian 98, Revision A.7, Gaussian, Inc., Pittsburgh PA, 1998.

[6] (a) M.C. Zerner, G. Loew, R. Kirchner and U. Mueller-Westerhoff, An Intermediate Neglect of Differential Overlap Technique for Spectroscopy of Transition-Metal Complexes, Ferrocene, *Journal of the American Chemical Society* **102** (1980), 589; (b) W.P. Anderson, D. Edwards and M.C. Zerner, Calculated Spectra of Hydrated Ions of the First Transition -Metal Series, *Inorganic Chemistry* **25** (1986), 2728–2732.

[7] J.F. Ward, Calculation of Nonlinear Optical Susceptibilities Using Diagrammatic Perturbation Theory, *Reviews of Modern Physics* **37** (1965), 1–18.

[8] D.R. Kanis, M.A. Ratner and T.J. Marks, Design and Construction of Molecular Assemblies with Large Second Order Optical Nonlinearities, *Chemical Review* **94** (1994), 195–242.

[9] ZINDO, 96.0/4.0.0., Molecular Simulations Inc., 1996.

[10] J.K. Feng, X.Y. Sun, A.M. Ren, K.Q. Yu and C.C. Sun, Comparative Study of the Second-Order Optical Properties of Spiroconjugated Compounds versus the Corresponding Polyenes and Polyphenylenes, *Journal of Molecular Structure (Theochem)* **489** (1999), 247–254.

[11] L.T. Cheng, W. Tam, S. Stevenson, G.R. Mederith, G. Rikken and S.R. Marder, Experimental investigations of organic molecular nonlinear optical polarizabilities, 1 Methods and results on benzene and stilbene derivatives, *Journal of Physical Chemistry* **95** (1991), 10631–10643. y

[12] S.M. Lecours, H.-W. Guan, S.G. DiMagno, C.H. Wang and M.J. Therien, Push-Pull Arylethynyl Porphyrins: New Chromophores that Exhibit Large Molecular First-Order Hyperpolarizabilities, *Journal of the American Chemical Society* **118** (1996), 1797–1503.

[13] C.B. Gorman, SR. Marder, An investigation of the interrelationships between linear and non-linear polarizabilities and bond-length alternation in conjugated organic molecules, *Proceedings of the National Academy of Science USA* **90** (1993), 11297–11301.

[14] (a) C. Godard, C. Lepetit and R. Chauvin, DFT Exploration of Structural and Magnetic Properties of [N]Annulene Ring Carbomers, *Chemical Communications*, 2000, pp. 1833–1834. (b) C. Lepetit, C. Godard and R. Chauvin, Aromaticity and Homoaraomaticity of Annulene Ring Carbomers, *New Journal of Chemistry* **25** (2001), 572–580. (c) C. Lepetit, B. Sivi and R. Chauvin, ELF analysis of Out-of-Plane Aromaticity and In-Plane Homoaromaticity in Carbo[N]annulenes and [N]Pericyclynes, *Journal of Physical Chemistry A* **107** (2003), 464–473.

[15] See for example: E.M. Breitung, C.-F. Shu and R.J. MacMahon, Thiazole and Thiophene Analogues of Donor-Acceptors Stilbenes: Molecular Hyperpolarizabilities and Structure-Property Relationships, *Journal of the American Chemical Society* **122** (2000), 1154–1160.

[16] (a) Y. Kuwatani, N. Watanabe and I. Ueda, Synthesis of the First 3,6,9,12,15,18-Hexasubstituted-1,2,4,5,7,8,11,13, 14,16,17-Dodecadehydro[18]annulenes with $D_{6h}$ Symmetry, *Tetrahedron Letters* **36** (1995), 119–122. (b) R. Suzuki, H. Tsukuda, N. Watanabe, Y. Kuwatani and I. Ueda, Synthesis Structure and Properties of 3,9,15-Tri- and 3,6,9,12,15,18-Hexasubstituted Dodecadehydro [18] annulenes ($C_{18}H_{33}$ and $C_{18}R_6$) with $D_{6h}$ Symmetry, *Tetrahedron* **54** (1998), 2477–2496.

[17] R. Chauvin, Carbomers. II. En Route to [C,C]$_6$Carbobenzene, *Tetrahedron Letters* **36** (1995), 401–404.

[18] L. Maurette, C. Sui-Seng, M. Soleilhavoup and R. Chauvin, in: *Carbocarbocycles à Cinq et Six Chaînons*, L. Maurette, ed., Ph.D thesis, Université Paul Sabatier, sept. 27th, Toulouse, 2002.

[19] C. Tedeschi, C. Saccavini, L. Maurette, M. Soleilhavoup and R. Chauvin, 1,4-Diynes from Alkynyl-Propargyl Coupling Reactions, *Journal of Organometallic Chemistry* **670** (2003), 151–169.

[20] L. Maurette, C. Godard, S. Frau, C. Lepetit, M. Soleihavoup and R. Chauvin, On Ring Carbomers of Cyclobutane, Cyclopentane and Cyclodecane, and Cyclizing Bis(alkynyl-propargyl) Coupling, *Chemistry, a European Journal* (2001), 1165–1170.

[21] C. Saccavini, C. Tedeschi and R. Chauvin, unpublished results.

[22] (a) A. Dulcic and C. Sauteret, The Regularities Observed in the Second Order Hyperpolarizabilities of Variously Disubstituted Benzenes, *Journal of Chemical Physics* **69** (1978), 3453–3457; (b) S.R. Marder, C.B. Gorman, B.G. Tiemann and L-T. Cheng, Stronger Acceptors Can Diminish Nonlinear Optical Response in Simple Donor-Acceptor Polyenes, *Journal of the American Chemical Society* **115** (1993), 3006–3007.

[23] M. Blanchard-Desce, V. Alain, P.V. Bedworth, S.R. Marder, A. Fort, C. Runser, M. Barzoukas, S. Lebus and R. Wortmann, Large Quadratic Hyperpolarizabilities with Donor-Acceptor Polyenes Exhibiting Optimum Bond Length Alternation: Correlation between Structure and Hyperpolarizability, *Chemistry, a European Journal* **3** (1997), 1091–1104.

[24] B.G. Tiemann, L.T. Cheng and S.R. Marder, The Effect of Varying Ground-state Aromaticity on the First Molecular Electronic Hyperpolarizabilities of Organic Donor-Acceptor Molecules, Journal of the Chemical Society, *Chemical Communications* (1993), 735–737.

[25] (a) H. Choi and M. Kertesz, Do Localized Structures of [14]-and [18]Annulenes exist? *Journal of the American Chemical Society* **119** (1997), 11994–11995; (b) M.K. Cyranski and T.M. Krygowski, Two Sources of the Decrease of Aromaticity: Bond Length Alternation and Bond Elongation. Part I. An Analysis Based on Benzene Ring Deformations, *Tetrahedron* **55** (1999), 6205–6210.

[26] T.M. Krygowski and M.K. Cyranski, Structural Aspects of Aromaticity, *Chemical Review* **101** (2001), 1385–1419.

[27] (a) A. Dulcic, C. Flytzanis, C.L. Tang, D. Pépin, M. Fétizon and Y. Hoppilliard, Length Dependence of the Second-Order Optical Nonlinearity in Conjugated Hydrocarbons, *Journal of Chemical Physics* **74** (1981), 1559–1563; (b) M. Blanchard-Desce, R. Wortmann, S. Lebus, J.-M. Lehn and P. Krämer, Intramolecular Charge Transfer in Elongated Donor-Acceptor Conjugated Polyenes, *Chemical Physics Letters* **243** (1995), 526–532; (c) M. Blanchard-Desce, C. Runser, A. Fort, M. Barzoukas, J.-M. Lehn, V. Bloy and V. Alain, Large Hyperpolarizabilities with Donor-Acceptor Polyenes Functionalized with Strong Donors. Comparison with Donor-Acceptor Diphenylpolyenes, *Chemical Physics* **199** (1995), 253–261; (d) M. Blanchard-Desce, J.-M. Lehn, M. Barzoukas, C. Runser, A. Fort, G. Puccetti, I. Ledoux and J. Zyss, Functionalized Polyenes and Carotenoids with Enhanced Nonlinear Optical Responses, *Nonlinear Optics* **10** (1995), 23–36; (e) M. Barzoukas, C. Runser, A. Fort and M. Blanchard-Desce, A Two-State Description of (Hyper)polarizabilities of Push-Pull Molecules Based on a Two Form Model, *Chemical Physics Letters* **257** (1996), 531–537.

[28] M. Blanchard-Desce, V. Alain, L. Midrier, R. Wortmann, S. Lebus, C. Glania, P. Krämer, A. Fort, J. Muller and M. Barzoukas, Intramolecular Charge Transfer and Enhanced Quadratic Optical Non-Linearities in Push-Pull Polyenes, *Journal of Photochemistry and Photobiology A: Chemistry* **105** (1997), 115–121.

[29] (a) Y.-G.K. Shin, B.S. Brunschwig, C. Creutz and N. Sutin, Toward a Quantitative Understanding of Dipole Moment Changes in Charge-Transfer Transitions: Electroabsorption Spectroscopy of Transition-Metal Complexes, *Journal of the American Chemical Society* **117** (1995), 8668–8669; (b) G.U. Bublitz, R. Ortiz, C. Runser, A. Fort, M. Barzoukas, S.R. Marder and S.G. Boxer, Stark Spectroscopy of Donor-Acceptor Polyenes: Correlation with Nonlinear Optical Measurements, *Journal of the American Chemical Society* **119** (1997), 2311–2312.

[30] B.V.V. S.N. Prabhakara Rao, S.C. Mathur, D.P. Tewari and M. Banerjee, Theoretical Study on Naphthyl-Phenylacetylenes for Second Order Nonlinear Optical Applications, *Canadian Journal of Chemistry* **75** (1997), 1041–1046.

[31] B. Kirtman and B. Champagne, Nonlinear Optical Properties of Quasilinear Conjugated Oligomers, Polymers and Organic Molecules, *International Reviews in Physical Chemistry* **16** (1997), 389–420.

[32] P.C. Hiberty and G. Ohanessian, Valence-Bond Description of Conjugated Molecules. 3. The Through-Resonance Concept in Para-Substituted Nitrobenzenes, *Journal of the American Chemical Society* **106** (1984), 6963–6968.

[33] M. Barzoukas, A. Fort and M. Blanchard-Desce, Chain-Length Dependence of the Quadratic Hyperpolarizability in Two Series of Push-Pull Polyenes: a Two-Form Model, *New Journal of Chemistry* **21** (1997), 309–316.

[34] L.T. Cheng, W. Tam, S.H. Stevenson, G.R. Mederith, G. Rikken and S.R. Marder, Experimental Investigation of Organic Molecular Nonlinear Optical Polarizabilities. 1. Methods and Results on Benzene and Stilbene Derivatives, *Journal of Physical Chemistry* **95** (1991), 10631–10643.
[35] (a) D.R. Kanis, P.G. Lacroix, M.A. Ratner and T.J. Marks, Electronic Structure and Quadratic Hyperpolarizabilities in Organotransition-Metal Chromophores Having Weakly Coupled p-Networks. Unusual Mechanisms for Second-Order Response, *Journal of the American Chemical Society* **116** (1994), 10089–10102; (b) D.R. Kanis, M.A. Ratner and T.J. Marks, Calculation and Electronic Description of Quadratic Hyperpolarizabilities. Toward a Molecular Understanding of NLO Responses in Organotransition Metal Chromophores, *Journal of the American Chemical Society* **114** (1992), 10338–10357.
[36] S. Di Bella and I. Fragala, Two-Dimensional Characteristics of the Second-Order Nonlinear Optical Response in Dipolar Donor-Acceptor Coordination Complexes, *New J. Chem.* **26** (2002), 285–290.
[37] S. Di Bella, I. Fragala, I. Ledoux, M.A. Diaz-Garcia and T.J. Marks, Synthesis, Characterization, Optical Spectroscopy, Electronic Structure and Second-Order Nonlinear Optical (NLO) Properties of a Novel Class of Donor-Acceptor Bis(salicylaldiminate)nickel Schiff Base NLO Chromophores, *J. Am. Chem. Soc.* **119** (1997), 9550–9557.
[38] D.L. Albert, T.J. Marks and M.A. Ratner, Rational Design of Molecules with Large Hyperpolarizabilities. Electric Field, Solvent Polarity, and Bond Length Alternation Effects on Merocyanine Dye Linear and Nonlinear Optical Properties, *Journal of Physical Chemistry* **100** (1996), 9714–9725.

Journal of Computational Methods in Sciences and Engineering 4 (2004) 589–596
IOS Press


# On the theoretical calculation of the static dipole polarizabilities of $Li_n$, $Cu_n$ ($n \leqslant 8$) and $Li_nCu_m$ ($n, m \leqslant 4$)

Patricio Fuentealba[a,*], Luis Padilla-Campos[b] and Orfa Reyes[a]
[a]*Departamento de Física, Facultad de Ciencias, Universidad de Chile, Casilla 653, Santiago, Chile*
[b]*Departamento de Quimica, Facultad de Ciencias Basicas, Universidad de Antofagasta, Casilla 170, Antofagasta, Chile*

**Abstract.** The static dipole polarizabilities of three series of clusters have been calculated and compared. They are the bare lithium clusters, the copper clusters and the mixed lithium – copper clusters. The calculations have been done solving the Kohn – Sham equations with an hybrid exchange – correlation functional. Some discrepancies with the experimental values for lithium clusters have been discussed. It has been found that the dipole polarizabilities of the copper cluters are smaller than the ones of lithium clusters, and the mixed ones follow close the values found for the copper clusters.

## 1. Introduction

The study of small metal clusters is an active area of research. Their properties may show both similarities and differences to their bulk counterparts. Clusters containing from a few to several thousands of atoms have been investigated in order to study the evolution of properties of bulk matter from properties of atoms [1]. In general, for metallic clusters, the chemical bonding is profoundly changed in the formation of a cluster whereas the solid – state-like properties like the dielectric constant are of no significance in a cluster of few atoms. The dipole polarizability, however, represents a suitable molecular property that can take on the role of the dielectric constant and can be measured without touching the cluster [2]. The dipole polarizability comprises the information of the bonding character in the cluster, and on the other hand, through the Classius – Mosotti relation it yields information about the dielectric constant of the solid – like macroscopic particle formed with the cluster. The experimental measurement of cluster polarizabilities is relatively easy in comparison to other properties of neutral clusters. In presence of an electric field the neutral clusters are deflected in linear order due to their dipole polarizabilities [3], allowing their measurement. Hence, the polarizabilities can be one of the most important piece of information about the nature of the bonding and the geometrical structure of the neutral clusters. To fulfill this goal it is necessary to have a reliable theoretical model to understand the principal factors influencing in the polarizabilities. The jellium model has been successfully applied to study the polarizability of large clusters approaching the bulk properties. For small clusters, however, one expects more bonding and geometrical effects than the incorporated in the jellium model. Therefore,

*Corresponding author. E-mail: pfuentea@uchile.cl.

ab initio calculations are necessary, and the experience gained in calculating the dipole polarizabilities of atoms and molecules is useful. The importance of polarizability data to rationalize experimental observations has been recently stressed [2], justifying the effort of looking for reliable theoretical models to calculate it and make valuable suggestions for further experiments.

It is also important to mention the relationship and importance of the dipole polarizability to chemical reactivity. It has been found that the dipole polarizability is proportional to the cubic power of the softness [4], a quantity defined in chemical reactivity to understand the propensity of a molecule to react. This fact paves the way to propose a "minimum polarizability principle" which states that the natural direction of evolution of any system is toward a state of minimum polarizability [5]. The range of validity of this principle has been investigated in many works [6]. Hence, the dipole polarizability could be one important cluster property to understand the reactivity of the clusters, i.e. how they are formed and which are the most stable one. An example of this line of work can be found in [7].

In this paper, the static dipole polarizabilities of lithium and copper clusters have been calculated and discussed in comparison with the dipole polarizabilities of mixed copper – lithium clusters which have been also calculated. Copper metal atoms are in some sense similar to the alkali metal atoms. The bare copper clusters are similar in structure to the alkali metal atom clusters. The most stable structures are planar until the cluster of six atoms, the cluster of seven atoms is a pentagonal bipyramids and the one formed by eight atoms is a tetrahedron. It is known that in copper clusters the d – electron do not participate directly in the bonding. Although they are important to polarize the electron cloud. On the other hand, the copper metal atoms are more electronegative than the lithium atoms and they have also a bigger ionic radii, but the atomic radii is smaller. Therefore, one can expect similarities and differences in the structure and electronic properties of the mixed clusters in comparison to the copper metal clusters or the lithium metal clusters. In fact, the geometrical structures of the mixed clusters are similar to the ones of the bare metals clusters.

The bimetallic Li – Cu systems are important in order to understand the promotion effects of the alkali atoms on the copper surface. It is known that the catalytic properties of copper surfaces are dramatically changed in presence of alkali metal atoms [8]. The alkali metal atoms donate charge to the surface increasing the reactivity of electrophilic species on the surface. The exact mechanism is, however, not well understood. Therefore, the study of small $Li_nCu_m$ clusters is important in order to understand the electronic structure of these systems.

The theoretical study of electronic properties of clusters formed by a small amount of atoms have been markedly dominated by the density functional based methods, specially the Kohn – Sham method. However, contrary to the situation in the metals where the local approximation (LDA) performs well, for clusters it is necessary to incorporate corrections to the local models. The gradient corrected versions of the LDA have been widely applied to clusters confirming the reliability of the results for energetic properties of the ground state. For the calculation of the dipole polarizability the use of density functional based methods could be more questionable. It is known that the actual exchange – correlation functionals are not able to yield accurate results for van der Waals interactions [9]. Therefore, the same should occurr in regard of polarizability calculations. In fact, it has been demostrated that the exchange – correlation potential derived from the commonly used functionals do not have the correct asymptotic behaviour [10]. This has a strong influence in the calculated polarizabilities. However, while this is certainly the case for atoms, it seems that for a conglomerate of atoms the bonding structures are more important than the asymptotic behaviour of the potential [11]. Hence, we are in a situation where a property, the dipole polarizability, it is calculated with more accuracy for the more complicated systems (molecules and clusters) than it is for the simpler systems (atoms). Another important point to consider in the calculation of the dipole polarizability is the basis set dependence. It is well known that this property is highly sensitive to the quality of the basis set. Therefore, the use of diffuse functions is necessary.

## 2. Methodology and computational details

The electronic energy and properties have been calculated by solving the Kohn – Sham equations in an atomic basis set formed by gaussian functions. For the copper atoms the Stuttgart pseudopotential [12] has been used together with the corresponding basis set which has been denoted basis set A. Then an augmented basis set has been constructed adding one diffuse function of each symmetry. The exponents have been chosen to be the smallest one present in the 6-311++G basis set. It will be denoted basis set B. The pseudopotential replaces ten core electrons thus nineteen valence electrons are considered. For the lithium atoms the D95 basis set [13] (basis set A) has been used. As in copper atom a second basis set (basis set B) has been constructed on the same way. The calculations have been done using the B3LYP [14] exchange correlation functionals. All the geometries were fully optimized without symmetry restrictions. The started geometries have been changed three or four times in order to have some confidence that the final geometry is the one of lowest energy. The mean polarizability has been calculated from the diagonal components of the tensor as:

$$\langle\alpha\rangle = \frac{1}{3}(\alpha_{xx} + \alpha_{yy} + \alpha_{zz}) \tag{1}$$

and the polarizability anisotropy as:

$$|\Delta\alpha| = \sqrt{\frac{1}{2}[(\alpha_{xx} - \alpha_{yy})^2 + (\alpha_{xx} - \alpha_{zz})^2 + (\alpha_{yy} - \alpha_{zz})^2]} \tag{2}$$

The clusters are on its principal axes. All the calculations have been done using the GAUSSIAN 98 program [15].

## 3. Results and discussion

The dipole polarizability of the clusters $Li_n$, $Cu_n$ ($n \leqslant 8$) and $LinCu_m$ ($n.m \leqslant 4$) has been calculated at the optimized geometries. For the bare lithium and copper clusters the geometries are similar to the ones already informed in the literature [7,16,17]. The trimers are triangular in a $^2B_2$ state. The tetramers are rhombus in a $^1A_g$ state. For $n = 5$ and 6 the ground state geometries are planar of $C_{2v}$ and $D_{3h}$ symmetry, respectively. The ground state is $^2A_1$ and $^1A_1$, respectively. The heptamers are pentagonal bypiramidal in a $^2A_2$ state. For $n = 8$ the optimized geometries for the ground state result to be a tetrahedron which is the accepted structure for $Li_8$ but not for $Cu_8$ [7,16] where there is also a $C_{2v}$ structure of very similar energy. The optimized structures of the lithium – copper mixed clusters are similar. In general, the lithium atoms tend to break the copper – copper bonds forming the more polar lithium – copper bonds. The only exception is the $Cu_4Li_3$ cluster where it was impossible to find a pentagonal bypiramidal structure. A tetrahedral structure with a vacancy appears to be the ground state. They will be discussed in conjunction with other properties of these clusters elsewhere [18].

In Tables 1 and 2 the dipole polarizabilities of the lithium clusters have been tabulated. The mean polarizability increases slowly with the number of atoms and it is interesting to note the small increases in going from $Li_6$ to $Li_7$, although there is a geometrical change from a planar structure to a three dimensional one. The reason could be find in the compact structure of $Li_7$. The anisotropy shows a maximum in $Li_6$ going to zero for the highly symmetric $Li_8$. The values calculated with basis set A or B are very similar. The augmented basis set B presents slightly higher values but the differences are

Table 1
Static Dipole Polarizabilities (in a.u.) of Lithium clusters calculated with basis set A

| | $\alpha_{xx}$ | $\alpha_{yy}$ | $\alpha_{zz}$ | $\langle\alpha\rangle$ | $\lvert\Delta\alpha\rvert$ |
|---|---|---|---|---|---|
| Li | 140.3 | 140.3 | 140.3 | 140.3 | 0 |
| $Li_2$ | 161.1 | 161.1 | 264.5 | 195.6 | 103.4 |
| $Li_3$ | 190.9 | 499.9 | 264.7 | 318.5 | 279.5 |
| $Li_4$ | 221.2 | 277.2 | 563.5 | 354.0 | 318.0 |
| $Li_5$ | 250.4 | 629.2 | 474.3 | 451.3 | 329.9 |
| $Li_6$ | 287.2 | 639.9 | 640.0 | 522.4 | 352.8 |
| $Li_7$ | 568.4 | 568.4 | 452.0 | 529.6 | 116.4 |
| $Li_8$ | 603.5 | 603.5 | 603.5 | 603.5 | 0 |

Table 2
Static Dipole Polarizabilities (in a.u.) of Lithium clusters calculated with basis set B

| | $\alpha_{xx}$ | $\alpha_{yy}$ | $\alpha_{zz}$ | $\langle\alpha\rangle$ | $\lvert\Delta\alpha\rvert$ |
|---|---|---|---|---|---|
| Li | 142.3 | 142.3 | 142.3 | 142.3 | 0 |
| $Li_2$ | 165.7 | 165.7 | 266.2 | 199.2 | 100.5 |
| $Li_3$ | 193.3 | 500.4 | 266.7 | 320.1 | 277.8 |
| $Li_4$ | 223.3 | 279.4 | 563.7 | 355.5 | 316.1 |
| $Li_5$ | 254.1 | 625.5 | 476.5 | 452.0 | 323.7 |
| $Li_6$ | 293.6 | 642.3 | 642.0 | 526.0 | 348.6 |
| $Li_7$ | 567.4 | 567.4 | 458.1 | 531.0 | 109.3 |
| $Li_8$ | 604.0 | 604.0 | 604.0 | 604.0 | 0 |

always less than 4.0 a.u.. It is gratifying to see that the anisotropies are also very similar, a quantity very sensitive to the quality of the basis set. Hence, these results give some confidence in the accuracy of our calculations in regards to the basis set. In Table 3, the values calculated with basis set B are compared with other theoretical values and with the experimental results. Among other theoretical values [19] the newest ones have been tabulated. They have been calculated using a non-hybrid gradient corrected exchange correlation functional and the basis set designed by Sadlej and Urban [20]. As mentioned in the introduction one can not expect a good concordance with the experimental value for the atom but one can see that the relative difference get smaller for the clusters with more atoms. However, it is important to remark that for $Li_3$ and $Li_6$ there are a significative difference between the theoretical values and the experimental ones. There are also other theoretical values not quoted in the table which show a great difference with respect to the experimental one. In the special case of $Li_3$ pilot calculations with other functionals, different basis sets, a more fine grid and other more expensive methods like MP2 and CCSD(T) have been done. In all cases, the theoretical calculations are significatively higher than the experimental ones. For instance, a CCSD(T) calculation with the 6-311G** basis set gives a value of 333.5 a.u.. A much more exhaustive investigation have been done for the dipole polarizability of $Li_4$ [21] coming to the same conclusion. The theoretical values are systematically higher than the experimental ones. The point is important because all the known effects which are not considered in the theoretical calculations like temperature, vibrations and thermal expansion can only increase the values. A similar discrepancy has been recently discussed for the case of silicon clusters [22]. Note that for $Li_8$ the two theoretical values quoted in Table 3 are similar, although they correspond to clusters of different geometry. [19] informs a capped centered trigonal prism as the ground state.

In Tables 4 and 5 the dipole polarizabilities of the copper clusters have been tabulated. Again, the mean polarizability increases slowly with the number of atoms and there is also a small increases in going from $Cu_6$ to $Cu_7$. The values calculated with basis set A and B are very similar with the exception

Table 3
Comparison of calculated and experimental dipole polarizabilities (in a.u.) for Lithium clusters

| | $\langle\alpha\rangle^{a}$ | $\lvert\Delta\alpha\rvert^{a}$ | $\langle\alpha\rangle^{b}$ | $\lvert\Delta\alpha\rvert^{b}$ | $\langle\alpha\rangle^{c}$ |
|---|---|---|---|---|---|
| Li | 142.3 | 0 | 151.2 | | 164.0 |
| $Li_2$ | 199.2 | 100.5 | 221.4 | 126.9 | 221.1 |
| $Li_3$ | 320.1 | 277.8 | 340.1 | | 233.6 |
| $Li_4$ | 355.5 | 316.1 | 394.1 | | 327.2 |
| $Li_5$ | 452.0 | 323.7 | 465.6 | | 427.5 |
| $Li_6$ | 526.0 | 348.6 | 502.1 | 267.2 | 359.2 |
| $Li_7$ | 531.0 | 109.3 | 562.2 | | 539.0 |
| $Li_8$ | 581.1 | 0 | 599.2 | 0 | 559.3 |

[a]This work ,basis set B.
[b]Theoretical values [19].
[c]Experimental values [19].

Table 4
Static Dipole Polarizabilities (in a.u.) of Copper clusters calculated with basis set A

| | $\alpha_{xx}$ | $\alpha_{yy}$ | $\alpha_{zz}$ | $\langle\alpha\rangle$ | $\lvert\Delta\alpha\rvert$ |
|---|---|---|---|---|---|
| Cu | 39.3 | 39.3 | 39.3 | 39.3 | 0 |
| $Cu_2$ | 57.5 | 57.5 | 103.5 | 72.8 | 46.0 |
| $Cu_3$ | 78.2 | 212.3 | 92.6 | 127.7 | 127.5 |
| $Cu_4$ | 89.1 | 118.7 | 233.8 | 147.2 | 132.4 |
| $Cu_5$ | 103.0 | 270.4 | 188.0 | 187.1 | 145.0 |
| $Cu_6$ | 115.0 | 260.9 | 260.9 | 212.3 | 145.9 |
| $Cu_7$ | 256.0 | 256.0 | 185.7 | 232.6 | 70.3 |
| $Cu_8$ | 262.6 | 262.6 | 262.6 | 262.6 | 0 |

of $Cu_5$ where the difference reaches 10.0 a.u., an error of 5%. In contrast to the case of lithium clusters where there are many theoretical calculations and experimental values, for the copper clusters only two recently calculations of the dipole polarizabilities have been found [7,16]. To our knowledge there is no measurement of the dipole polarizability of copper clusters. In Table 6, the mean polarizability and the anisotropy are compared with the other theoretical calculations. In [16] the calculations are all – electron with a non hybrid gradient corrected exchange correlation functional whereas in [7] the Los Alamos pseudopotential with the respective basis set have been used. Note that the results presented in this work are the only ones which include in the pseudopotential the relativistic effects. For copper atom the influence of the relativistic effects in the polarizability has been estimated to reduce it in around 7.0 a.u. [23] which is almost the difference between the value reported here and the other ones. Very recently, for $Cu_2$ the relativistic effects on the dipole polarizabilitiy have been estimated to reduce it also in around 7.0 a.u. [24]. Nothing is known about the relativistic effect in the dipole polarizability of larger systems. The geometry of all clusters in the different works are similar exception made of $Cu_8$ where this work found the tetrahedron structure as the ground state whereas the other two works found a $C_{2v}$ structure. In general, the agreement among the three different calculations is good, with the values reported here slightly lower. It is interesting to note that for $Cu_8$ the values are similar demonstrating once again that different geometries do not affect significatively the mean polarizability as long as the structures are compact.

In Table 7 the dipole moments and polarizabilities of the mixed clusters $Cu_nLi_m$ are presented. In Fig. 1, the dipole polarizabilities per atom for the three series of clusters studied are compared. One can see that the values for the mixed clusters follow close the values obtained for the copper clusters which

Table 5
Static Dipole Polarizabilities (in a.u.) of Copper clusters calculated with basis set B

| | $\alpha_{xx}$ | $\alpha_{yy}$ | $\alpha_{zz}$ | $\langle\alpha\rangle$ | $\|\Delta\alpha\|$ |
|---|---|---|---|---|---|
| Cu | 39.4 | 39.4 | 39.4 | 39.4 | 0 |
| $Cu_2$ | 58.5 | 58.5 | 104.4 | 73.8 | 45.9 |
| $Cu_3$ | 79.3 | 214.9 | 93.1 | 129.1 | 129.3 |
| $Cu_4$ | 90.1 | 120.1 | 235.0 | 148.4 | 132.5 |
| $Cu_5$ | 108.6 | 264.1 | 217.9 | 196.9 | 138.3 |
| $Cu_6$ | 116.7 | 261.6 | 261.8 | 213.4 | 145.0 |
| $Cu_7$ | 256.7 | 256.7 | 186.6 | 233.3 | 70.1 |
| $Cu_8$ | 263.4 | 263.4 | 263.4 | 263.4 | 0 |

Table 6
Comparison of calculated dipole polarizabilities (in a.u.) for Copper clusters with other theoretical values

| | $\langle\alpha\rangle^a$ | $\|\Delta\alpha\|^a$ | $\langle\alpha\rangle^b$ | $\|\Delta\alpha\|^b$ | $\langle\alpha\rangle^c$ |
|---|---|---|---|---|---|
| Cu | 39.4 | 0 | 47.0 | 0 | 52.6 |
| $Cu_2$ | 73.8 | 45.9 | 78.5 | 39.6 | 77.6 |
| $Cu_3$ | 129.1 | 129.3 | 130.1 | 90.0 | 137.1 |
| $Cu_4$ | 148.4 | 132.5 | 151.5 | 126.2 | 153.9 |
| $Cu_5$ | 196.9 | 138.3 | 192.1 | 139.3 | 196.5 |
| $Cu_6$ | 213.4 | 145.0 | 217.6 | 141.4 | 221.2 |
| $Cu_7$ | 233.3 | 70.1 | 233.1 | 68.7 | 242.9 |
| $Cu_8$ | 263.4 | 0 | 256.8 | 47.4 | 267.6 |

[a] This work ,basis set B.
[b] [19].
[c] [7].

Table 7
Dipole moment (in Debye Units) and static dipole polarizabilities (in a.u.) of mixed Lithium-Copper clusters

| | $\alpha_{xx}$ | $\alpha_{yy}$ | $\alpha_{zz}$ | $\langle\alpha\rangle$ | $\|\Delta\alpha\|$ | $\|\mu\|$ $(D)$ |
|---|---|---|---|---|---|---|
| $CuLi(^1\Sigma_g)$ | 78.8 | 78.8 | 111.1 | 89.6 | 32.3 | 1.9 |
| $Cu_2Li(^2B_2)$ | 81.3 | 169.6 | 109.4 | 120.1 | 78.1 | 1.5 |
| $Cu_2Li_2(^1A_g)$ | 101.1 | 124.8 | 167.1 | 131.0 | 57.9 | 0 |
| $Cu_3Li(^1A_1)$ | 91.8 | 227.1 | 115.1 | 144.7 | 125.3 | 1.2 |
| $Cu_3Li_2(^2B_2)$ | 136.2 | 187.6 | 201.0 | 174.9 | 59.2 | 3.0 |
| $Cu_3Li_3(^1A_1)$ | 141.0 | 218.8 | 218.8 | 192.9 | 77.8 | 0 |
| $Cu_4Li(^2A_1)$ | 109.0 | 285.3 | 181.5 | 191.9 | 153.5 | 0.5 |
| $Cu_4Li_2(^1A_g)$ | 221.1 | 221.1 | 145.5 | 195.9 | 75.6 | 0 |
| $Cu_4Li_3(A')$ | 302.1 | 302.1 | 247.7 | 284.0 | 54.4 | 1.0 |
| $Cu_4Li_4(^1A_1)$ | 249.9 | 249.9 | 249.9 | 249.9 | 0 | 0 |

are much lower than the ones of the lithium clusters. The three series presents an oscillatory behavior but only the lithium clusters show an important decreases at the beginning of the series. This can be due to the abnormal high value of the dipole polarizability of the lithium atom, i.e. it is higher than the dipole polarizability of the sodium atom. The small value of the polarizabilities of the mixed clusters can be due to the more polar bonds which points out to more localized electrons. The $Cu_4Li_3$ system tends to have a value higher in the series which can be due to the change in the geometry in comparison to the respective $Cu_7$ cluster.

Concluding, the static dipole polarizabilities of the mixed lithium – copper clusters have been calculated

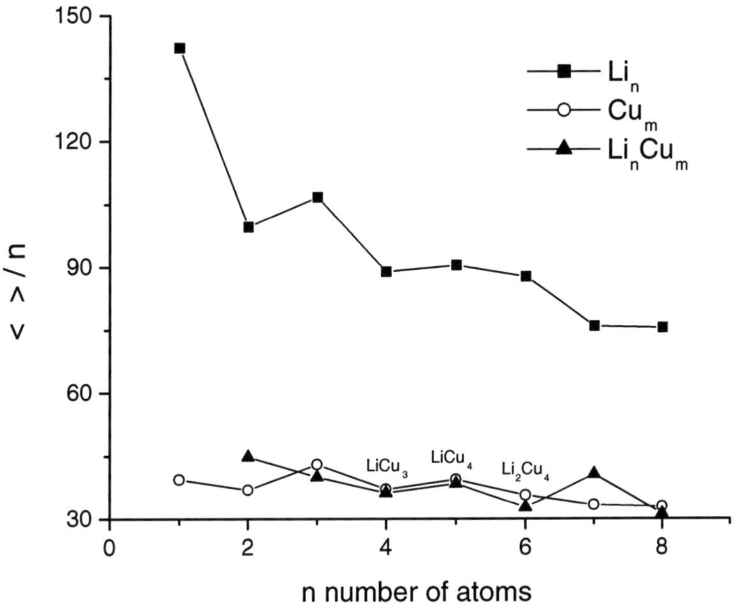


Fig. 1.

and compared with analogous calculations for the bare lithium and copper clusters. The calculations have been done solving the Kohn – Sham equations with an hybrid exchange – correlation functional. It has been found that the copper clusters present values much smaller than the lithium clusters, and the mixed ones follow very close the trend of the copper clusters. The copper clusters are more compact than the lithium clusters and in the mixed ones the electrons are more localized.

## Akcnowledgements

This work has been supported by FONDECYT (Fondo Nacional de Desarrollo Científico y Tecnológico), grant 1010649.

## References

[1] W. Ekardt, *Metal Clusters*, Wiley: USA, 1999.

[2] J.A. Becker, *Angew. Chem. Int. Ed.* **36** (1997), 1390.

[3] W.A. de Heer, *Rev. Mod. Phys.* **65** (1993), 611; M. Brack, *Rev. Mod. Phys.* **65** (1993), 677.

[4] P.J. Politzer, *J. Chem. Phys.* **86** (1987), 1072; T.K. Ghanty and S. Ghosh, *J. Phys. Chem.* **97** (1993), 4951; P. Fuentealba and O. Reyes, *Theochem* **282** (1993), 65; Y. Simon and P. Fuentealba, *J. Phys. Chem.* **A 102** (1998), 2029.

[5] P.K. Chattaraj and S. Sengupta, *J. Phys. Chem.* **A 100** (1996), 16126.

[6] P.K. Chattaraj, P. Fuentealba, P. Jaque and A. Toro, *J. Phys. Chem.* **A 103** (1999), 9307; P. Fuentealba, Y. Simon and P.K. Chattaraj, *J. Phys. Chem.* **A 104** (2000), 3185; T. Mineva, E. Sicilia and N. Russo, *J. Am. Chem. Soc.* **120** (1998), 9053; U. Hohm, *J. Phys. Chem.* **A 104** (2000), 8418; M. Torrent, J. Luis, M. Duran and M. Sola, *J. Chem. Phys.* **117** (2002), 10561.

[7] P. Jaque and A. Toro, *J. Chem. Phys.* **117** (2002), 3208.

[8] R.D. Diehl and R. McGrath, *Surf. Sci. Report* **23** (1996), 43.

[9] D.J. Lacks and R.G. Gordon, *Phys. Rev.* **A47** (1993), 4681.

[10] E. Engel, J.A. Chevary, L.D. Macdonald and S.H. Vosko, *Z. Phys.* **D23** (1992), 7.

[11] P. Fuentealba and Y. Simon, *J. Phys. Chem.* **101** (1997), 4231.
[12] M. Dolg, U. Wedir, H. Stoll and H. Preuss, *J. Chem. Phys.* **86** (1986), 866.
[13] T.H. Dunning and P. Hay, in: *Modern Theoretical Chemistry*, (Vol. 3), H.H. Schaefer, ed., Plenum: New York, 1976, p. 1.
[14] A.D. Becke, *J. Chem. Phys.* **98** (1993), 5648.
[15] Gaussian 98, Revision A.9; M.J. Frisch, G.W. Trucks, H.B. Schlegel, G.E. Scuseria, M.A. Robb, J.R. Cheeseman, V.G. Zakrzewski, J.A. Montgomery, Jr., R.E. Stratmann, J.C. Burant, S. Dapprich, J.M. Millam, A.D. Daniels, K.N. Kudin, M.C. Strain, O. Farkas, J. Tomasi, V. Barone, M. Cossi, R. Cammi, B. Mennucci, C. Pomelli, C. Adamo, S. Clifford, J. Ochterski, G.A. Petersson, P.Y. Ayala, Q. Cui, K. Morokuma, D.K. Malick, A.D. Rabuck, K. Raghavachari, J.B. Foresman, J. Cioslowski, J.V. Ortiz, A.G. Baboul, B.B. Stefanov, G. Liu, A. Liashenko, P. Piskorz, I. Komaromi, R. Gomperts, R.L. Martin, D.J. Fox, T. Keith, M.A. Al-Laham, C.Y. Peng, A. Nanayakkara, M. Challacombe, P.M.W. Gill, B. Johnson, W. Chen, M.W. Wong, J.L. Andres, C. Gonzalez, M. Head-Gordon, E.S. Replogle and J.A. Pople, Gaussian, Inc, Pittsburgh PA, 1998.
[16] K. Jug, B. Zimmermann, P. Calaminici and A. Koester, *J. Chem.Phys.* **116** (2002), 4497.
[17] R.O. Jones, A.I. Lichtenstein and J. Hutter, *J. Chem. Phys.* **106** (1997), 4566.
[18] L. Padilla and P. Fuentealba, manuscript in preparation.
[19] D. Rayane, A. Allouche, E. benichou, R. Antoine, M. Aubert – Frecon, Ph. Dugourd, M. Broyer, C. Ristori, F. Chandezon, B. Huber and C. Guet, *Eur. Phys. J.* **D 9** (1999), 243.
[20] A. Sadlej and M. Urban, *J. Mol. Struc.* (*THEOCHEM*) **80** (1991), 234.
[21] G. Maroulis and D. Xenides, *J. Phys. Chem.* **A103** (1999), 4590.
[22] V.E. Bazterra, M.C. Caputo, M. Ferraro and P. Fuentealba, *J. Chem. Phys.* **117** (2002), 11158.
[23] P. Fuentealba, H. Stoll, L.V. Szentpály, P. Schwerdtfegery and H. Preuss, *J. Phys.* **B16** (1983), L323.
[24] T. Saue and H. Jensen, *J. Chem. Phys.* **118** (2003), 522.

Journal of Computational Methods in Sciences and Engineering 4 (2004) 597–609
IOS Press 

# Computation and interpretation of Raman scattering intensities

Kathleen M. Gough*, Concetta Lupinetti and Richard Dawes
*Department of Chemistry, University of Manitoba, Winnipeg, Manitoba, Canada*

**Abstract.** The experimental measurement and the theoretical calculation of Raman scattering intensities present a number of challenges. We evaluate the currently available options with regard to our on-going studies of intensity in hydrocarbons and silanes. In the treatment of experimental data, anharmonic force fields permit the calculation of combination and overtone intensities as well as the resolution of anharmonic resonances. The structural dependence of the magnitude of the mean molecular polarizability derivative associated with the stretch of a single bond is confirmed at the B3LYP/aug-cc-pVTZ level of calculation. Although they do not exhibit nonlinear optical properties, derivatives for the straight chain alkanes show nonlinear trends in common with silanes, polyenes and polyaromatic hydrocarbons. The dynamic correction has been evaluated using time dependent DFT with the same basis set (TDDFT). The best calculations yield results that agree with the measured values to within experimental error.

Keywords: Raman scattering intensity, polarizability derivative, alkanes, polyaromatic hydrocarbons, Density functional theory, DFT, TDDFT, anharmonic force field, inverse eigenvalue method

## 1. Introduction

Vibrational intensities provide information about the molecular charge distribution and its change during a vibration. Early experimental results indicated that the Raman trace scattering intensity associated with the stretch of a CH bond was the same, for a range of hydrocarbons including alkanes, alkenes, alkynes and benzene [1]. From a study of CH-stretching vibrations in perdeuterated molecules, more accurate data showed that this conclusion was incorrect [2]. Prompted by these results, we began an investigation of Raman trace scattering intensities in hydrocarbons that has included both the measurement of absolute Raman trace scattering intensities in small hydrocarbons, and the complementary calculation of intensities in terms of the polarizability derivatives, $\partial\alpha/\partial r$, from *ab initio* molecular orbital (MO) methods [3–5]. Our goals have been to provide reliable *ab initio* calculations of molecular polarizability derivatives and to seek general relationships between intensity (associated with a particular internal coordinate) and structure (the position and orientation of a bond within a given molecular conformation).

Intensity patterns have been examined in a large selection of hydrocarbons [4,5], including the all-trans straight chain hydrocarbons to C25, rings C3 to C6, including both boat and chair conformations of cyclohexane, propellanes, bicycloalkanes and a tetracycloalkane [5], as well as a series of primary models for conducting polymers: unsaturated chains and rings, and several silanes [5]. Patterns identified

*Corresponding author: Kathleen M. Gough, Ph.D., Department of Chemistry, Room 350 Parker Building, 144 Dysart Road, University of Manitoba, Winnipeg, Manitoba, Canada R2T 2N2. Tel. +1 204 474 6262; Fax: +1 204 474 7608; E-mail: kmgough@ms.umanitoba.ca.

for the CH bond stretch in straight chain hydrocarbons continue to appear in the latter molecules. Our results have shown that, for the X-H bond stretches (X = C, Si), the polarizability is most sensitive to changes at the end of a chain. Hence, the derivative is greatest when the terminal, in-plane (Hip) CH or SiH bonds in long chains are stretched, while it is smaller when methylenic (Hm) bonds are stretched in chains of any length. The pattern extends to rings, where intensity is greatest for CH bonds at the further extremities. The polarizability derivatives associated with the C-C, C = C and Si-Si backbone segments exhibit different, but definable, patterns [4,5]; however, the intensity is greatest for bonds located in the interior of the structure. Steric hindrance is invariably associated with a significant decrease in calculated intensities.

Wavefunctions from MO calculations have been analyzed with the theory of Atoms In Molecules (AIM) [6,7], in order to correlate redistribution of the charge density in the vibrating molecules with calculated and observed intensities [4]. AIM theory provides a unique method for obtaining the electron population associated with an atom in a molecule, with or without an applied field. To date, all molecules studied with AIM, including non-conductors, have been found to have a dielectric nature: there is a transfer of charge from one end of the molecule to the other to create a dipole which opposes the applied field, while the atomic dipoles of the interior atoms oppose the surface dipole created by the charge transfer.

The rapid developments in computational capability in the last decade necessitate a reconsideration of the methodology we have employed in the past, and the improvements that we may incorporate, given the size and type of molecules that we are investigating. Numerous basis sets have been developed for the calculation of electronic properties, and in particular, the molecular polarizability [8–20]. The accuracy of calculations can be further refined through incorporation of electron correlation, configuration interaction, nuclear polarizability, and dynamic rather than static systems.

Electronic properties are normally obtained from derivatives (analytical or finite difference) calculated at the optimized geometry; however, it has been argued that experimental geometries yield better results [16,21]. A more rigorous solution to this problem is to employ a higher-order anharmonic force field to obtain geometries within the more realistic anharmonic potential [22–24]. The position expectation values for the nuclei no longer coincide with the potential minimum and the molecular electronic properties are slightly altered [25–27]. Of far greater experimental importance are the extreme effects of various resonances that frequently complicate recorded spectra. As will be discussed under treatment of experimental data, calculated higher-order force fields, as well as dipole and polarizability fields, yield corrections for resonant "intensity borrowing" from fundamental transitions by higher-order transitions [23]. This correction will be shown to be important for the comparison of calculated intensities with those recorded experimentally.

We have three goals for this paper. First, we will examine the computational options and improvements that are now routinely possible, to determine the cost/benefits of incorporating them into our analyses. Secondly, we will outline the additional improvements possible in the analysis of experimental data, specifically the extension to higher order force fields to resolve anharmonic resonances. Finally, we will compare some of our experimental and theoretical data with recent computational results for model compounds possessing non-linear optical properties, to gain more insight into the patterns that we have found in Raman trace scattering intensities.

## 2. Experimental and theoretical approach

Our combined experimental and theoretical approach is described in detail elsewhere [3–5], and only a brief overview is presented here. Trace (isotropic) spectra are chosen because i) there are fewer bands:

the derivative of the mean molecular polarizability, $\partial\overline{\alpha}/\partial q_i$, is non-zero only for totally symmetric normal modes; ii) bands are narrow and well resolved: Q branches only; and iii) trace scattering intensity parameters obey sum rules upon isotopic substitution: no rotational correction is required [3]. Absolute intensities are then obtained by scaling relative to $N_2$ as an internal standard. Since the exciting line is far removed from any electronic or fundamental vibrational transition energies, the scattering formalism developed by Placzek [28] applies. We have previously assumed the double harmonic approximation, i.e. absence of both mechanical and electrical anharmonicity. Under these conditions, the Raman trace scattering cross-section is given by:

$$\frac{\partial\sigma}{\partial\Omega} = \left(\frac{\pi^2}{2\varepsilon_0^2}\right)(\nu_0 - \nu)^4 \left[1 - \exp\left(\frac{-hc\nu_i}{kT}\right)\right]^{-1} \left(\frac{\partial\overline{\alpha}}{\partial q_i}\right)^2 \tag{1}$$

The polarizability derivatives have been converted to symmetry coordinates via the L matrix obtained from a harmonic force field analysis, and then to internal coordinates, $\partial\overline{\alpha}/\partial r$. The spectra of all isotopomers can be fit by the intensity parameters of the parent molecule. For the MO calculations, although some larger bases were tested [4], Dunning's split valence basis set with polarization functions (D95(d,p)) has been used for the majority of our work, as it yielded consistent relative parameters. The wavefunction from the MO calculation has been analyzed with the AIM software [6,7], to yield bond critical points and define the surfaces of atomic subspaces. Atomic properties are found by integration over these subspaces and molecular properties obtained by summation over all atoms. The calculations have been performed on wavefunctions obtained at equilibrium and at vibrationally extended or compressed geometries, with and without an applied static electric field. This method has enabled us to draw an insightful picture of charge redistribution in a molecule during a vibration, within the limitations of the clamped-nuclei and static field approximations [4].

## 3. Discussion

### *3.1. Treatment of experimental data*

Insight into the origin of large intensity variations observed even within a group of similar compounds requires that results be placed into a common framework. Much of the work in our laboratory has been concentrated on saturated hydrocarbons. Force-field methods are employed to transform the data into intensity parameters with respect to internal coordinates. This allows intensity parameters to be described in terms of the same coordinates used to describe the geometries and greatly facilitates the recognition and development of structure/intensity relationships [3].

The trace scattering spectrum is composed only of contributions arising from totally symmetric modes ($A_{1g}$). In a best-case scenario, the peaks will be well separated and free from anharmonic resonant interactions, and the integrated intensities can be assigned to the corresponding dimensionless normal modes. A sign ambiguity exists, since these intensities are related to the squares of the intensity parameters. Previously, the ambiguity has been resolved through numerical optimization of intensity data over a set of molecular isotopomers, through which the character of the normal modes varies. We have found that polarizability calculations with at least a moderate level of theory will always predict signs for the intensity parameters in agreement with these optimizations. If other isotopic data are lacking, we may be guided by theory. Using the best available force-field, the L-matrix is used to transform these normal mode intensities into parameters with respect to internal coordinates. Thus in this best case, using the L-matrix and the signs determined by theory, it is only necessary to solve a small, linear system of equations to obtain the desired internal coordinate intensity parameters.

The next level of complexity is encountered when the spectrum is disturbed by anharmonic resonances. If there is only one, totally symmetric fundamental mode in each region of the spectrum, then the treatment of splitting due to resonances is still relatively straightforward. This is true for ethane in which the C-H stretching mode ($A_{1g}$) is split into four bands [3]. The first level of approximation is to assign any trace scattering in the region to the fundamental. This approximation neglects second derivatives of the polarizability that would provide inherent intensity to the second-order transitions involved in the resonance. Theoretical papers describing calculated Raman intensities often compare their results with experimental numbers obtained in this way without comment [29]. Montero has published formulae derived from perturbation theory that may be used to correct intensities for one fundamental mode involved in weak or strong resonances. These formulae require a cubic force field, and a quadratic polarizability field [30–32]. Often, as with ethane, the inherent intensity of higher-order transitions is found to be small, indicating that much of the recorded intensity should be attributed to the fundamental. In either method, the L-matrix transformation to internal coordinates is performed as before.

A worst-case scenario exists when two or more fundamentals of similar energy are disturbed by multiple resonances. The resulting resonant polyad must be resolved through diagonalization of a matrix describing all of the resonant interactions through fourth order. The off-diagonal elements (the Fermi and Darling-Dennison interaction constants) are derived from the cubic and quartic force fields [33–35], and are known to be accurate even when calculated at a relatively moderate level of theory. The diagonal elements are the deperturbed anharmonic frequencies of the involved states. These are less accurately known, since even the initial harmonic frequencies usually require scaling [25]. Furthermore, in complex cases, many resonant denominators must be removed. Ultimately, this causes a breakdown in the perturbative method, and limits the size of system to which it may be applied. The eigenvalues of this resonance interaction matrix should yield the experimentally observed transition energies, while the eigenvectors provide the composition of each transition in terms of the fundamentals and combinations or overtones thereof. This completes the goal of distributing the observed intensity amongst the fundamental modes. In practice, if the diagonal elements are of poor quality, diagonalization will generally not yield particularly good agreement with the observed transition frequencies. Given accurate interaction constants and experimental eigenvalues, the inverse-eigenvalue problem may be solved to obtain a set of diagonal elements that will produce the observed transitions [36]. Efficient algorithms exist for this procedure, and although there is the potential for a multiplicity of solutions, in practice only one physically reasonable set of values will be produced [37]. Furthermore, one can test the sensitivity of the eigenvectors to the smaller errors in the interaction constants. This approach allows the less accurate diagonal elements to be improved empirically in order to obtain improved eigenvectors. Finally, using the derived eigenvectors, the intensity may be distributed amongst the fundamentals for transformation into the internal coordinate intensity parameters as before.

## 4. General considerations on theoretical options

There are many options for improving our theoretical computations beyond the level employed in the past. Choices include the size of the basis set, addition of diffuse and/or polarization functions, inclusion of electron correlation, configuration interaction, vibrational responses to the applied field, and the choice between dynamic and static calculations.

The importance of adding diffuse and/or polarization functions to basis sets is well known and has been explored in many studies [8–13,17]. In general, large bases [8–13] with split valence and exponent optimization [9–16], diffuse functions [8,19,20,38], and polarization functions [14,38,39] are known to improve the accuracy of the results. Diffuse functions appear to be more important than polarization functions; however, if the latter are optimized for the calculation of electronic properties, the results can be as good as those for much larger bases with diffuse functions. For example, Van Caillie and Amos [11] found that, for numerous small molecules, Sadlej's polarization functions [40] yielded results that were just as good as those from an augmented basis set (aug-cc-pVTZ) whether calculated as SCF, or with Density Functional Theory (DFT) [Generalized Gradient Approximation (GGA), such as B3LYP; or Local Density Approximation (LDA)]. Inclusion of electron correlation explicitly, through Moeller-Plesset perturbation theory [9,10,12–15,19,20,41] or implicitly, with DFT [11,19,20,42–50], is desirable, but can lead to unexpected anomalies. In the case of DFT, local field problems have been identified and improved functionals are now being developed [42,43]. Configuration interaction may be important in some cases, but may be computationally too expensive for molecules with more than a few atoms [10,11,13,49,50].

Nuclear coordinates may enter at several levels. The polarizability values may be averaged over the zero point nuclear displacements. Additionally, nuclear displacements in the presence of an electric field, called vibrational relaxation [51] or vibrational polarizability [52], may make a significant contribution to the polarizability in some cases, but the reported values were far from convergence with respect to basis set. The contribution to the total polarizability is not likely to exceed 2% for our molecules, and the effect on the derivative is likely minimal.

Finally, dynamic calculations incorporating the excitation wavelength can improve the accuracy of results by a few percent [11,47–50,53–58]. This aspect is discussed with examples in the next section.

## 5. Comparison with recent theoretical work

### *5.1. Dynamic calculation of polarizability derivatives*

While theoretical methodologies continue to advance with regard to calculating Raman intensities [29], reliable experimental results are often lacking, or are accompanied by large uncertainties. In other cases, only modest corrections have been applied to problems with anharmonic resonances. Results are reported in various formats and force-field transformations are required in order to make useful comparisons. For example, theoretical calculations that included the influence of dynamic polarizabilities on Raman intensities [29] could not be verified due to uncertainties in the published experimental data. From the data listed in that paper, it is not clear that the dynamic polarizability correction is improving agreement with experiment; rather, the agreement appears to be worsened for some compounds. We have transformed our most accurate results for ethane into the format and units reported by Van Caillie and Amos [29]. As can be seen in Table 1, any of the better functionals, along with the dynamic correction, produce an excellent agreement with experiment. These results provide an important affirmation of work in this direction. Our latest experimental data for bicyclo-[1.1.1]-pentane also point towards an improvement from dynamic polarizability corrections (Table 2). The data have not yet been fully corrected for the effects of anharmonic resonances but the final values are expected to be within a few percent of the dynamic values.

In our previous studies, we had used the D95(d,p) basis, noting its consistency for comparison with experiment; however, the results were typically overestimating the derivative for a CC stretch, and

Table 1
Raman scattering activity of ethane (Å$^4$ amu$^{-1}$)

| $C_2H_6$ | Static Raman scattering activity[a] | | | | Dynamic Raman scattering activity[a] | | | | Expt[b] | Expt[c] |
|---|---|---|---|---|---|---|---|---|---|---|
| | SCF | B3LYP | PBEØ | B97-1 | SCF | B3LYP | PBEØ | B97-1 | 514.5 nm | 514.5 nm |
| $\nu_1$ | 354.11 | 372.17 | 368.02 | 370.21 | 392.06 | 420.27 | 413.81 | 417.06 | 360.1 | 410.37 |
| $\nu_2$ | 0.34 | 0.02 | 0.02 | 0.02 | 0.48 | 0.04 | 0.02 | 0.04 | 2.2 | 0.0 |
| $\nu_3$ | 15.33 | 11.78 | 10.76 | 11.78 | 10.95 | 13.09 | 11.83 | 13.09 | 16.9 | 14.55 |

[a]From [29].
[b]From [59].
[c]From [3], and [Murphy, private communication], calculated for this work.

Table 2
Raman trace scattering parameters for bicyclo-[1.1.1]-pentane, propane, and cyclohexane ($\partial\bar{\alpha}/\partial r \times 10^{-30}$ Cm/V), from theory and experiment

| $C_5H_8$ | SCF[a] | B3LYP[b] | B3LYP[b]+ Dynamic[c] |
|---|---|---|---|
| Bridgehead C-H | 1.386 | 1.698 | 1.826 |
| Methylene C-H | 1.124 | 1.347 | 1.436 |
| $C_3H_8$ | | | Expt[d] |
| In-Plane C-H | 1.181 | 1.401 | 1.438 |
| Out-of-Plane C-H | 1.053 | 1.240 | 1.252 |
| Methylene C-H | 1.126 | 1.329 | 1.366 |
| C-C | 1.329 | 1.186 | 1.212 |
| $C_6H_{12}$ | | | Expt[e] |
| Equatorial C-H | 1.259 | 1.454 | 1.33(F) |
| Axial C-H | 1.021 | 1.222 | 1.17(F) |
| C-C | 1.217 | 1.110 | 1.03(F) |

[a]HF/D95(d,p), reference 5a.
[b]B3LYP/aug-cc-pVTZ, (CADPAC6) [60], calculated for this work.
[c]TDDFT (CADPAC6) [60], calculated for this work.
[d]Reference 3c.
[e]Reference 3b. (F) Strongly affected by Fermi resonances (see discussion).

underestimating that for the CH stretch. The B3LYP functionals, coupled with a large basis set (aug-cc-pVTZ), produce $\partial\bar{\alpha}/\partial r$ values that are in excellent agreement with experiment for the CH and CC stretching modes of bicyclopentane and propane (Table 2). The results for cyclohexane are also much improved, but agreement with experiment is less good. The latter bands are strongly perturbed by Fermi resonance and the isotopic sum rules are not obeyed well. To resolve this problem, it will be necessary to employ the higher order force field methods described above. Overall, the derivative for the CH stretch was increased by about 18–20%, while that for the CC was lowered by about 10%, compared to the HF method with the smaller basis set. The dynamic correction tends to increase the values by a few percent, and this would produce nearly exact agreement within the uncertainty limits of the best experimental data.

### 5.2. *Structural dependence of polarizability derivatives*

Recent work by Castiglioni et al. [61–64] on anomalous Raman bands in polyaromatic hydrocarbons (PAH) and graphitic samples reveals interesting similarities in trends observed for polarizabilities and derivatives among both saturated and unsaturated hydrocarbons. The authors classified their PAHs as either benzenoid (a lattice of aromatic rings joined by C-C single bonds) or non-benzenoid, on the basis of bond length alternation (BLA). The BLA was less pronounced in the "non-benzenoid" group. Regardless of their classification, the value of $\partial\overline{\alpha}/\partial r_{cc}$ was found to switch from + to − for the alternating shorter and longer bonds, while the absolute magnitude of the derivative decreased from the central to the peripheral bonds. This behavior is also seen in the quasi-one-dimensional PAHs, the rylenes, where the central CC bonds aligned with the longer dimension of the molecule give rise to $\partial\overline{\alpha}/\partial r_{cc}$ that is more than double that of their counterparts on the $D_{6h}$ PAH, C114. The $D_{2h}$ PAH, C78, with intermediate anisotropy, exhibits intermediate behavior. The sign alternation, coupled with the decreasing value of the derivative from interior to exterior of the molecule, is also evident in the straight chain alkenes [5, 65,66].

In the three PAHs, whether classified as benzenoid or not, the $\partial\overline{\alpha}/\partial r_{cc}$ for the bonds at the periphery do not follow the patterns for the interior [64]. In C114, bonds 13 through 16 are all classified as aromatic, but their derivatives are lower than those of the interior aromatic-type bonds. In C78, the bonds designated as C-C type exhibit an orientational dependence. The two bonds with the greatest absolute magnitude of $\partial\overline{\alpha}/\partial r_{cc}$ are bonds 2 and 9, and they are oriented parallel to the longer molecular axis. The PAH data were obtained at a relatively low level of theory (BLYP/6-31G), but the pattern seems to be real. For the straight chain alkanes, we also observed a decrease in the derivative for the terminal CC bonds (Fig. 1, bottom). Our original data were obtained with the D95(d,p) basis [5], but we now have repeated the calculations with the B3LYP/aug-cc-pVDZ basis and find that the pattern persists. The variations and anisotropy of the individual terms warrant closer investigation.

The non-linear optical (NLO) properties of polyacetylenes and polysilanes have been extensively investigated with *ab initio* methods for many years [14,16,18,63,65–69]. Common conclusions include the observation that the polarizability (among other properties) increases non-linearly with chain length, up to a point, but increases only linearly beyond that point. This is sometimes represented as an asymptotic limit to the average value of the longitudinal polarizability per chain sub-unit. The non-linearity is attributed variously to the delocalization of $\pi$ electrons (long chain alkenes, etc) or $\sigma$ electrons (polysilanes) along the chain backbone. Long chain alkanes (of the polyethylene family) such as those we have studied with regard to Raman intensities are deemed linear in their behavior [5]. We note here the non-linear properties of the polarizability *derivatives* of the alkanes and the similarity of these values to those of the NLO compounds. The value of $\partial\overline{\alpha}/\partial r_{Hip}$, that is, the derivative of the polarizability associated with the stretch of the methyl CH bond lying in the plane of an all-trans alkane, exhibits the same non-linearity as the longitudinal polarizability of an unsaturated chain. The value increases non-linearly with chain length but approaches an asymptotic limit as chain length increases beyond 15 carbons. All of the structural dependencies that we have noted in the different ring and cage hydrocarbons, as well as the length and orientation dependencies noted for the alkenes and silanes [5] fit within the overall picture of nonlinearity. The values of $\partial\overline{\alpha}/\partial r_{CC \text{ or s Si-Si}}$ (the bonds that comprise the backbone of the long chain molecules we have studied) also exhibit behaviors similar to those of the polarizabilities of the NLO.

In Fig. 1, we have compared the derivatives for the CC stretching modes in three of the PAHs reported [64]: the C78 and C114, classified as benzenoid, and the C138, denoted as non-benzenoid, with those of three straight chain alkanes (C9, C15 and C25). The straight chain alkanes are not NLO compounds: they exhibit only the slightest bond length alternation (± 0.001 Å) in the C15 chain both at the HF/D95(d,p) [5] and B3LYP/aug-cc-pVDZ levels (this work). The $\sigma$ conjugation in the alkanes is minimal compared with that of the silanes, or with the $\pi$ conjugation of the aromatic systems, while the HOMO-LUMO gap is large. However, the same type of behavior found in the PAHs is seen in the derivatives of the alkanes. Most intriguing to us, therefore, are the common patterns to be found among the derivatives of the mean molecular polarizability with respect to the internal coordinates. AIM

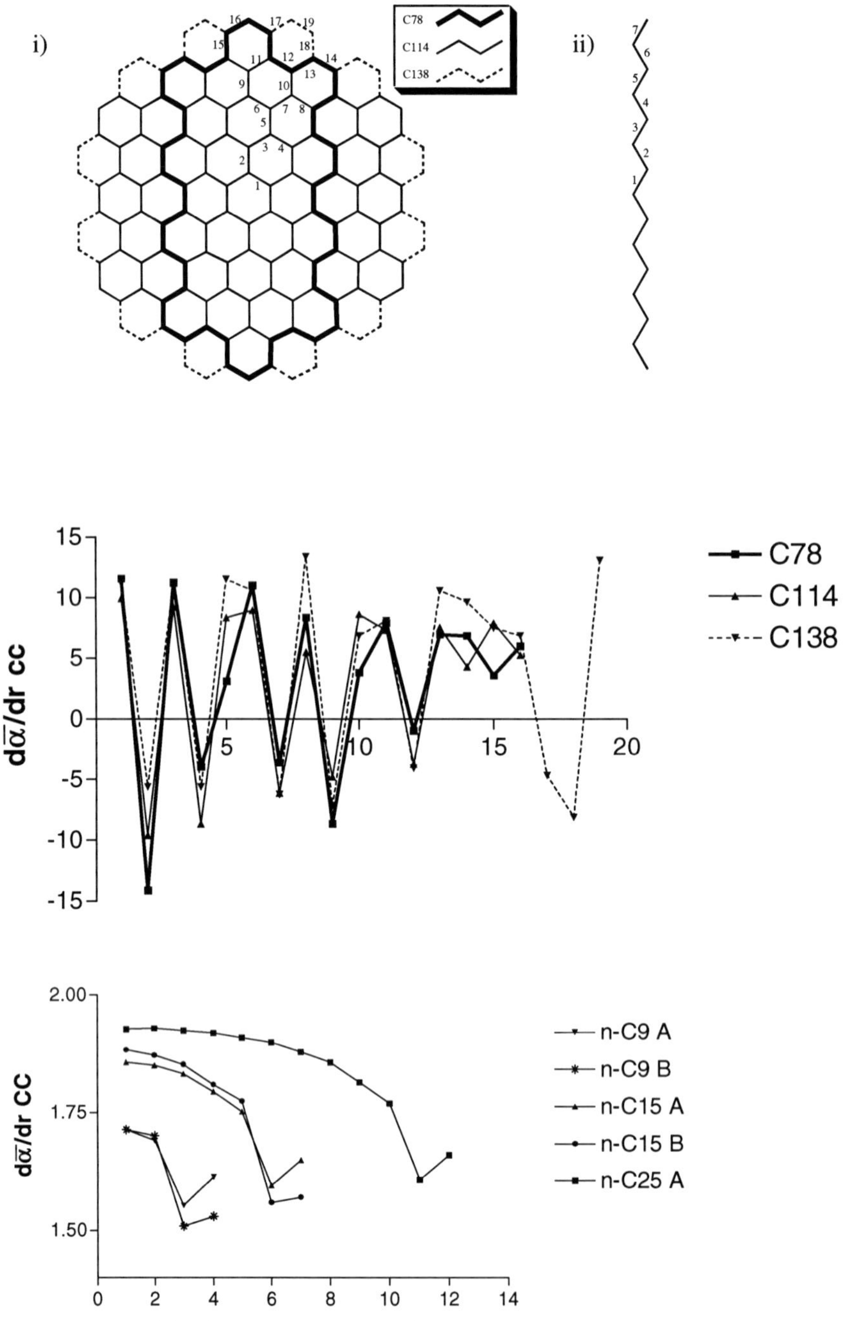


Fig. 1. Derivatives of the mean molecular polarizability for CC stretching modes. Top: i) Structures and bond numbering of PAHs C78 (bold outline), C114 (solid outline) and C138 (entire structure, including solid and dashed outer lines); ii) Structure and bond numbers for straight chain alkanes, C9, C15 and C25 Middle: Graph depicting variation of $(\partial\overline{\alpha}/\partial r$ $(10 \times 10^{-30}$ cm/V) Cm/V) for molecules in (i), data from reference 64. Bottom: Graph depicting variation of $(\partial\overline{\alpha}/\partial r$ $(\times\, 10^{-30}$ cm/V) for molecules in (ii). Data labeled (A) calculated as HF/D95(d,p) [5]; data labeled (B) calculated as B3LYP/aug-cc-pVDZ (this work).

analysis of the charge redistribution within the alkanes and alkenes, and, within computational limits, the silanes and PAHs, is seen to be an important direction for further insight into these patterns.

In a future paper, we will present a detailed analysis of bicyclo-[1.1.1]-pentane. This is an ideal candidate for testing the present limits in high level *ab initio* calculations, as it is small, highly symmetrical, and possesses only one type of CC and two very distinct CH bonds. For the bridgehead CH, the intensity parameter $\partial\alpha/\partial r_{CH}$ is predicted to be larger than that for any saturated hydrocarbon studied to date, while that for the methylene is predicted to be similar to any normal methylene in a linear alkane. Experimental verification of this prediction and AIM analysis are important steps in the development of our model.

## 6. Summary

Possible improvements to our theoretical methods cover a range of possibilities. For the computation of polarizability and polarizability derivatives, some considerations are more important than others. Clearly, large basis sets with diffuse and polarization functions should always be used, since the improvements are systematic and do not simply result in a scalable increase in property values. We have found that for small to medium molecules, the aug-cc-pVTZ basis set provides excellent results. Furthermore, our basis set tests, up to aug-cc-pV5Z on small, saturated hydrocarbons, have shown that the vibrational eigenvectors and Raman intensities are essentially converged with respect to basis set at the triple zeta level. The dynamic correction is an exciting development that appears to consistently increase the static values by a few percent. This often produces accuracy within experimental error. However, this additional correction comes at great computational expense at this time. For all but small test cases, given the predictable increase, it seems reasonable to accept that the static values may be slightly low. It would certainly not be practical to employ this method for some of the extended systems discussed in this paper.

Vibrational relaxation merits some investigation. While we have not found it necessary to include this effect in our calculations to date, there is certainly evidence that it could be important in some cases. Since geometry optimizations and nuclear forces can now be routinely computed in the presence of an applied field (e.g.: [70]), it would be useful to perform survey calculations to find a molecule exhibiting anomalously large relaxation effects, as a candidate for experimental confirmation.

In terms of the level of theory appropriate for Raman intensity calculations, we feel that most systems of interest to us are best suited to DFT calculations at the GGA level. While CCSD(T) and similar methods are important for benchmarking and comparison purposes for functional development, the types of systems in which the most unusual electronic behaviors are provoked, anomalous intensities, non-linearities, etc., are simply too large to treat this way.

For the experimental perspective, it has become important to use theoretical methods more extensively for the treatment of recorded spectra. The use of isotopomers has been valuable but anharmonicities limit the transferability of intensity parameters and accuracy is reduced when isotope sum rules are poorly obeyed. Sign ambiguities and anharmonic resonances can often be resolved through theory. The next level of accuracy demands that higher order force field methods be employed.

All of the systems considered in this paper, including saturated and unsaturated long chain hydrocarbons, as well as extended sheets such as the PAH's, exhibit some remarkably similar trends. Bond length alternation along the system, accompanied by non-linear increases in polarizability derivatives towards the central bonds, finally reaching a linear asymptote is a common theme. Another consistent observation is an "end effect", where stretches of bonds at the end of a chain or at the edge of a sheet produce unusual polarizability derivative values. These effects have been predicted by a wide range of theoretical methods. The general trends are observed at the HF level with small basis sets (HF/6-31G*), and are maintained at various higher levels (e.g.: B3LYP/aug-cc-pVTZ). A combined theoretical and experimental probe into the similarities between apparently categorically different systems is required. Understanding the mechanisms behind the non-linear regime and the end effect is the current challenge.

## Acknowledgements

This work was supported by a grant from the Natural Sciences and Engineering Research Council of Canada. We are grateful to W.F. Murphy for useful discussions.

## References

[1] T. Yoshino and H.J. Bernstein, *Spectrochim. Acta* **14** (1959), 127–.

[2] R.G. Snyder, A.L. Aljibury, H.L. Strauss, H.L. Casal, K.M. Gough and W.F. Murphy, Isolated C-H stretching vibrations of n-alkanes: Assignments and relation to structure, *J. Chem. Phys.* **81** (1984), 5352–5361.

[3] (a) K.M. Gough and W.F. Murphy, Intensity parameters for the Raman trace scattering of ethane, *J. Chem. Phys.* **85** (1986), 4290–4296; (b) K.M. Gough and W.F. Murphy, Harmonic force field and Raman trace scattering intensity parameters for cyclohexane, *J. Chem. Phys.* **87** (1987), 1509–1519; (c) K.M. Gough and W.F. Murphy, Raman trace scattering intensity parameters for propane, *J. Chem. Phys.* **87** 3341–3346; (d) K.M. Gough and W.F. Murphy, The Raman scattering intensity parameters of acetylene, *J. Mol. Struct.* **224** (1990), 73–88.

[4] (a) K.M. Gough, Theoretical analysis of molecular polarizabilities and polarizability derivatives in hydrocarbons, *J. Chem. Phys.* **91** (1989) , 2424–2432; (b) K.M. Gough, H.K. Srivastava and K. Belohorcová, Analysis of Raman-trace-scattering intensities in alkanes with the theory of atoms in molecules, *J. Chem. Phys.* **98** (1993), 9669–; (c) K.M. Gough, H.K. Srivastava and K. Belohorcová, Molecular polarizability and polarizability derivatives in cyclohexane analysed with the theory of atoms in molecules, *J. Phys. Chem.* **98** (1994), 771; (d) K.M. Gough and H.K. Srivastava, Electronic charge flow and Raman trace scattering intensities for CH stretching vibrations in n-pentane, *J. Phys. Chem.* **100** (1996), 5210.

[5] (a) K.M. Gough and J.R. Dwyer, Effect of structure and conformation on Raman trace scattering intensities in hydrocarbons, *J. Phys. Chem.* **A 102** (1998), 2723; (b) K.M. Gough, J.R. Dwyer and R. Dawes, *Ab initio* analysis of C-H and C-C stretching intensities in Raman spectra of hydrocarbons, *Can. J. Chem.* **78** (2000), 1035; (c) C. Lupinetti and K.M. Gough, *Ab initio* analysis of Raman trance scattering intensities in alkenes and silanes, *J. Raman Spectrosc.* **33** (2002), 147–154.

[6] R.F.W. Bader, *Atoms in Molecules: a Quantum Theory; The International Series of Monographs on Chemistry 22*; Clarendon Press: Oxford, 1990.

[7] F. Biegler-Konig, J. Schonbohm and D. Bayles, AIM2000-a program to analyze and visualize atoms in molecules, *J. Comput. Chem.* **22** (2001), 545.

[8] M. Dory, L. Beudels, J.G. Fripiat, J. Delhalle, J.M. André and M. Dupuis, *Ab Initio* CPHF calculations of the static polarizability and second hyperpolarizability of small molecules: Comparisons between standard and moderately large basis sets augmented with diffuse functions, *International Journal of Quantum Chemistry* **42**, (1992), 1577–1594.

[9] G. Maroulis, A study of basis set and electron correlation effects in the *ab initio* calculation of the electric dipole hyperpolarizability of ethene ($H_2C = CH_2$), *Journal of Chemical Physics* **97** (1992), 4188–4194.

[10] G. Maroulis, Electric polarizability and Hyperpolarizability of carbon monoxide, *Journal of Physical Chemistry* **100** (1996), 13466–13473.

[11] C. Van Caillie and R.D. Amos, Static and dynamic polarisabilities, Cauchy coefficients and their anisotropies: a comparison of standard methods, *Chemical Physics Letters* **291** (1998), 71–77.

[12] G. Maroulis, D. Xenides, U. Hohm and A. Loose, Dipole, dipole-quadrupole, and dipole-octopole polarizability of adamantane, $C_{10}H_{16}$, from refractive index measurements, depolarized collision-induced light scattering, conventional *ab initio* and density functional theory calculations, *Journal of Chemical Physics* **115** (2001), 7957–7967.

[13] G. Maroulis, C. Makris, U. Hohm and W. Wachsmuth, Determination of the complete polarizability tensor of 1,3-butadiene by combination of refractive index and light scattering measurements and accurate quantum chemical *ab Initio* calculations, *Journal of Physical Chemistry A* **103** (1999), 4359–4367.

[14] G.J.B. Hurst, M. Dupuis and E. Clementi, *Ab initio* analytic polarizability, first and second hyperpolarizabilities of large conjugated organic molecules: Applications of polyenes $C_4H_6$ to $C_{22}H_{24}$, *Journal of Chemical Phyiscs* **89** (1988), 385–395.

[15] M.A. Spackman, Accurate Prediction of Static Dipole Polarizabilities with Moderately Sized Basis Sets, *Journal of Physical Chemistry* **93** (1989), 7594–7603.
[16] B. Kirtman and M. Hasan, Linear and nonlinear polarisabilities of trans-polysilane from *ab initio* oligomer calculations, *Journal of Chemical Physics* **96** (1992), 470–479.
[17] H.-J. Werner and W. Meyer, PNO-CI and PNO-CEPA studies of electron correlation effects V. Static dipole polarizabilities of small molecules, *Molecular Physics* **31** (1976), 855–872.
[18] B. Champagne, J.G. Fripiat and J.M. André, *Ab initio* Quantum Chemical Calculations of Polarizabilities of Polymeric Systems, *Nonlinear Optics* **6** (1994), 229–261.
[19] A. Hinchliffe and H.J. Soscún-M, Density functional studies of molecular polarizabilities. Part 3; ethene, buta-1,3-diene and hexa-1,3,5-triene, *Electronic Journal of Theoretical Chemistry* **2** (1997), 315–324.
[20] S. Millefiori and A. Alparone, *Ab initio* and density functional theory calculations of the dipole polarizabilities of ethene, benzene and naphthalene, *Journal of Molecular Structure (Theochem)* **422** (1998), 179–190.
[21] B. H. Lengsfield III., S. Havriliak, M.E. Colvin and H.F. Schaefer III., Analytic MC SCF infrared intensities: Application to formaldehyde, *Chemical Physics Letters* **135** (1987), 340–345.
[22] W.H. Green Jr., A. Willetts, D. Jayatilaka and N.C. Handy, *Ab initio* prediction of fundamental, overtone and combination band infrared intensities, *Chem. Phys. Lett.* **169** (1990), 127–137.
[23] A. Willetts, N.C. Handy, W.H. Green Jr. and D. Jayatilaka, Anharmonic corrections to vibrational transition intensities, *J. Phys Chem.* **94** (1990), 5608–5616.
[24] W. Schneider and W. Theil, Anharmonic force fields from analytic second derivatives: method and application to methyl bromide, *Chem. Phys. Lett.* **157**(4) (1989), 367–373.
[25] P.E. Maslen, N.C. Handy, R.D. Amos and D. Jayatilaka, Higher analytic derivatives IV. Anharmonic effects in the benzene spectrum, *J. Chem. Phys.* **97**(6) (1992), 4233–4254.
[26] J.M.L. Martin, T.J. Lee and P.R. Taylor, A purely *ab initio* spectroscopic quality quartic force field for acetylene, *J. Chem. Phys.* **108** (1998), 676–691.
[27] D.A. Clabo Jr., W.D. Allen, R.B. Remington, Y. Yamaguchi and H.F. Schaefer III, A systematic study of molecular vibrational anharmonicity and vibration-rotation interaction by self-consistent-field higher-derivative methods. Asymmetric top molecules, *Chemical Physics* **123** (1988), 187–239.
[28] G. Placzek, *US Atomic Energy Commission, UCRL-Trans-524(L), translated from Handbuch der Radiologie*, (2nd, ed.), edited by E. Marks (Academische Leipzig, 1934), Vol. 6, part II, 1962, pp. 204–374.
[29] C. Van Caillie and R. Amos, Raman intensities using time dependent density functional theory, *Phys. Chem. Chem. Phys.* **2** (2000), 2123–2129.
[30] S. Montero, Raman intensities of Fermi diads. I. Overtones in resonance with nondegenerate fundamentals, *J. Chem. Phys.* **79**(9) (1983), 4091–4100.
[31] D. Bermejo and S. Montero, Anharmonic potential functions as derived from Raman intensities: Methane, *J. Chem. Phys.* **81**(9) (1984), 3835–3841.
[32] S. Montero, Anharmonic Raman intensities of overtone, combination and difference bands, *J. Chem. Phys.* **77**(1) (1982), 23–29.
[33] A.F. Borro, I.M. Mills and E. Venuti, Quartic anharmonic resonances in acetylenes and haloacetylenes, *J. Chem. Phys.* **102** (1995), 3938–3944.
[34] I.M. Mills and A.G. Robiette, On the relationship of normal modes to local modes in molecular vibrations, *Molecular Physics* **56** (1985), 743–765.
[35] M.J. Bramley, S. Carter, N.C. Handy and I.M. Mills, A refined quartic force field for Acetylene: Accurate calculation of the vibrational spectrum, *J. Mol. Spec.* **157** (1993), 301–336.
[36] P. Palmieri, R. Tarroni, M.M. Huhn, N.C. Handy and A. Willetts, An improved anharmonic force field of $CHClF_2$, *Chemical Physics* **190** (1995), 327–344.
[37] A.S. Deakin and T.M. Luke, On the inverse eigenvalue problem for matrices, *J. Phys. A. Math. Gen.* **25** (1992), 635–648.
[38] M. Torrent-Sucarrat, M. Sola, M. Duran, J.M. Luis and B. Kirtman, Basis set and electron correlation effects on *ab initio* electronic and vibrational nonlinearoptical properties of conjugated organic molecules, *J. Chem. Phys.* **118** (2003), 711–718.
[39] P. Karamanis and G. Maroulis, Static electric dipole polarizability and hyperpolarizability of fluorodiacetylene, *Theochem* **621** (2003), 157–162.
[40] A.J. Sadlej, Medium-size polarized basis sets for high-level-correlated calculations of molecular electric properties. II. Second-row atoms: silicon through chlorine, *Theoret. Chim. Acta* **79** (1991), 123–140.
[41] A. Hinchliffe and H.J. Soscún M., *Ab initio* studies of the dipole polarizabilities of conjugated molecules. Part 5. The five-membered heterocyclics $C_4H_4E$ (E = BH, AlH, $CH_2$, $SiH_2$, NH, PH, O and S), *Journal of Molecular Structure (Theochem)* **331** (1995), 109–125.

[42] M. van Faassen, P.L. de Boeij, R. van Leeuwen, J.A. Berger and J.G. Snijders, Application of time-dependent current-density-functional theory to nonlocal exchange-correlation effects in polymers, *Journal of Chemical Physics* **118** (2003), 1044–1053.

[43] Z.-L. Cai, K. Sendt and J.R. Reimers, Failure of density-functional theory and time-dependent density-functional theory for large extended $\pi$ systems, *Journal of Chemical Physics* **117** (2002), 5543–5549.

[44] S. Millefiori, A. Alparone and A. Millefiori, Structure and (Hyper)polarizabilities of Five-membered Heterocycles $C_4H_4XH_2$ (X = C, Si, Ge, Sn), *Journal of Chemical Research* (**S**) (1999), 238–239.

[45] B. Champagne, E.A. Perpète, S.J.A. van Gisbergen, E.-J Baerends, J.G. Snijders, C. Soubra-Ghaoui, K.A. Robins and B. Kirtman, Assessment of conventional density functional schemes for computing the polarizabilities and hyperpolarizabilities of conjugated oligomers: An *ab initio* investigation of polyacetylene chains, *Journal of Chemical Physics* **109** (1998), 10489–10498.

[46] M. Le Guennec, K. Evain and B. Illien, Calculation of static mean polarisability and polarisability anisotropy. Statistical comparison with the results of gases and influence of the geometrical parameters, *Journal of Molecular Structure (Theochem)* **542** (2001), 167–176.

[47] C. Van Caillie and R.D. Amos, Static and dynamic polarisabilities, Cauchy coefficients and their anisotropies: an evaluation of DFT functionals, *Chemical Physics Letters* **328** (2000), 446–452.

[48] H. Soscún, J. Hernández, R. Escobar and A. Hinchliffe, Ethyne in static and dynamic electric fields: an *ab initio* and DFT study of the linear and nonlinear dipole polarizability, *Asian Journal of Spectroscopy* **4** (2000), 119–126.

[49] M. Pecul and S. Coriani, The effect of triple excitations in coupled cluster calculations of Raman scattering cross-sections, *Chemical Physics Letters* **355** (2002), 327–338.

[50] M. Pecul and A. Rizzo, Linear response coupled cluster calculation of Raman scattering cross sections, *Journal of Chemical Physics* **116** (2002), 1259–1268.

[51] M. Gussoni, M. Rui and G. Zerbi, Electronic and relaxation contribution to linear molecular polarizability. An analysis of the experimental values, *Journal of Molecular Structure* **447** (1998), 163–215.

[52] D.M. Bishop, Molecular vibrational and rotational motion in static and dynamic electric fields, *Reviews of Modern Physics* **62** (1990), 343–374.

[53] U. Hohm and U. Trümper, Dispersion of Mean Molecular Dipole Polarizability and Second Density Virial Coefficients of 1,3-Butadiene, *Berichte der Bunsen-Gesellschaft für Physikalische Chemie* **96** (1992), 1061–1063.

[54] Y. Luo, H. Ågren, H. Koch, P. Jørgensen and T. Helgaker, Random-phase calculations of frequency-dependent polarizabilities and hyperpolarizabilities of long polyene chains, *Physical Review B* **51** (1995), 14949–14957.

[55] A. Kumar, W.J. Meath, P. Bündgen and A.J. Thakkar, Reliable anisotropic dipole properties, and dispersion energy coefficients, for $O_2$ evaluated using constrained dipole oscillator strength techniques, *Journal of Chemical Physics* **105** (1996), 4927–4937.

[56] D. Spelsberg and W. Meyer, Dynamic multipole polarizabilities, reduced spectra, and interaction coefficients for $N_2$ and CO, *Journal of Chemical Physics* **111** (1999), 9618–9624.

[57] A. Hu, D. M. York and T.K. Woo, Time-dependent density functional theory calculations of molecular static and dynamic polarizabilities, Cauchy coefficients and their anisotropies with atomic numerical basis functions, *Journal of Molecular Structure (Theochem)* **591** (2002), 255–266.

[58] M. Pecul and A. Rizzo, Relativistic effects on the electric polarizabilities and their geometric derivatives for hydrogen halides and dihalogens – a Dirac-Hartree-Fock study, *Chemical Physics Letters* **370** (2003), 578–588.

[59] J. Martin and S. Montero, Raman intensities of ethane and deuterated derivatives, *J. Chem. Phys.* **80** (1984), 4610–4619.

[60] CADPAC: The Cambridge Analytic Derivatives Package Issue 6, Cambridge, 1995. A suite of quantum chemistry programs developed by R.D. Amos with contributions from I.L. Alberts, J.S. Andrews, S.M. Colwell, N.C. Handy, D. Jayatilaka, P.J. Knowles, R. Kobayashi, K.E. Laidig, G. Laming, A.M. Lee, P.E. Maslen, C.W. Murray, J.E. Rice, E.D. Simandiras, A.J. Stone, M.-D. Su and D.J. Tozer.

[61] C. Castiglioni, C. Mapelli, F. Negri and G. Zerbi, Origin of the D line in the Raman spectrum of graphite: A study based on Raman frequencies and intensities of polycyclic aromatic hydrocarbon molecules, *Journal of Chemical Physics* **114** (2001), 963–974.

[62] M. Rigolio, C. Castiglioni, G. Zerbi and F. Negri, Density functional theory prediction of the vibrational spectra of polycyclic aromatic hydrocarbons: effect of molecular symmetry and size on Raman intensities, *Journal of Molecular Structure* **563–564** (2001), 79–87.

[63] C. Castiglioni, F. Negri, M. Rigolio and G. Zerbi, Raman activation in disordered graphites of the A' symmetry forbidden $k \neq 0$ phonon: The origin of the D line, *Journal of Chemical Physics* **115** (2001), 3769–3778.

[64] F. Negri, C. Castiglioni, M. Tommasini and G. Zerbi, A Computational Study of the Raman Spectra of Large Polycyclic Aromatic Hydrocarbons: Toward Molecularly Defined Subunits of Graphite, *Journal of Physical Chemistry A* **106** (2002), 3306–3317.

[65] S. Suhai, On the Structure dependence of the static longitudinal linear electric polarizability of infinite polyenes: An *ab-initio* uncoupled perturbed Hartree-Fock crystal orbital study, *Molecular Engineering* **1** (1991), 115–129.

[66] D.A. Morton-Blake and M.E. Mura, Polarizability calculations in conjugated hydrocarbon chains, *Synthetic Metals* **53** (1992), 37–46.

[67] B. Champagne and J.M. André, Towards the calculation of polarizabilities of infinite polymeric chains. Uncoupled and coupled Hartree-Fock schemes, *Nonlinear Optics* **9** (1995), 25–36.

[68] B. Jansik, B. Schimmelpfennig, P. Norman, Y. Mochizuki, Y. Luo and H. Ågren, Size, order, and dimensional relations for silicon cluster polarizabilities, *Journal of Physical Chemistry A* **106** (2002), 395–399.

[69] B. Kirtman, B. Champagne, F.L. Gu and D.M. Bishop, Coupled-perturbed Hartree-Fock treatment of infinite periodic systems: application to static polarizabilities and hyperpolarizabilities of polydiacetylene, polybutatriene, and interacting pairs of polyacetylene chains, *International Journal of Quantum Chemistry* **90** (2002), 709–718.

[70] Gaussian 03, Revision A.1, M.J. Frisch, G.W. Trucks, H.B. Schlegel, G.E. Scuseria, M.A. Robb, J.R. Cheeseman, J.A. Montgomery, Jr., T. Vreven, K.N. Kudin, J.C. Burant, J.M. Millam, S.S. Iyengar, J. Tomasi, V. Barone, B. Mennucci, M. Cossi, G. Scalmani, N. Rega, G.A. Petersson, H. Nakatsuji, M. Hada, M. Ehara, K. Toyota, R. Fukuda, J. Hasegawa, M. Ishida, T. Nakajima, Y. Honda, O. Kitao, H. Nakai, M. Klene, X. Li, J.E. Knox, H.P. Hratchian, J.B. Cross, C. Adamo, J. Jaramillo, R. Gomperts, R.E. Stratmann, O. Yazyev, A.J. Austin, R. Cammi, C. Pomelli, J.W. Ochterski, P.Y. Ayala, K. Morokuma, G.A. Voth, P. Salvador, J.J. Dannenberg, V.G. Zakrzewski, S. Dapprich, A.D. Daniels, M.C. Strain, O. Farkas, D.K. Malick, A.D. Rabuck, K. Raghavachari, J.B. Foresman, J.V. Ortiz, Q. Cui, A.G. Baboul, S. Clifford, J. Cioslowski, B.B. Stefanov, G. Liu, A. Liashenko, P. Piskorz, I. Komaromi, R.L. Martin, D.J. Fox, T. Keith, M.A. Al-Laham, C.Y. Peng, A. Nanayakkara, M. Challacombe, P.M.W. Gill, B. Johnson, W. Chen, M.W. Wong, C. Gonzalez and J.A. Pople, Gaussian, Inc., Pittsburgh PA, 2003.

Journal of Computational Methods in Sciences and Engineering 4 (2004) 611–640
IOS Press


# Electric properties of diatomic molecules: A comparison of finite basis set and finite difference Hartree-Fock calculations

J. Kobus[a,*], D. Moncrieff[b] and S. Wilson[c]
[a]*Instytut Fizyki, Uniwersytet Mikołaja Kopernika, ul. Grudziądzka 5/7, 87-100 Toruń, Poland*
[b]*School of Computational Science & Information Technology, Florida State University, Tallahassee, FL 32306, USA*
[c]*Rutherford Appleton Laboratory, Chilton, Oxfordshire OX11 0QX, UK*

**Abstract.** Recent work comparing the accuracy with which the electric properties of diatomic molecules can be obtained in the Hartree-Fock model from finite basis set (the algebraic approximation) and finite difference techniques is reviewed. The electric moments $\mu$, $\Theta$, $\Omega$ and $\Phi$ are considered together with the electric dipole polarizability $\alpha_{zz}$ and hyperpolarizability $\beta_{zzz}$. By comparing results obtained within the finite basis set approximation with those given by using finite difference techniques, accurate estimates of the basis set truncation errors corresponding to different electric properties of diatomic molecules can be gained. Accurate Hartree-Fock calculations of these electric properties are an essential prerequisite of precision correlation studies.

Keywords: Hartree-fock, algebraic approximation, even-tempered basis set, basis set truncation error, finite difference, multipole moment, polarizability, hyperpolarizability

## 1. Introduction

Recent years have witnessed a growing interest in the understanding of nonlinear optical processes at the molecular level. Such processes have technological applications in optoelectronics, for example, in the construction of optical computing devices and in high speed data transmission [1–9]. The nonlinearity in the response of molecules to electromagnetic fields facilitates a mechanism for amplification, for modulation and for frequency changes in optical signals in a manner analogous to the way in which the nonlinear characteristics of transistors support these functions in semiconductor-based electronics. In particular, molecular systems with large hyperpolarizabilities are potential candidates for applications in optoelectronics.

The response of molecules to external electric fields provides one of the principal sources of information about fundamental properties of matter. When exposed to an external electric field, a molecule undergoes a complicated polarization process which involves distortion of the electron distribution and a simultaneous perturbation of the vibrational and rotational motion. The theoretical description of the effects of electric fields on molecular systems provides a test both of the theories and the associated

*Corresponding author. E-mail: jacek.kobus@phys.uni.torun.pl.

approximation techniques and algorithms. The electronic and nuclear problems are usually decoupled within the Born-Oppenheimer approximation. The electronic problem is the more demanding computational problem and has therefore commanded greater attention. The vibrational and rotational problems contain some subtleties and the interested reader is referred to the review [10] by Bishop for further details. The present review is concerned with the electronic problem. Moreover, it is concerned with approximate solutions of the molecular electronic structure problem within the Hartree-Fock approximation, the ubiquitous independent particle model. Comparisons with experiment are not of primary interest in the studies reviewed here. Accurate Hartree-Fock calculations are an essential prerequisite of precision correlation studies. For diatomic molecules the finite difference method can be used to approximate the solutions of the Hartree-Fock equation [11–14]. Molecules containing two atoms therefore provide important "benchmarks" with which the corresponding finite basis set Hartree-Fock studies can be compared [15–17]. Since finite basis set methods can be routinely applied to polyatomic molecules and to studies which take account of electron correlation effects, such comparison provide important insight into the nature of the algebraic approximation; that is, the approximation resulting for the use of finite basis set expansions.

The outline of this review is as follow: In Section 2, we describe the various truncation errors which arise in (non-relativistic) molecular electronic structure calculations. This serves to place the present review in context. In Section 3, we describe the background to the present review giving an overview of the techniques developed for implementing both finite basis set and finite difference approximations. The theoretical apparatus employed in the study of the electric properties of molecular systems is given in Section 4. This serves to establish notation and the definitions of the quantities calculated. The computational methods employed are described in Section 5, whilst in Section 6 some prototypical results are presented and discussed. Section 7 contains conclusions.

## 2. Truncation errors in molecular electronic structure calculations

The molecular electronic structure problem is most often approached in two stages (see, for example, the monographs by Wilson [18], by Szabo and Ostlund [19], and by McWeeny [20], or volume 2 of the *Handbook of Molecular Physics and Quantum Chemistry* [21]). Firstly, an independent electron model is assumed. In non-relativistic studies this is almost invariably the Hartree-Fock model whilst within the framework of relativistic quantum mechanics the corresponding Dirac-Hartree-Fock model is employed. Secondly, the corrections to the independent electron model are determined – the so-called electron correlation effects. The total energy expectation value is then written

$$E = E_{\mathrm{HF}} + E_{\mathrm{corr}} \tag{1}$$

The correlation energy may be determined by configuration interaction [24], many-body perturbation theory [25] or a coupled cluster expansion [26]. In practice, the expansion for the correlation energy has to be truncated – the configuration interaction may be limited to single and double replacements with respect to some chosen reference function or functions, the many-body perturbation expansion is usually taken to some low order, or the cluster expansion includes only single and double excitations with perhaps perturbative estimates of the triple excitation component. The correlation energy can therefore be written

$$E_{\mathrm{corr}} = E_{\mathrm{corr}}^{\mathrm{est}} + \Delta E_{\mathrm{corr}} \tag{2}$$

where $E_{\mathrm{corr}}^{\mathrm{est}}$ is the estimate of the correlation energy supported by a given model and $\Delta E_{\mathrm{corr}}$ is the remainder term. Combining Eqs (1) and (2) gives

$$E = E_{\mathrm{HF}} + E_{\mathrm{corr}}^{\mathrm{est}} + \Delta E_{\mathrm{corr}} \tag{3}$$

For calculations carried out within the algebraic approximation, this equation becomes

$$E = (E_{\mathrm{mHF}} + \varepsilon_{\mathrm{mHF}}) + (E_{\mathrm{mcorr}} + \varepsilon_{\mathrm{mcorr}}) + (\Delta E_{\mathrm{mcorr}} + \varepsilon_{\mathrm{mcorr}}) \tag{4}$$

where $E_{\mathrm{mHF}}$ is the matrix Hartree-Fock energy – the estimate of the Hartree-Fock energy supported by the chosen basis set – and $\varepsilon_{\mathrm{mHF}}$ is the basis set truncation error associated with $E_{\mathrm{mHF}}$. Values of $\varepsilon_{\mathrm{mHF}}$ may be obtained by comparing the results of matrix Hartree-Fock calculations with finite difference (or finite element) Hartree-Fock calculations since the latter is near exact provided an adequate grid is chosen. In Eq. (4), $E_{\mathrm{mcorr}}$ is the estimate of the correlation energy that is usually reported; that is, the correlation energy obtained with a given model supported by a given basis set and $\varepsilon_{\mathrm{mcorr}}$ is the associated basis set truncation error. $E_{\mathrm{mcorr}}$ may, for example, represent the `mbpt2` (or `mp2`) energy supported by a double $\zeta$ basis set. $\varepsilon_{\mathrm{mcorr}}$ is then difference between the exact `mbpt2` energy, represented in Eq. (2) by $E_{\mathrm{corr}}^{\mathrm{est}}$, and the estimate supported by the double $\zeta$ basis set, $E_{\mathrm{mcorr}}$. For the full configuration interaction method both terms in the third parenthesis in Eq. (4) are zero since the description of electron correlation is exact within the chosen basis set. For limited configuration interaction, truncated many-body perturbation series or cluster expansions both terms in the third parenthesis in Eq. (4) are non-zero but neither are calculated.

In calculating some expectation value for a property $A(X)$, say, which depends on a parameter $X$, the following expression may be written

$$A(X) = A(X)_{\mathrm{mHF}} + \chi(X)_{\mathrm{mHF}} + A(X)_{\mathrm{mcorr}} + \chi(X)_{\mathrm{mcorr}} + \Delta A(X)_{\mathrm{mcorr}} + \Delta\chi(X)_{\mathrm{mcorr}}. \tag{5}$$

The various quantities in Eq. (5) are defined by analogy with those in Eq. (4) so that, for example, $A(X)_{\mathrm{mHF}}$ is the matrix Hartree-Fock expectation value of $A$ and $\chi_{\mathrm{mHF}}$ is the associated basis set truncation error.

Let us now consider a system for which the hamiltonian depends upon the parameter $X$. In the present work this will be an applied field. The total energy and other expectation values now depend on $X$ so, in general, the derivative of $A$ with respect to $X$ may be defined as

$$\left(\frac{\partial A}{\partial X}\right)_{X=0} = \lim_{\delta X \to 0} \frac{A(\delta X) - A(0)}{\delta X}. \tag{6}$$

Obviously, Eq. (4) allows us to decompose this derivative into the corresponding contributions.

Finite difference calculations with suitably chosen grids can afford solutions of the Hartree-Fock equations for diatomic molecules to an accuracy beyond the sub-$\mu$Hartree level. The results of such calculations can be compared with the corresponding results of matrix Hartree-Fock studies to provide values of $\varepsilon_{\mathrm{mHF}}$, of $\chi_{\mathrm{mHF}}$, and of $\left(\frac{\partial \chi_{\mathrm{mHF}}(X)}{\partial X}\right)_{X=0}$. More particularly, such comparisons encourage the design of sequences of basis sets with which the approach to the Hartree-Fock limit can be explored.

In this review, we compare matrix Hartree-Fock calculations using systematic sequences of distributed universal even-tempered basis sets with finite difference Hartree-Fock studies of the electric moments $\mu$, $\Theta$, $\Omega$ and $\Phi$ and the electric dipole polarizabilities $\alpha_{zz}$ and hyperpolarizabilities $\beta_{zzz}$ for a number of diatomic molecules. Recent studies [22,23] have provided numerical evidence that, with a suitable choice of Gaussian basis set, it is possible to achieve convergence of the electric moments $\mu$, $\Theta$, $\Omega$ and $\Phi$ and the molecular electric dipole polarizabilities $\alpha_{zz}$ and hyperpolarizabilities $\beta_{zzz}$ to their limiting Hartree-Fock values.

## 3. The algebraic approximation and finite difference techniques

### 3.1. Finite basis set expansions

The use of basis set expansions is one of the greatest strengths and, at the same time, perhaps one of the greatest weaknesses of modern *ab initio* quantum chemistry (see, for example, the reviews by Wilson [17, 27,28,31] or those by Davidson and Feller [29,30]). Finite basis set expansions facilitate tractable computations of molecular electronic structures for arbitrary polyatomic systems by centring a subset of basis functions on each atomic centre and, in some cases, other points such as bond midpoints. Finite basis set expansions transform the integro-differential Hartree-Fock equations into a set of algebraic equations for the expansion coefficients which determine the weight of each basis function in the approximation of a particular orbital. This algebraic approximation provides a representation of the occupied orbitals which comprise the Hartree-Fock description of the system under investigation. It also provides a representation of the unoccupied orbitals which span the virtual spectrum and which form an essential ingredient of the description of electron correlation effects, which are the corrections to the Hartree-Fock model within the framework of non-relativistic quantum mechanics [32]. These virtual orbitals are required when implementing, for example, configuration interaction, many-body perturbation theory and coupled cluster expansions to account for electron correlation. Finite basis set expansions are particularly useful in studies of the relativistic molecular electronic structure problem which begins with the relativistic generalization of the Hartree-Fock equations, the Dirac-Hartree-Fock problem. Here the algebraic approximation not only affords a representation of the positive energy branch of the spectrum, which includes both occupied and unoccupied single particle state functions, but also a representation of the negative energy branch of the spectrum [33–35]. This negative energy branch is taken to be filled with electrons in the Dirac model. The negative energy branch might appear to play no role in theories of molecular electronic structure implemented within the so-called no virtual pair approximation in which only single particle state functions within the positive energy branch are explicitly considered. However, the negative and positive energy branches together form a complete set of solutions of the Dirac-Hartree-Fock equations and hence the positive energy functions implicitly depend on the solutions in the negative energy branch.

The weakness of basis set expansions arises from the need to use sets of limited size in order to keep a particular computation tractable. This, as we shall explain below, is also one of its greatest strengths. The first stage in the implementation of the algebraic approximation is to evaluate the molecular integrals involving the components of the hamiltonian operator over the chosen basis functions. If $N$ denotes the number of basis functions then this computation formally scales as $\sim N^4$. From these integrals and an approximation for the occupied orbitals in terms of the chosen basis set, the matrix representation of the Fock operator, which describes the motion of a single particle in the averaged field of the other electrons in the system, is constructed. Fock matrix construction also scales as $N^4$. Diagonalization of the Fock matrix, which scales as $N^3$, yields a new approximation to the occupied orbitals which can be used to construct an improved Fock matrix. This process is continued until self-consistency is achieved. Having completed the self-consistent field calculation attention is turned to the correlation problem. Integrals over basis functions then have to be transformed to integrals over molecular orbitals. This transformation scales as $N^5$. The computational complexity of electron correlation energy calculations depends upon the method but, for example, the fourth order many-body perturbation theory energy component associated with triply excited intermediate states is among the most demanding and scales as $N^7$. When using limited computing resources it is therefore important to employ basis sets of limited size. This inevitably gives rise to a *basis set truncation error*; an error which cannot be removed by any refinement of theory – such as more extensive configuration interaction or higher order many-body perturbation expansions, but which can be reduced by extension of the basis set. In the vast majority of contemporary molecular electronic structure calculations, the errors in calculated energies associated with basis set truncation effects are much larger – perhaps by one or two orders of magnitude – than the energy *differences* which are important in the determination of the chemical properties of a particular chemical system. Nevertheless, by using suitably chosen basis sets energy differences can be obtained quite accurately in spite of the fact that the calculated total energies upon which they rest may contain large errors. The choice of basis set then becomes part of the *art* of quantum chemistry. Recipes are established in which the basis set becomes an integral part of the chosen theoretical model. For example, the determination of equilibrium bond lengths requires the determination of the difference in the energy associated with different bond distances. It is known that by using small basis sets of *double* $\zeta$ type for molecules containing first-row atoms within the self-consistent field approximation estimates of equilibrium bond lengths within ~0.1Å of experimental values can be obtained. Using a more extensive basis set, including functions to describe polarization functions, leads to equilibrium bond lengths which differ from the experimental values by ~0.02–0.04 Å. However, the addition of polarization functions is essential if bond angles are to be determined. This pragmatic approach to basis set selection can be regarded as one of the greatest strengths of basis set expansion techniques. The choice of basis set is regarded as part of the *theoretical model chemistry* [36]. A particular method using a particular basis set acquires a certain degree of predictive capability once has been applied to a range of problems and its accuracy and applicability established. Fundamentally, there is no rigorous theoretical justification for this approach and yet it has formed the basis of the vast majority of applied quantum chemistry over the past fifty years.

Basis set expansions were first used in the work of Heitler and London [37] in 1927. Their seminal work heralded not only the birth of quantum chemistry but also the introduction of the basis set *and* with it basis set truncation errors. The Hartree-Fock equations were first cast in a form suitable for calculations within the algebraic approximation by Hall [38] and independently by Roothaan [39] in the early 1950s. At that time the development of the first electronic digital computers raised the possibility of the automated calculation of molecular structure and properties. The *matrix* Hartree-Fock equations of Hall and Roothaan provide the foundation for much of modern computational quantum chemistry. At about the same time, the Gaussian basis function was introduced by McWeeny [40] and independently by Boys [41]. Although the disadvantages of this choice of function were initially thought to outweigh it advantages, it is today almost ubiquitous. By the 1960s, the idea that molecular basis sets should be constructed from atomic basis sets, supplemented by *polarization functions* to describe the distortion of the atoms in the molecular environment, was firmly established. Much effort was devoted to the construction of tables of optimized atomic basis sets which could be use as ingredients of molecular basis sets. During the 1960s and 1970s, the emphasis was on the development of basis sets for self-consistent field (matrix Hartree-Fock) calculations, but gradually the main thrust of work on basis set development turned to the electron correlation problem and, in more recent years, the relativistic electronic structure problem. In many contemporary quantum chemical studies the choice of basis set is regarded as an integral part of the assumed theoretical model. A knowledge of the "performance" of a given basis set then becomes vital but, as Davidson and Feller point out in a review [29] published in 1986

> *Occasionally, for reasons of economy or computational feasibility, a basis set is selected for which the computed results are nearly without significance.*

### 3.2. Systematic implementation of the algebraic approximation

In his seminal paper on the open-shell matrix Hartree-Fock approximation, Roothaan [42] writes of the finite basis set self-consistent field method

*It is... not necessary to restrict the process to... relatively crude applications; rather we wish to use it as a practical method to obtain the fully optimized Hartree-Fock orbitals by making the basis set... sufficiently large and flexible. That this is possible without having to take an impracticably large basis set has been shown in a number of recent calculations on atoms [43] and the $H_2$ molecule [44].*

Since 1960, progress in the design of basis sets capable of supporting high precision and in methods for exploiting them has been considerable. Forty years ago in 1963, Schwartz [45] emphasized that

*...an orderly plan of attack is... essential for any mathematical analysis of the convergence rate. The old habit of picking the "best" (chosen by art) choice of a fixed number of terms is to be discarded if one wants to see how the problem converges.*

He went on

*One must choose a set of basis functions and then gradually add more... in some systematic manner.*

Also in 1963, Reeves [46,47] advocated the idea of using basis sets with exponents forming a geometric progression[1]. Specifically, the exponents, $\zeta_k$, for an even-tempered basis set are generated by the formula

$$\zeta_k = \alpha\beta^k, \beta > 1, k = 1, 2, \ldots, N \tag{7}$$

Ten years later in 1973, Ruedenberg and his co-workers [48–54] discovered that even-tempered basis sets afforded a systematic method for generating atomic basis sets. Initially, the parameters $\alpha$ and $\beta$ were optimized for specific atoms and indeed specific atomic shells. Then, Silver and his coworkers [55–60] recognized that since large and flexible basis sets are required for accurate calculations the optimization of $\alpha$ and $\beta$ for each atom can be avoided[2]. They introduced the *universal* basis set and, in particular, the *universal even-tempered* basis set[3].

The even-tempered generating formula is not unique. Equation (7) may be written in the form

$$\ln \zeta_k = \ln \alpha + k \ln \beta, k = 1, 2, \ldots, N \tag{8}$$

or alternatively as the recursion

$$\zeta_{k+1} = \zeta_k \beta, \quad k = 1, 2, \ldots, N-1 \tag{9}$$

In 1981, Wilson [76] suggested

$$\zeta_{k+1} = \zeta_k \left(\beta + k\gamma\right), \quad k = 1, 2, ..., N-1 \tag{10}$$

in which $\gamma$ is an additional parameter, as a useful generalization of the recursion Eq. (9). Huzinaga and his coworkers [77–79] introduced a further parameter in their well-tempered basis sets with exponents defined by

$$\zeta_k = \alpha\beta^k \left[1 + \gamma\left(\frac{k-1}{N}\right)^\delta\right], \quad k = 1, 2, \ldots, N \tag{11}$$

---

[1] These even-tempered Gaussian basis sets were first suggested by McWeeny [40] in his doctoral dissertation in 1948.

[2] See also the work of Mezey [61].

[3] Universal even-tempered basis sets have been employed in a wide variety of calculations [62–75] at both the matrix Hartree-Fock level and including correlation effects, in non-relativistic and relativistic studies.

Klobukowski [80,81] generalized the expansion Eq. (8) to a polynomial in $k$

$$\ln \zeta_k = \sum_{j=0}^{j=j_{max}} A_j k^j, k = 1, 2, \ldots, N \tag{12}$$

Putting $j_{max} = 1$ recovers the even-tempered generating formula. Putting $j_{max} = N - 1$ can be equivalent to setting each of the exponents equal to its variationally optimized value. Recently, Petersson et al. [82] have pointed out that Eq. (12) becomes ill-conditioned for higher values of $j_{max}$ and have proposed an alternative form based on an expansion in Legendre polynomials, $P_j(x)$

$$\ln \zeta_k = \sum_{j=0}^{j=j_{max}} B_j P_j \left( \frac{2(k-1)}{N-1} - 1 \right), \quad k = 1, 2, \ldots, N \tag{13}$$

By 1979, Ruedenberg and his coworkers [83,84] had demonstrated that sequences of even-tempered basis sets can be constructed that systematically approach a complete set. As the number of basis functions, $N$, is increased, the even-tempered basis sets will become complete sets provided that $\alpha$ and $\beta$ are functions of $N$, i.e.

$$\begin{aligned} \alpha &= \alpha(N) \\ \beta &= \beta(N) \end{aligned} \tag{14}$$

such that

$$\lim_{N \to \infty} \alpha(N) = 0 \tag{15}$$

$$\lim_{N \to \infty} \beta(N) = 1 \tag{16}$$

$$\lim_{N \to \infty} \beta(N)^N = \infty \tag{17}$$

For atoms these systematic sequences of even-tempered basis sets of Gaussian functions were shown to approach the Hartree-Fock limit. By 1980, the concept of the universal basis set had been combined with the use of systematic sequences of even-tempered basis sets to give universal systematic sequences of even-tempered basis sets [85]. These sequences were generated by the following recursions

$$\begin{aligned} \alpha(N) &= \left[ \frac{\beta(N) - 1}{\beta(N-1) - 1} \right]^a \alpha(N-1), a > 0 \\ \ln \beta(N) &= \left[ \frac{N}{N-1} \right]^b \ln \beta(N-1), -1 < b < 0 \end{aligned} \tag{18}$$

For atoms such sequences were shown to support an accuracy comparable with that achieved in the finite difference Hartree-Fock method [84] and also later with that of the finite difference Dirac-Hartree-Fock method [86]. The empirical formulae of Schmidt and Ruedenberg [84] for generating sequences of even-tempered basis sets are not unique. Recently, Kryachko and Wilson [87] have considered alternative schemes.

For molecular systems basis sets are distributed; that is, they are centred on a number of different points in space. Most often these expansion centres are taken to coincide with the atomic nuclei, but they may correspond to other points, such as the bond centre so as to describe the accumulation of charge associated with bonding. Further details of the distributions employed in the calculations reviewed here will be given in Section 5.1.

### 3.3. Molecular finite difference Hartree-Fock calculations

Finite difference methods have been employed in atomic structure calculations since the pioneering work of Hartree [88] in the 1930s. In spherical polar co-ordinates $r$, $\theta$ and $\phi$, the central field problem can be factored into an angular part depending on $\theta$ and $\phi$ which can be handled analytically and an $r$-dependent radial part which can be approximated by one-dimensional finite difference techniques [89]. The year 1983 saw the first application of finite difference methods to a molecular problem. Laaksonsen, Pyykkö and Sundholm [90] used the elliptical coordinates (prolate spheroidal coordinates) $\xi$, $\eta$ and $\phi$ with

$$\xi = \frac{r_a + r_b}{R} \tag{19}$$

$$\eta = \frac{r_a - r_b}{R} \tag{20}$$

in which $r_a$ and $r_b$ are the distances from the points $(0, 0, -R/2)$ and $(0, 0, +R/2)$ to develop a two-dimensional finite difference Hartree-Fock method for diatomic molecules. The angular part of the problem depending on $\phi$ can be handled analytically. The $(\xi, \eta)$-dependent part of the problem is solved on a two-dimensional grid.

For diatomic molecules, the use of prolate spheroidal coordinates leads to the Hartree-Fock equations in the form of Poisson-like two-dimensional partial differential equations for the molecular orbitals and potentials. Upon high order discretization the corresponding sets of linear equations are solved by the successive overrelaxation method which proved to be both stable and efficient [91,92,94]. A detailed description of the finite difference approach to the Hartree-Fock problem for diatomic molecules may be found elsewhere [93–95].

The finite difference equations are usually defined on a rectangular region

$$[1, \xi_\infty] \times [-1, 1] \tag{21}$$

with $\xi_\infty$ sufficiently large. The coordinate system actually employed in the present finite difference calculations is defined through the transformation

$$\mu = \cosh^{-1} \xi, \quad (0 \leqslant \mu \leqslant \infty) \tag{22}$$

and

$$\nu = \cos^{-1} \eta, \quad (0 \leqslant \nu \leqslant \pi) \tag{23}$$

In these coordinates the grid points were distributed uniformly according to

$$\mu_i = \mu_{i-1} + \delta_\mu, \; i = 1, 2, \ldots, n_\mu, \quad \mu_0 = 0 \tag{24}$$

with a step given by $\delta_\mu = \cosh^{-1} \xi_\infty / n_\mu$, where $\xi_\infty = 2R_\infty / R$ and $R_\infty$ is an approximation to infinity and $n_\mu$ is the number of points, and

$$\nu_j = \nu_{j-1} + \delta_\nu, \; j = 1, 2, \ldots, n_\nu, \qquad \nu_0 = 0 \tag{25}$$

with a step given by $\delta_\nu = 2\pi / n_\nu$ where $n_\nu$ is the number of points. These grids are designated

$$[n_\nu \times n_\mu; R_\infty] \tag{26}$$

### 3.4. A comparison of finite basis set and finite difference Hartree-Fock calculations for molecules

An important application of finite difference Hartree-Fock studies for diatomic molecules [12–14] is in providing a measure of the basis set truncation error in calculations carried out within the algebraic approximation, i.e. the matrix Hartree-Fock method, which, as is well known, can be readily applied to polyatomic molecules. Most frequently, such comparisons have involved total energies and orbital energies [92,95–97,99–101], but visualization techniques have also been used [102] to examine orbital amplitude difference functions and also the local orbital energy function [103].

By 1993, it had been demonstrated [15,16] that basis sets for many-electron molecular systems can be systematically constructed so as to support total Hartree-Fock energies of an accuracy approaching that achieved in finite difference calculations [11]; namely, $\sim$1 $\mu$Hartree, and basis set truncation errors effectively eliminated for practical purposes.

In more recent years, improvements in finite difference algorithms have supported higher accuracy and facilitated applications to closed-shell diatomic systems containing a heavy atom [98,99], to open-shell species [100] and, very recently, to open-shell systems containing a heavy atom [101]. During the same period, there have been corresponding improvements in the effective implementation of the algebraic approximation [97,98].

In recent work [22], a comparison of the multipole moments obtained from finite basis set calculations in which the algebraic approximation is systematically implemented with the corresponding finite difference Hartree-Fock moments for diatomic molecules afforded a more detailed measure of the quality of the finite basis set results than comparisons of total energy values alone.

## 4. Electric properties of diatomic molecules

Our purpose in this section is to establish notation and to give the definitions of the multipole moments, the electric dipole polarizability and hyperpolarizability in the case of diatomic molecules.

The hamiltonian operator for a molecule in weak interaction with fixed external field may be written as a Taylor expansion about some chosen point [104–111]

$$\mathcal{H} = \mathcal{H}_0 - \sum_\alpha \mu_\alpha F_\alpha - \frac{1}{3}\sum_{\alpha,\beta} \Theta_{\alpha\beta} F_{\alpha\beta} - \frac{1}{15}\sum_{\alpha,\beta,\gamma} \Omega_{\alpha\beta\gamma} F_{\alpha\beta\gamma} - \frac{1}{105}\sum \Phi_{\alpha\beta\gamma\delta} F_{\alpha\beta\gamma\delta} \cdots, \qquad (\alpha,\beta = x,y,z) \tag{27}$$

where $\mathcal{H}_\prime$ is the hamiltonian of the free molecule. $F_\alpha$ is the field at the origin, $F_{\alpha\beta}$ is the field gradient at the chosen expansion centre, *etc.* $\mu_\alpha$ is the dipole moment, $\Theta_{\alpha\beta}$ is the quadrupole moment, $\Omega_{\alpha\beta\gamma}$ is the octopole moment and $\Phi_{\alpha\beta\gamma\delta}$ is the hexadecapole moment. The energy of the molecule is

$$\begin{aligned}E &= \langle\Psi|\mathcal{H}|\Psi\rangle\\ &= E_0 - \sum_{\alpha}\mu_\alpha^{(0)}F_\alpha - \frac{1}{2!}\sum_{\alpha,\beta}\alpha_{\alpha\beta}F_\alpha F_\beta - \frac{1}{3!}\sum_{\alpha,\beta,\gamma}\beta_{\alpha\beta\gamma}F_\alpha F_\beta F_\gamma - \frac{1}{4!}\sum_{\alpha,\beta,\gamma,\delta}\gamma_{\alpha\beta\gamma\delta}F_\alpha F_\beta F_\gamma F_\delta \ldots\\ &\quad -\frac{1}{3}\sum_{\alpha,\beta}\Theta_{\alpha\beta}^{(0)}F_{\alpha\beta} - \frac{1}{3}\sum_{\alpha,\beta,\gamma}A_{\gamma,\alpha\beta}F_\gamma F_{\alpha\beta} - \frac{1}{6}\sum_{\alpha,\beta,\gamma,\delta}B_{\alpha\beta,\gamma\delta}F_\alpha F_\beta F_{\gamma\delta}\ldots - \frac{1}{15}\sum_{\alpha,\beta,\gamma}\Omega_{\alpha\beta\gamma}^{(0)}F_{\alpha\beta\gamma}\\ &\quad -\frac{1}{6}\sum C_{\alpha\beta,\gamma\delta}F_{\alpha\beta}F_{\gamma\delta}\ldots - \frac{1}{105}\sum_{\alpha,\beta,\gamma,\delta}\Phi_{\alpha\beta\gamma\delta}^{(0)}F_{\alpha\beta\gamma\delta} - \frac{1}{15}\sum_{\alpha,\beta,\gamma,\delta}E_{\alpha,\beta\gamma\delta}F_\alpha F_{\beta\gamma\delta} - \ldots \end{aligned} \quad (28)$$

where $\mu_\alpha^{(0)}$ is the permanent dipole moment, $\Theta_{\alpha\beta}^{(0)}$ is the permanent quadrupole moment, $\Omega_{\alpha\beta\gamma}^{(0)}$ is the permanent octopole moment and $\Phi_{\alpha\beta\gamma\delta}^{(0)}$ is the permanent hexadecapole moment. $\alpha_{\alpha\beta}$ is the dipole polarizability and $\beta_{\alpha\beta\gamma}$ and $\gamma_{\alpha\beta\gamma\delta}$ are the hyperpolarizabilities. $A_{\gamma,\alpha\beta}$ is the dipole-quadrupole polarizability, $E_{\alpha,\beta\gamma\delta}$ is the dipole-octopole polarizability, $C_{\alpha\beta,\gamma\delta}$ is the quadrupole polarizability and $B_{\alpha\beta,\gamma\delta}$ is the dipole-dipole-quadrupole hyperpolarizability.

The dipole moment may be written as the expansion

$$\begin{aligned}\mu_\alpha &= \mu_\alpha^{(0)} + \sum_{\beta}\alpha_{\alpha\beta}F_\beta + \frac{1}{2}\sum_{\beta,\gamma}\beta_{\alpha\beta\gamma}F_\beta F_\gamma + \frac{1}{3}\sum_{\beta,\gamma,\delta}\gamma_{\alpha\beta\gamma\delta}F_\beta F_\gamma F_\delta + \ldots + \frac{1}{3}\sum_{\beta,\gamma}A_{\alpha,\beta\gamma}F_{\beta\gamma}\\ &\quad + \frac{1}{6}\sum_{\beta,\gamma,\delta}B_{\alpha\beta,\gamma\delta}F_\beta F_{\gamma\delta} + \ldots,\end{aligned} \quad (29)$$

the quadrupole moment as the expansion

$$\Theta_{\alpha\beta} = \Theta_{\alpha\beta}^{(0)} + \sum_{\gamma}A_{\gamma,\alpha\beta}F_\gamma + \frac{1}{2}\sum_{\gamma,\delta}B_{\gamma\delta,\alpha\beta}F_\gamma F_\delta + \ldots + \sum_{\gamma,\delta}C_{\alpha\beta,\gamma\delta}F_{\gamma\delta} + \ldots, \quad (30)$$

and the octopole moment may be expressed in the form

$$\Omega_{\alpha\beta\gamma} = \Omega_{\alpha\beta\gamma}^{(0)} + \sum_{\delta}E_{\delta,\alpha\beta\gamma}F_\delta + \ldots \quad (31)$$

$\alpha_{\alpha\beta}$ may be written

$$\alpha_{\alpha\beta} = -\left(\frac{\partial^2 E}{\partial F_\alpha \partial F_\beta}\right)_{F_\alpha=0,\,F_\beta=0} \quad (32)$$

whilst for $\beta_{\alpha\beta\gamma}$ we have

$$\beta_{\alpha\beta\gamma} = -\left(\frac{\partial^3 E}{\partial F_\alpha \partial F_\beta \partial F_\gamma}\right)_{F_\alpha=0,\,F_\beta=0,\,F_\gamma=0} \quad (33)$$

The dipole moment may be written as the expansion Eq. (30) and so the dipole polarizability may also be written in the form

$$\alpha_{\alpha\beta} = \left(\frac{\partial \mu_\alpha}{\partial F_\beta}\right)_{F_\beta=0} \quad (34)$$

and the hyperpolarizability as

$$\beta_{\alpha\beta\gamma} = \left( \frac{\partial^2 \mu_\alpha}{\partial F_\beta \partial F_\gamma} \right)_{F_\beta=0,\, F_\gamma=0} \tag{35}$$

In the matrix Hartree-Fock calculations considered in the present review, the total multipole moments for a given molecule were determined as tensor quantities, the $2^n$-pole being an $n$-rank tensor which has the following explicit form for the first five values of $n$:

$$n = 0: \quad q = \sum_A Z_A - \int \rho dv \tag{36}$$

$$n = 1: \quad \mu_\alpha = \sum_A Z_A R_{A\alpha} - \int r_\alpha \rho dv \tag{37}$$

$$n = 2: \quad Q_{\alpha\beta} = \sum_A Z_A R_{A\alpha} R_{A\beta} - \int r_\alpha r_\beta \rho dv \tag{38}$$

$$n = 3: \quad R_{\alpha\beta\gamma} = \sum_A Z_A R_{A\alpha} R_{A\beta} R_{A\gamma} - \int r_\alpha r_\beta r_\gamma \rho dv \tag{39}$$

$$n = 4: \quad S_{\alpha\beta\gamma\delta} = \sum_A Z_A R_{A\alpha} R_{A\beta} R_{A\gamma} R_{A\delta} - \int r_\alpha r_\beta r_\gamma r_\delta \rho dv \tag{40}$$

In these expressions, $Z_A$ is the charge of nucleus $A$ and $(R_{Ax}, R_{Ay}, R_{Az})$ are its cartesian coordinates. $\rho$ is the total electronic charge distribution and $r_\alpha$ is the $\alpha$-coordinate of the electronic charge. The multipole moments are symmetric in all indices. Only the lowest non-zero moment is independent of the choice of origin. In the present matrix Hartree-Fock calculations, the moments were calculated with respect to the centre of nuclear charge in cases where only atom-centred basis functions were used and with respect to an arbitrarily chosen centre when bond-centred functions were employed. However, for quadrupole moments and higher moments it is often convenient to use alternative definitions.

Diatomic molecules possess an axis of symmetry and thus it is convenient to define the multipole moments as irreducible tensors [107]. In such a case, there is only one independent component for each electric multipole moment and this can be defined as follows:

$$M^{(k)} = \sum_A Z_A R_{Az}^k - \int r^k P_k(z/r) \rho dv. \tag{41}$$

where $P_k$ are Legendre polynomials of degree $k$. Values of $M^{(k)},\ k = 1, 2, 3, 4$ were evaluated from the orbitals determined by the finite difference Hartree-Fock program. Specifically, the first four moments defined in this way are

$$M^{(1)} \equiv \mu = \mu_z \tag{42}$$

$$M^{(2)} \equiv \Theta = \Theta_{zz} \tag{43}$$

$$M^{(3)} \equiv \Omega = \Omega_{zzz} \tag{44}$$

$$M^{(4)} \equiv \Phi = \Phi_{zzzz} \tag{45}$$

These moments were determined with respect to the geometrical centre, i.e. the bond mid-point, in the finite difference Hartree-Fock calculations.

The multipole moments determined from the matrix Hartree-Fock calculations can be related to irreducible tensor form evaluated by the finite difference Hartree-Fock program as follows:

$$\begin{aligned}
\Theta_{zz} &= Q_{zz} - \frac{1}{2}(Q_{xx} + Q_{yy}) \\
\Omega_{zzz} &= R_{zzz} - \frac{3}{2}(R_{xxz} + R_{yyz}) \\
\Phi_{zzzz} &= S_{zzzz} - 3(S_{xxzz} + S_{yyzz}) + \frac{3}{8}(S_{xxxx} + 2S_{xxyy} + S_{yyyy})
\end{aligned} \tag{46}$$

in which all quantities are transformed so as to refer to the same origin, the bond mid-point.

For a diatomic species with its nuclei on the $z$ axis, the dipole moment is given by

$$\mu_z = -\left(\frac{\partial E}{\partial F_z}\right)_{F_z=0}, \tag{47}$$

the polarizability is

$$\alpha_{zz} = -\left(\frac{\partial^2 E}{\partial F_z^2}\right)_{F_z=0} = -\left(\frac{\partial \mu_z}{\partial F_z}\right)_{F_z=0}, \tag{48}$$

with $\alpha_{xx} = \alpha_{yy}$ similarly defined, and the hyperpolarizability is

$$\beta_{zzz} = -\left(\frac{\partial^3 E}{\partial F_z^3}\right)_{F_z=0} = -\left(\frac{\partial^2 \mu_z}{\partial F_z^2}\right)_{F_z=0}, \tag{49}$$

with $\beta_{xxx} = \beta_{yyy}$ defined in a similar way.

## 5. Computational methods

### 5.1. Finite basis set calculations

Most contemporary molecular electronic structure codes employing the algebraic approximation evaluate polarizabilities and hyperpolarizabilities analytically and thus avoid the use of finite field methods. For a given theoretical approach the accurate implementation of the algebraic approximation is crucial. The finite basis sets of distributed universal even-tempered, spherical harmonic Gaussian functions described in the present review of the Hartree-Fock model were constructed following the systematic procedure which we have established in our previous publications [15,16,98,97,74,112,113]. Basis

| | | | | |
|---|---|---|---|---|
| [X:$sp$;H:$s$] | $\rightarrow$ | [X:$spd$;H:$sp$] | $\rightarrow$ | [X:$spdf$;H:$spd$] |
| $\downarrow$ | | $\downarrow$ | | $\downarrow$ |
| [X:$sp$;H:$s$;bc:$s$] | $\rightarrow$ | [X:$spd$;H:$sp$;bc:$sp$] | $\rightarrow$ | [X:$spdf$;H:$spd$;bc:$spd$] |
| $\downarrow$ | | $\downarrow$ | | $\downarrow$ |
| [X:$sp$;H:$s$;bc:$sp$] | $\rightarrow$ | [X:$spd$;H:$sp$;bc:$spd$] | $\rightarrow$ | [X:$spdf$;H:$spd$;bc:$spdf$] |

Fig. 1. Sequences of distributed universal even-tempered basis sets used in molecular electronic structure calculations for diatomic hydrides, XH.

subsets were centred on each atomic centre, designated X. A subset of functions centred on the bond midpoint, designated bc, was also included. Functions of $s$, $p$, $d$ and $f$ symmetry were located on the X nucleus in a sequence designated $sp$, $spd$, $spdf$. For diatomic hydrides, the highest symmetry type centred on H nucleus was $\ell - 1$, where $\ell$ is the highest symmetry type centred on other nucleus X. The $s$ subset consists of the $30s$ even-tempered set taken from a systematic sequence of such basis sets [84] and defined in our previous publications whilst the $p$, $d$ and $f$ subsets were the $15p$, $15d$ and $15f$ subsets also defined previously. The functions in the subset centred on the bond mid-point were of $s$, $p$, $d$ and $f$ symmetry with the restriction that the highest symmetry included was the same as that included on X. The sequence of distributed universal even-tempered basis sets employed in the study of diatomic hydrides can be represented by Fig. 1.

The largest basis set considered can therefore be represented by

$$[\mathbf{X} : spdf; \mathrm{H} : spd; \mathrm{bc} : spdf],$$

which can be written more fully as

$$[\mathbf{X} : 30s15p15d15f; \mathrm{H} : 30s15p15d; \mathrm{bc} : 30s15p15d15f].$$

Now for this basis set and for some of the smaller basis sets in the sequence above, computational linear dependence can make the matrix Hartree-Fock calculations intractable and it is therefore necessary to delete some functions to control this. Some of the more contracted functions centred on the bond midpoint were also deleted on the grounds of computational efficiency. In previous work, we have used the notation $(p + n + q)\ell$ to denote an even-tempered basis set of $n$ functions of $\ell$-symmetry which has been supplemented by $p$ more diffuse functions generated with the same even-tempered parameters and by $q$ more contracted functions generated in a similar fashion. By admitting negative values of $p$ and $q$ we can employ the same notation to define the basis sets used in the present calculations. Thus, for example, $[\mathrm{bc} : (-3 + 30 - 18)s]$ denotes a bond centred subset of $s$ functions which is truncated, both to control computational linear dependence and in the interests of efficiency, by deleting the 3 most diffuse functions and the 18 most contracted functions from the original $30s$ set.

The sequence of distributed universal even-tempered basis sets employed in the study of the diatomic molecules XY, in which neither X nor Y are H, can be represented by Fig. 2. Now the highest symmetry types included on both atomic centres are the same.

$$\begin{array}{ccccc} [\mathrm{X}:sp;\mathrm{Y}:sp] & \rightarrow & [\mathrm{X}:spd;\mathrm{Y}:spd] & \rightarrow & [\mathrm{X}:spdf;\mathrm{Y}:spdf] \\ \downarrow & & \downarrow & & \downarrow \\ [\mathrm{X}:sp;\mathrm{Y}:sp;\mathrm{bc}:sp] & \rightarrow & [\mathrm{X}:spd;\mathrm{Y}:spd;\mathrm{bc}:spd] & \rightarrow & [\mathrm{X}:spdf;\mathrm{Y}:spdf;\mathrm{bc}:spdf] \end{array}$$

Fig. 2. Sequences of distributed universal even-tempered basis sets used in molecular electronic structure calculations for the diatomic molecules XY, (X,Y $\neq$ H).

The finite basis set calculations summarized in this review were performed with the commercially available GAUSSIAN programs of Frisch et al. [114]. However, it should be emphasized that the basis sets described here could be employed with other quantum chemical codes.

### 5.2. Finite difference calculations

The finite difference Hartree-Fock calculations were carried out on a two-dimensional grid designated $[n_\nu \times n_\mu; R_\infty]$, where $n_\nu$ and $n_\mu$ are the numbers of points associated with the prolate spheroidal coordinates $\nu$ and $\mu$ and $R_\infty$ is a practical approximation to infinity which is used to establish boundary conditions. All finite difference Hartree-Fock calculations reported in this paper were carried out with the program of Kobus et al. [93,94].[4] With this program, total energies and dipole moments can be evaluated as an expectation values. Polarizabilities and hyperpolarizabilities can be evaluated by the finite field approximation. In order to assess the accuracy of the finite field approach in finite difference Hartree-Fock calculations we also determined the dipole moment from energy expectation values corresponding to an applied field by using central difference approximations. In particular, we used the 3 point central difference approximation (with errors of order $F_z^2$):

$$\mu_z = -\left(\frac{\partial E}{\partial F_z}\right)_{F_z=0} \approx -\frac{1}{2F_z}\left(E(+F_z) - E(-F_z)\right) \tag{50}$$

and its 5 point analogue (with errors of order $F_z^4$):

$$\mu_z \approx -\frac{1}{12F_z}\left(-E(+2F_z) + 8E(+F_z) - 8E(-F_z) + E(-2F_z)\right). \tag{51}$$

Dipole moments determined by using Eqs (50) and (51) can be compared with those evaluated as an expectation value. The dipole polarizability is determined either by numerical differentiation of the energy, i.e., the numerical second derivative with respect to the applied field, or by numerical differentiation of the dipole moment, i.e., the numerical first derivative with respect to the field. The 3 point central difference approximations in these two cases are:

$$\alpha_{zz} = -\left(\frac{\partial^2 E}{\partial F_z^2}\right)_{F_z=0} \approx -\frac{1}{F_z^2}\left(E(F_z) - 2E(0) + E(-F_z)\right), \tag{52}$$

[4]The finite difference Hartree-Fock program of Kobus et al. is available from *http://staff.csc.fi/~laaksone/Num2d.html.*

$$\alpha_{zz} = -\left(\frac{\partial \mu_z}{\partial F_z}\right)_{F_z=0} \approx \frac{1}{2F_z}\left(\mu_z(F_z) - \mu_z(-F_z)\right), \qquad (53)$$

whilst their 5 point analogs are:

$$\alpha_{zz} \approx -\frac{1}{12F_z^2}\left(-E(2F_z) + 16E(F_z) - 30E(0) + 16E(-F_z) - E(-2F_z)\right), \qquad (54)$$

$$\alpha_{zz} \approx \frac{1}{12F_z}\left(-\mu_z(2F_z) + 8\mu_z(F_z) - 8\mu_z(-F_z) + \mu_z(-2F_z)\right). \qquad (55)$$

The values of the polarizability obtained by taking numerical first derivatives of the dipole moment are more accurate than those obtained by taking second derivatives of the energy. The hyperpolarizability is found by numerical differentiation of the dipole moment. The numerical second derivative for the 3 point central difference approximation is:

$$\beta_{zzz} = -\left(\frac{\partial^2 \mu_z}{\partial F_z^2}\right)_{F_z=0} \approx \frac{1}{F_z^2}\left(\mu_z(F_z) - 2\mu_z(0) + \mu_z(-F_z)\right), \qquad (56)$$

and the 5 point analogue is:

$$\beta_{zzz} \approx \frac{1}{12F_z^2}\left(-\mu_z(2F_z) + 16\mu_z(F_z) - 30\mu_z(0) + 16\mu_z(-F_z) - \mu_z(-2F_z)\right). \qquad (57)$$

Because its approximation involves a numerical second derivative, the hyperpolarizability is given to a lower accuracy than the polarizability. Since numerical differentiation is known to be prone to round-off error, careful choice of both the particular approximation employed and the value of the finite field $F_z$ is required.

## 6. Some prototypical results

In this section, we present some prototypical results demonstrating comparisons of finite basis set and finite difference Hartree-Fock studies. We begin, in Section 6.1, with a study of the ground state of the FH molecule at its equilibrium geometry which concentrates on the total ground state energy and the dipole moment. This comparison is extended to the multipole moments $\mu$, $\Theta$, $\Omega$ and $\Phi$ in Section 6.2 where we employ the open-shell BeF molecule as a prototype. In Section 6.3, we compare finite basis set and finite difference multipole moments for a range of diatomic fluorides. We turn our attention to polarizabilities and hyperpolarizabilities in Section 6.4 where we again use the FH molecule as a prototype. In Section 6.5, we consider the energy, dipole moment and polarizability of the FH molecule as a function of nuclear geometry.

### *6.1.* FH *ground state at the equilibrium nuclear geometry*

Matrix Hartree-Fock energies for the ground state of the FH molecule for a nuclear separation of 1.7328 bohr, the experimentally determined equilibrium value [115], are given in Table 1 as a function of the basis set employed. These values are compared with the Hartree-Fock energy obtained from a two-dimensional finite difference calculation using a grid of $77,827$ points, designated $[223 \times 349; 65]$,

Table 1
Comparison of approximations to the Hartree-Fock energy of the FH molecule at its equilibrium geometry. Total energies are in Hartree. Energy differences are in $\mu$Hartree. $\Delta_E$ records the change in the energy with respect to the previous row. $\delta_E$ is the difference between the energy expectation value supported by a given basis set and the finite difference value obtained with the [223 × 379;100] grid

| Basis set/Grid | Notes | $E_{mHF} + 100$ | $\Delta_E$ | $\delta_E$ |
|---|---|---|---|---|
| F:$sp$;H:$s$ | | −0.043 659 7 | | 27 142.7 |
| F:$sp$;H:$s$;bc:$s$ | [a] | −0.056 662 0 | 13 002.3 | 14 140.5 |
| F:$sp$;H:$s$;bc:$sp$ | [b] | −0.069 722 4 | 13 060.4 | 1 080.1 |
| F:$spd$;H:$sp$ | | −0.069 475 4 | | 1 327.1 |
| F:$spd$;H:$sp$;bc:$sp$ | [c] | −0.070 757 8 | 1 282.4 | 44.7 |
| F:$spd$;H:$sp$;bc:$spd$ | [d] | −0.070 768 5 | 10.7 | 33.9 |
| F:$spdf$;H:$spd$ | | −0.070 760 7 | | 41.8 |
| F:$spdf$;H:$spd$;bc:$spd$ | [e] | −0.070 801 0 | 40.3 | 1.5 |
| F:$spdf$;H:$spd$;bc:$spdf$ | [f] | −0.070 801 1 | 0.2 | 1.4 |
| F:$sp$;H:$s$ | | −0.043 659 7 | | 27 142.7 |
| F:$spd$;H:$sp$ | | −0.069 475 4 | 25 815.6 | 1 327.1 |
| F:$spdf$;H:$spd$ | | −0.070 760 7 | 1 285.3 | 41.8 |
| F:$sp$;H:$s$;bc:$s$ | [a] | −0.056 662 0 | | 14 140.5 |
| F:$spd$;H:$sp$;bc:$sp$ | [c] | −0.070 757 8 | 14 095.8 | 44.7 |
| F:$spdf$;H:$spd$;bc:$spd$ | [e] | −0.070 801 0 | 43.2 | 1.5 |
| F:$sp$;H:$s$;bc:$sp$ | [b] | −0.069 722 4 | | 1 080.1 |
| F:$spd$;H:$sp$;bc:$spd$ | [d] | −0.070 768 5 | 1 046.1 | 33.9 |
| F:$spdf$;H:$spd$;bc:$spdf$ | [f] | −0.070 801 1 | 32.6 | 1.4 |
| [223 × 349; 65] | | −0.070 802 483 | | |
| [223 × 379; 100] | | −0.070 802 492 | | |

[a] [F:30$s$15$p$; H:30$s$; bc:(0+30-18)$s$]. The energy tabulated should be compared with the energy of −100.056 662 2 Hartree obtained before this truncation of the basis set.
[b] [F:30$s$15$p$; H:30$s$; bc:(0+30-18)$s$(0+15-5)$p$]. The energy tabulated should be compared with the energy of −100.069 722 8 Hartree obtained before this truncation of the basis set.
[c] [F:30$s$15$p$15$d$; H:30$s$15$p$; bc:(-3+30-18)$s$(-3+15-5)$p$].
[d] [F:30$s$15$p$15$d$; H:30$s$15$p$; bc:(-3+30-18)$s$(-3+15-5)$p$(-3+15-5)$d$].
[e] [F:30$s$15$p$15$d$15$f$; H:30$s$15$p$15$d$; bc:(-4+30-18)$s$(-4+15-5)$p$(-4+15-5)$d$].
[f] [F:30$s$15$p$15$d$15$f$; H:30$s$15$p$15$d$; bc:(-4+30-18)$s$(-4+15-5)$p$(-4+15-5)$d$(-4+15-5)$f$].

and a second calculation using $84,517$ points arranged on a grid designated $[223 \times 379; 100]$. The finite difference Hartree-Fock energies differ by $0.009$ $\mu$Hartree and this difference demonstrates the sensitivity of finite difference calculations to the choice of grid.

The first three energies recorded in Table 1 correspond to the basis sets in the first column of the sequence defined in Fig. 1; specifically

$$[\mathrm{F} : sp; \mathrm{H} : s] \rightarrow [\mathrm{F} : sp; \mathrm{H} : s; \mathrm{bc} : s] \rightarrow [\mathrm{F} : sp; \mathrm{H} : s; \mathrm{bc} : sp].$$

The first calculation uses an atom-centred basis set containing only functions of the symmetries required to describe the component ground state atoms and yields a total matrix Hartree-Fock energy which is in error by ∼27,143 $\mu$Hartree. The remaining energies presented in this set of three correspond to basis sets which also contain subsets centred on the bond mid-point. Some functions were deleted from the bond centred subset in order to control computational linear dependence and to render the computations less demanding of computer resources. However, the associated basis set truncation effects were monitored and found to be small; details are given in the notes to Table 1. Addition of a bond-centred set of $s$ symmetry to the [F:$sp$;H:$s$] basis set lowers the energy by some 13,002 $\mu$Hartree, nearly half the error in the smallest atom-centred set. Including a subset of $p$ functions on the bond midpoint lowers the energy by a further 13,061 $\mu$Hartree, just over 92% of the remaining basis set truncation error. However, the basis set truncation error associated with the largest basis set of this sub-sequence is 1080 $\mu$Hartree.

The second set of three energies given in Table 1 correspond to the second column of the sequence defined in Fig. 1:

$$[\mathrm{F} : spd; \mathrm{H} : sp] \rightarrow [\mathrm{F} : spd; \mathrm{H} : sp; \mathrm{bc} : sp] \rightarrow [\mathrm{F} : spd; \mathrm{H} : sp; \mathrm{bc} : spd].$$

This sub-sequence explores the addition of bond centred functions to the atom-centred set [F:$spd$;H:$sp$]; first by adding functions of the same symmetries as those centred on the H atom and then by including the symmetries associated with the F nucleus. The third set of three energies in Table 1 correspond to the final column of the sequence defined in Section 5.1, i.e.

$$[\mathrm{F} : spdf; \mathrm{H} : spd] \rightarrow [\mathrm{F} : spdf; \mathrm{H} : spd; \mathrm{bc} : spd] \rightarrow [\mathrm{F} : spdf; \mathrm{H} : spd; \mathrm{bc} : spdf]$$

and explores the addition of bond-centred functions to the atom-centred basis set designated [F:$spdf$;H:$spd$].

The remaining matrix Hartree-Fock energies presented in Table 1 correspond to the rows of the sequence of universal basis sets defined in Fig. 1. The first set of three energies correspond to the sequence

$$[\mathrm{F} : sp; \mathrm{H} : sp] \rightarrow [\mathrm{F} : spd; \mathrm{H} : sp] \rightarrow [\mathrm{F} : spdf; \mathrm{H} : spd].$$

This sub-sequence of basis sets is more commonly employed in molecular studies. The atom-centred basis set containing only functions of the symmetries required to describe the component ground state atoms yields a total matrix Hartree-Fock energy which is in error by $\sim$27,143 $\mu$Hartree. The addition of the primary set of polarization functions, i.e. $p$ and $d$ functions on H and F, respectively, reduces the total energy by $\sim$25,816 $\mu$Hartree and reduces the basis set truncation error in the first reported energy by over 95%. The secondary set of polarization functions, $d$- and $f$-functions on the H and F nuclei, respectively, lead to a further reduction in the energy of $\sim$1285$\mu$Hartree, which represents about 5% of the energy lowering associated with the primary set but almost 97% of the error in the energy supported by the [F:spd;H:sp] basis set.

Approximately half of the energy lowering obtained by addition of the primary polarization sets in the atom-centred calculations is obtained by addition of a bond-centred set of $s$ symmetry to the [F:sp;H:s] basis set. However, including a subset of $p$ functions on the bond midpoint supports a total energy lying some 247 $\mu$Hartree below that achieved with the atom centred basis set containing primary sets of polarization functions and only 38 $\mu$Hartree above that achieved with that with atom-centred secondary polarization sets.

The penultimate sub-sequence of basis sets corresponds to the second row of the complete sequence defined in Fig. 1, i.e.

$$[\mathrm{F} : sp; \mathrm{H} : s; \mathrm{bc} : s] \rightarrow [\mathrm{F} : spd; \mathrm{H} : sp; \mathrm{bc} : sp] \rightarrow [\mathrm{F} : spdf; \mathrm{H} : spd; \mathrm{bc} : spd].$$

These basis sets contain functions centred on the bond centre of the same symmetries as those centred on the H atom. Adding the primary set of polarization functions on all centres to the [F:$sp$;H:$s$;bc:$s$] set recovers $99.7\%$ of the basis set truncation error and $96.6\%$ of the remaining error is removed on adding the secondary polarization functions to all centres.

The final sub-sequence of basis sets corresponds to the third row of the complete sequence defined in Fig. 1, i.e.

$$[\mathrm{F}:sp;\mathrm{H}:s;\mathrm{bc}:sp] \rightarrow [\mathrm{F}:spd;\mathrm{H}:sp;\mathrm{bc}:spd] \rightarrow [\mathrm{F}:spdf;\mathrm{H}:spd;\mathrm{bc}:spdf].$$

Each set contains functions centred on the bond centre of the same symmetries as those centred on the F atom. Addition of primary polarization functions to all centres in the [F:$sp$;H:$s$;bc:$sp$] set recovers 96.9$\%$ of the error and 96.2$\%$ of the remaining error is recovered on addition of secondary polarization functions on all centres.

The total matrix Hartree-Fock energy supported by the largest basis set defined in Table 1 differs from the lower finite difference energy by 1.4 $\mu$Hartree. The accuracy achieved for the total Hartree-Fock energy for the FH ground state is thus comparable with that observed in our other studies and the improved convergence of this energy with respect to the basis set chosen by the inclusion of subsets centred on the bond midpoint is conclusively demonstrated.

Matrix Hartree-Fock dipole moments for the ground state of the FH molecule are given in Table 2 as a function of the basis set employed. The corresponding finite difference Hartree-Fock dipole moment was calculated using the grids designated $[223 \times 349; 65]$ and $[223 \times 379; 100]$. The two finite difference Hartree-Fock dipole moments agree to all figures given (i.e. $10^{-8}$ atomic units.) The dipole moment expectation values obtained from matrix Hartree-Fock calculations are arranged according to the sequence of basis sets defined in Section 5.1 and following the presentation of the energy expectation values given in Table 1. The convergence pattern seen for the dipole moment mirrors that for the total energy. For the atom-centred basis sets the largest change in the dipole moment is seen on the addition of the primary polarization functions, accounting for over 99.5$\%$ of the basis set truncation error measured in the dipole moment expectation value supported by the [F:$sp$;H:$s$] basis set. For the basis sets containing both atom- and bond-centred subsets the addition of bond-centred functions of $p$ symmetry to the [F:$sp$;H:$s$;bc:$s$] set has the most significant impact on the calculated expectation value recovering 98.8$\%$ of the basis set truncation error. The dipole moment supported by the largest basis set considered in Table 2 agrees with the finite difference value to 7 decimal places. The efficacy of even-tempered basis sets containing both atom-centred and bond-centred functions in supporting the Hartree-Fock dipole moment is demonstrated.

The accuracy achieved in the calculation of the dipole moment of FH summarized in Table 2 is comparable with that achieved in our other studies of a number of diatomic systems [22].

Another purpose of this review is to examine the accuracy with which Hartree-Fock electric dipole polarizabilities and hyperpolarizabilities can be determined by comparing matrix Hartree-Fock results with those obtained from finite difference studies. The determination of dipole polarizabilities and hyperpolarizabilities from finite difference Hartree-Fock calculations involves the application of a *finite field* and the numerical evaluation of first and second derivatives. To provide some insight into the accuracy of the finite field procedure in finite difference Hartree-Fock studies, we calculated dipole moments from energy expectation values according to the central difference approximation formulae Eqs (50) and (51) given above. A field $F_z$ of 0.0001 a.u. was applied along the $z$ axis. Finite difference Hartree-Fock calculations were carried out for $E(\pm F_z)$ and $E(\pm 2F_z)$ The results are given in the final two rows of Table 2. It can be seen the numerical determination of first derivatives of the energy with respect to applied field results in a loss of accuracy of about 2 significant figures.

Table 2
Comparison of approximations to the Hartree-Fock dipole moment of the FH molecule at its equilibrium geometry. Dipole moments are in a.u. Differences between calculated dipole moments are in a.u. $\times 10^{-3}$. $\Delta_{\mu_z}$ records the change in the dipole moment with respect to the previous row. $\delta_{\mu_z}$ is the difference between the dipole moment expectation value supported by a given basis set and the finite difference value obtained with the [223 × 379; 100] grid

| Basis set/Grid | Notes | $\mu_z$ | $\Delta_{\mu_z}$ | $\delta_{\mu_z}$ |
|---|---|---|---|---|
| F:*sp*;H:*s* | | 0.930 631 3 | | 174.555 |
| F:*sp*;H:*s*;bc:*s* | [a] [g] | 0.904 214 1 | 26.417 | 148.138 |
| F:*sp*;H:*s*;bc:*sp* | [b] [h] | 0.757 806 3 | 146.408 | 1.730 |
| F:*spd*;H:*sp* | | 0.756 864 7 | | 0.789 |
| F:*spd*;H:*sp*;bc:*sp* | [c] | 0.756 177 5 | 0.687 | 0.102 |
| F:*spd*;H:*sp*;bc:*spd* | [d] | 0.756 153 3 | 0.024 | 0.077 |
| F:*spdf*;H:*spd* | | 0.756 087 2 | | 0.011 |
| F:*spdf*;H:*spd*;bc:*spd* | [e] | 0.756 076 1 | 0.011 | 0.000 |
| F:*spdf*;H:*spd*;bc:*spdf* | [f] | 0.756 075 9 | 0.000 | 0.000 |
| F:*sp*;H:*s* | | 0.930 631 3 | | 174.555 |
| F:*spd*;H:*sp* | | 0.756 864 7 | 173.767 | 0.789 |
| F:*spdf*;H:*spd* | | 0.756 087 2 | 0.778 | 0.011 |
| F:*sp*;H:*s*;bc:*s* | [a] [g] | 0.904 214 1 | | 148.138 |
| F:*spd*;H:*sp*;bc:*sp* | [c] | 0.756 177 5 | 148.037 | 0.102 |
| F:*spdf*;H:*spd*;bc:*spd* | [e] | 0.756 076 1 | 0.101 | 0.000 |
| F:*sp*;H:*s*;bc:*sp* | [b] [h] | 0.757 806 3 | | 1.730 |
| F:*spd*;H:*sp*;bc:*spd* | [d] | 0.756 153 3 | 1.653 | 0.077 |
| F:*spdf*;H:*spd*;bc:*spdf* | [f] | 0.756 075 9 | 0.077 | 0.000 |
| [223 × 349; 65] | | 0.756 075 86 | | |
| [223 × 379; 100] | | 0.756 075 86 | | |
| [223 × 379; 100] | [i] | 0.756 075 7 | | |
| [223 × 379; 100] | [j] | 0.756 075 4 | | |

[a] [F:30$s$15$p$; H:30$s$; bc:(0+30-18)$s$]. The energy tabulated should be compared with the energy of −100.056 662 2 Hartree obtained before this truncation of the basis set.
[b] [F:30$s$15$p$; H:30$s$; bc:(0+30-18)$s$(0+15-5)$p$]. The energy tabulated should be compared with the energy of −100.069 722 8 Hartree obtained before this truncation of the basis set.
[c] [F:30$s$15$p$15$d$; H:30$s$15$p$; bc:(-3+30-18)$s$(-3+15-5)$p$].
[d] [F:30$s$15$p$15$d$; H:30$s$15$p$; bc:(-3+30-18)$s$(-3+15-5)$p$(-3+15-5)$d$].
[e] [F:30$s$15$p$15$d$15$f$; H:30$s$15$p$15$d$; bc:(-4+30-18)$s$(-4+15-5)$p$(-4+15-5)$d$].
[f] [F:30$s$15$p$15$d$15$f$; H:30$s$15$p$15$d$; bc:(-4+30-18)$s$(-4+15-5)$p$(-4+15-5)$d$(-4+15-5)$f$].
[g] The dipole moment tabulated should be compared with that of 0.904 213 8 a.u. obtained before the truncation of the basis set described in Note [a].
[h] The dipole moment tabulated should be compared with that of 0.757 805 6 a.u. obtained before the truncation of the basis set described in Note [b].
[i] First derivative of the energy with respect to applied field calculated by 3 point central difference approximation according to Eq. (50).
[j] First derivative of the energy with respect to applied field calculated by 5 point central difference approximation according to Eq. (51).

### 6.2. *Convergence of* $\mu$, $\Theta$, $\Omega$ *and* $\Phi$ *for* BeF *in the algebraic approximation*

In Table 3, the multipole moments and total Hartree-Fock energies for the ground state of the open-shell BeF molecule are displayed for a sequence of even-tempered basis sets and for two grids.

Table 3
Multipole moments and total Hartree-Fock energies for the BeF ground state for a sequence of even-tempered basis sets and for two grids[a]

| Basis set/Grid | $\mu$ | $\Theta$ | $\Omega$ | $\Phi$ | $E^{\mathrm{b}}$ |
|---|---|---|---|---|---|
| [Be,F:30$s$15$p$] | 1.7471 | −4.7600 | 5.2936 | −7.3890 | 49797.3 |
| [Be:32$s$17$p$;F:30$s$15$p$] | 1.7338 | −4.7080 | 5.0393 | −6.7973 | 50167.8 |
| [Be,F:30$s$15$p$15$d$] | 1.3024 | −4.0364 | 5.5505 | −6.2023 | 71281.7 |
| [Be:32$s$17$p$17$d$;F:30$s$15$p$15$d$] | 1.2744 | −4.1313 | 5.9385 | −7.4481 | 71508.2 |
| [Be,F:30$s$15$p$15$d$15$f$] | 1.2889 | −4.1133 | 5.7757 | −7.5722 | 72525.5 |
| [Be:32$s$17$p$17$d$17$f$;F:30$s$15$p$15$d$15$f$] | 1.2728 | −4.1623 | 5.8712 | −7.5922 | 72615.5 |
| [Be:32$s$17$p$;F:30$s$15$p$;bc:30$s$15$p$] | 1.3180 | −4.2176 | 5.7366 | −7.5686 | 71274.2 |
| [Be:32$s$17$p$;F:30$s$15$p$;bc:20$s$15$p$] | 1.3180 | −4.2176 | 5.7366 | −7.5686 | 71274.2 |
| [Be:32$s$17$p$;F:30$s$15$p$;bc:18$s$14$p$] | 1.3180 | −4.2176 | 5.7366 | −7.5686 | 71274.2 |
| [Be:32$s$17$p$;F:30$s$15$p$;bc:16$s$13$p$] | 1.3180 | −4.2176 | 5.7366 | −7.5686 | 71274.2 |
| [Be:32$s$17$p$;F:30$s$15$p$;bc:14$s$12$p$] | 1.3180 | −4.2176 | 5.7366 | −7.5686 | 71274.2 |
| [Be:32$s$17$p$;F:30$s$15$p$;bc:12$s$11$p$] | 1.3180 | −4.2176 | 5.7366 | −7.5686 | 71274.0 |
| [Be:32$s$17$p$17$d$;F:30$s$15$p$15$d$;bc:9$s$9$p$10$d$] | 1.2727 | −4.1616 | 5.8769 | −7.5818 | 72667.7 |
| [Be:32$s$17$p$17$d$17$f$;F:30$s$15$p$15$d$15$f$;bc:9$s$8$p$9$d$8$f$] | 1.2726 | −4.1624 | 5.8727 | −7.5843 | 72683.8 |
| $[169 \times 193, 40]$ | 1.2726 | −4.1622 | 5.8728 | −7.5857 | 72685.0 |
| $[319 \times 415, 40]$ | 1.2726 | −4.1622 | 5.8728 | −7.5857 | 72685.0 |

[a]The basis sets and grids employed are defined by the functions of $s$, $p$, $d$ and $f$ symmetry centred on the Be atom, those centred on the F atom, and those centred on the bond centre (bc). $\mu$ in Debye, $\Theta$ in Debye Ångstrom, $\Omega$ in Debye Ångstrom$^2$, $\Phi$ in Debye Ångstrom$^3$.
[b]$(E_{mHF} + 114.1\ E_h)$ in $E_h \times 10^{-6}$.

The two finite difference Hartree-Fock calculations employed the grids: [169 $\times$ 193,40], i.e. 32,617 points, and [319 $\times$ 415,40], i.e. 132,385 points. The total Hartree-Fock energies supported by these grids are −114.172 684 996 Hartree and −114.172 685 040 Hartree, respectively; a difference of 0.044 $\mu$Hartree. There is no difference between the dipole, quadrupole, octopole and hexadecapole moments supported by these grids to the accuracy quoted in Table 3.

In Table 3, the results of calculations for basis sets containing only atom-centred functions are recorded for both the 30s15p... sets and the 32s17p... sets. For basis sets containing both atom-centred and bond-centred functions only the 32s17p... sets are considered on the Be nucleus. For the isotropic basis set the dipole moment is in error by 36.2%; an error which is reduced markedly by the introduction of atom-centred polarization functions, first to 0.1% on the addition of $d$ functions and still further on adding $f$ functions. On the other hand, adding functions of $s$ and $p$ symmetry centred at the mid-point of the bond to the isotropic basis set reduces the error to ~3.6% of the exact value. Supplementing this $sp$ basis set by functions of $d$ symmetry centred both on the atomic nuclei and on the bond mid-point yields a dipole moment expectation value which is in error by less than 0.01%.

### 6.3. *Comparison of finite difference and finite basis set moments for diatomic fluorides in the Hartree-Fock approximation*

Multipole moments for some closed-shell diatomic fluorides containing an increasingly heavy atom and calculated relative to the geometrical centre are displayed in the upper portion of Table 4. In this Table, *fdHF* labels the finite difference Hartree-Fock results, *mHF* labels the matrix Hartree-Fock results and rows labelled $\delta$ give the differences between the finite difference and matrix Hartree-Fock values. The average absolute difference between the dipole moments values obtained from the finite difference and finite basis set calculations is 0.0015 Debye. The largest difference in the dipole moment values is for TlF for which a difference of 0.0030 Debye was observed which represents ~0.1% of the exact value.

For the quadrupole moments the average absolute difference is 0.0050 Debye Ångstrom with the largest difference being the 0.0112 Debye Ångstrom measured for TlF, some ~0.13% of the exact value. For the octopole moments the average difference is 0.0114 Debye Ångstrom $^2$, the largest difference is 0.0275 Debye Ångstrom$^2$, again for TlF and amounting to ~0.3% of the exact value. The average difference between the hexadecapole moments obtained from finite difference and finite basis set Hartree-Fock calculations is 0.0223 Debye Ångstrom$^3$ with the largest being 0.0490 Debye Ångstrom$^3$, some 0.24% of the exact (finite difference) value.

In the lower part of Table 4, multipole moments are presented for some open-shell diatomic fluorides containing an increasingly heavy atom calculated relative to the geometrical centre. The average difference between the non-zero dipole moments values obtained from the finite difference and finite basis set calculations is 0.0001 Debye. For the quadrupole moments this average difference is 0.0002 Debye Ångstrom and for the octopole moments the average difference is 0.0012 Debye Ångstrom $^2$ with the largest being 0.0028 Debye Ångstrom$^2$, some ~0.02% of the exact value, in the case of the CaF ground state. The average difference between the hexadecapole moments obtained from finite difference and finite basis set Hartree-Fock calculations is 0.0145 Debye Ångstrom$^3$. The largest difference is for the CaF molecules but at 0.0338 Debye Ångstrom$^3$ this corresponds to ~0.13% of the exact, finite difference value.

### 6.4. *Polarizabilities and hyperpolarizabilities*

We turn now to the polarizability of FH and Table 5 which records the matrix Hartree-Fock dipole polarizability as a function of the basis set employed. Values of $\alpha_{zz}$ are determined from the finite basis set calculations and compared with the corresponding value obtained from finite difference calculations. The first two finite difference polarizabilities, which are given in the lower part of Table 5, differ by $2.3 \times 10^{-6}$ a.u. These values were obtained from finite field dipole moment expectation values by means of the 5 point central difference approximation given in Eq. (55). They are compared with the corresponding finite field finite difference values obtained by using the 3 point central difference approximation to the first derivative of the dipole moment expectation values with respect to the applied field and by using both the 3 and the 5 point approximation to determine second derivatives of the energy expectation values with respect to the applied field. There is little difference between the polarizability estimates determined by the 3 point and the 5 point formulae for evaluating the first derivative of the dipole moment. Numerical differentiation of calculated dipole moments means that the quote polarizabilities have at most 7 or 8 significant figures. However, there is considerable loss of accuracy when the estimates of the polarizability are obtained by evaluating second derivatives of the energy expectation values obtained in the applied field.

The polarizability components obtained from matrix Hartree-Fock calculations are arranged according to the sequence of basis sets defined in Figure 1 and following the presentation of the energy expectation values given in Table 1. The convergence pattern for the polarizability components mirrors to a large extent that of the energy and the dipole moment. The value of the component $\alpha_{zz}$ supported by the [F:$sp$;H:$s$] basis set is in error by some 1.2 a.u. For the atom-centred sets, the error is reduced by 99.1% upon adding the primary set of polarization functions and a further 0.9% by the secondary set leading to an error of 155 $\times 10^{-6}$ a.u. for the largest atom-centred basis set considered. The basis set truncation error for the polarizability component $\alpha_{zz}$ measured with the smallest of the sets containing bond centred subsets is, at 1.400 a.u., 0.212 a.u. greater than that for the smallest atom-centred set. Adding functions of $p$-symmetry at the bond midpoint reduces this error by 93.5%, whilst adding a primary polarization

Table 4
Multipole moments for some diatomic fluroides containing an increasingly heavy atom calculated relative to the geometrical centre[a]

| Molecule | Method | $\mu$ | $\Theta$ | $\Omega$ | $\Phi$ | $E_{HF}$ |
|---|---|---|---|---|---|---|
| BF | fdHF[b] | 0.8713 | −3.9523 | 4.5293 | −3.6677 | −124.168 780 0 |
| | mHF[c] | 0.8713 | −3.9522 | 4.5290 | −3.6732 | −124.168 777 0 |
| | $\delta$ | 0.0000 | −0.0001 | 0.0003 | 0.0055 | 2.2 |
| AlF | fdHF[d] | −1.3227 | −5.8863 | 5.9769 | −7.2501 | −341.488 382 8 |
| | mHF[e] | −1.3228 | −5.8859 | 5.9754 | −7.2482 | −341.488 369 5 |
| | $\delta$ | 0.0001 | −0.0004 | 0.0015 | −0.0019 | 13.3 |
| GaF | fdHF[f] | −2.2235 | −5.9097 | 6.1715 | −9.7471 | −2022.836 882 2 |
| | mHF[g] | −2.2255 | −5.9048 | 6.1630 | −9.7719 | −2022.836 862 3 |
| | $\delta$ | 0.0020 | −0.0049 | 0.0085 | 0.0248 | 19.9 |
| InF | fdHF[f] | −2.8551 | −7.8301 | 8.2079 | −16.3040 | −5839.729 193 5 |
| | mHF[h] | −2.8574 | −7.8218 | 8.1887 | −16.2744 | −5839.729 178 0 |
| | $\delta$ | 0.0023 | −0.0083 | 0.0192 | −0.0296 | 15.5 |
| TlF | fdHF[i] | −3.2138 | −8.6314 | 9.0890 | −20.3563 | −19061.376 349 |
| | mHF[j] | −3.2168 | −8.6202 | 9.0615 | −20.3067 | −19061.376 288 |
| | $\delta$ | 0.0030 | −0.0112 | 0.0275 | −0.0496 | 61 |
| BeF | fdHF[k] | −1.2727 | −4.1622 | 5.8728 | −7.5857 | −114.172 685 040 |
| | mHF[l] | −1.2726 | −4.1624 | 5.8727 | −7.5843 | −114.172 683 8 |
| | $\delta$ | −0.0001 | 0.0002 | 0.0001 | −0.0014 | 1.2 |
| MgF | fdHF[k] | −3.1005 | −6.7716 | 9.3090 | −17.1685 | −299.159 119 304 |
| | mHF[m] | −3.1005 | −6.7715 | 9.3074 | −17.1509 | −299.159 112 9 |
| | $\delta$ | 0.0000 | −0.0001 | 0.0016 | −0.0176 | 6.4 |
| CaF | fdHF[n] | −2.6450 | −10.6872 | 14.2460 | −26.7662 | −776.329 958 335 |
| | mHF[o] | −2.6451 | −10.6871 | 14.2432 | −26.7324 | −776.329 956 0 |
| | $\delta$ | 0.0001 | −0.0001 | 0.0028 | −0.0338 | 2.3 |
| SrF | fdHF[p] | −2.5759 | −11.8026 | 14.1228 | −26.1455 | −3231.119 630 232 |
| | mHF[q] | −2.5760 | −11.8024 | 14.1225 | −26.1506 | −3231.119 629 2 |
| | $\delta$ | 0.0001 | −0.0002 | 0.0003 | 0.0051 | 1.1 |

[a] $\mu$ in Debye; $\Theta$ in Debye Ångstrom; $\Omega$ in Debye Ångstrom$^2$. $\Phi$ in Debye Ångstrom$^3$, $E_{HF}$ in Hartree, $\delta E$ in $\mu$Hartree.
[b] [169 × 193; 40],
[c] [B:26$s$18$p$18$d$18$f$ F: 26$s$18$p$18$d$8$f$;bc:23$s$13$p$13$d$14$f$],
[d] [169 × 193; 40] [98],
[e] [Al:28$s$20$p$20$d$20$f$;F:26$s$18$p$18$d$18$f$;bc:22$s$15$p$17$d$17$f$] [98].
[f] [295 × 295; 40] [98,99].
[g] [Ga:48$s$24$p$24$d$24$f$;F:30$s$15$p$15$d$15$f$;bc:27$s$12$p$14$d$15$f$] [98,99].
[h] [In:56$s$28$p$28$d$28$f$28$g$;F:30$s$15$p$15$d$15$f$;bc:51$s$23$p$24$d$25$f$25$g$] [99].
[i] [595 × 595; 45; ] [99].
[j] [Tl:58$s$29$p$29$d$29$f$29$g$;F:30$s$15$p$15$d$15$f$;bc:53$s$24$p$26$d$25$f$25$g$] [99].
[k] [319 × 415; 40] [101].
[l] [Be:32$s$17$p$17$d$17$f$;F:30$s$15$p$15$d$15$f$;bc:9$s$8$p$9$d$8$f$] [101].
[m] [Mg:32$s$17$p$17$d$17$f$;F30$s$15$p$15$d$15$f$;bc:9$s$9$p$10$d$8$f$] [101].
[n] [349 × 499; 80] [101].
[o] [Ca:48$s$27$p$27$d$27$f$;F:30$s$15$p$15$d$15$f$;bc:10$s$9$p$9$d$11$f$] [101].
[p] [391 × 547; 80] [101].
[q] [Sr:56$s$31$p$31$d$31$f$31$g$;F:30$s$15$p$15$d$15$f$;bc:9$s$9$p$9$d$9$g$] [101].

sets on the atom centres accounts for another 6.3%. The secondary set of polarization functions recovers ∼0.15% of the error in the smallest set containing bond-centred functions. It can be seen that the largest basis set considered in Table 5 supports a value of $\alpha_{zz}$ which lies within $23 \times 10^{-6}$ a.u. of the finite

Table 5

Comparison of approximations to the Hartree-Fock dipole polarizability of the FH molecule at its equilibrium geometry. Dipole polarizabilities are in a.u. Differences between calculated dipole polarizabilities are in a.u. $\times 10^{-3}$. $\Delta_{\alpha_{xx}}$ and $\Delta_{\alpha_{zz}}$ record the changes in the dipole polarizability components $\alpha_{xx}$ and $\alpha_{zz}$ with respect to the previous row. $\delta_{\alpha_{zz}}$ is the difference between the dipole polarizability component $\alpha_{zz}$ supported by a given basis set and the finite difference value obtained with the [223 × 379; 100] grid using a 5 point central difference approximation applied to finite field dipole moment expectation values

| Basis set/Grid | Notes | $\alpha_{zz}$ | $\Delta_{\alpha_{zz}}$ | $\delta_{\alpha_{zz}}$ |
|---|---|---|---|---|
| F:*sp*;H:*s* | | 4.565 125 | | 1189.13 |
| F:*sp*;H:*s*;bc:*s* | [a] [g] | 4.353 277 | 211.85 | 1400.98 |
| F:*sp*;H:*s*;bc:*sp* | [b] [h] | 5.663 546 | −1310.27 | 90.71 |
| F:*spd*;H:*sp* | | 5.743 414 | | 10.84 |
| F:*spd*;H:*sp*;bc:*sp* | [c] | 5.751 749 | −8.34 | 2.50 |
| F:*spd*;H:*sp*;bc:*spd* | [d] | 5.752 072 | −0.32 | 2.18 |
| F:*spdf*;H:*spd* | | 5.754 100 | | 0.16 |
| F:*spdf*;H:*spd*;bc:spd | [e] | 5.754 233 | −0.13 | 0.02 |
| F:*spdf*;H:*spd*;bc:*spdf* | [f] | 5.754 232 | 0.00 | 0.02 |
| F:*sp*;;H:*s* | | 4.565 125 | | 1189.13 |
| F:*spd*;H:*sp* | | 5.743 414 | −1178.29 | 10.84 |
| F:*spdf*;H:*spd* | | 5.754 100 | −10.69 | 0.16 |
| F:*sp*;H:*s*;bc:*s* | [a] [g] | 4.353 277 | | 1400.98 |
| F:*spd*;H:*sp*;bc:*sp* | [c] | 5.751 749 | -1398.47 | 2.50 |
| F:*spdf*;H:*spd*;bc:*spd* | [e] | 5.754 233 | -2.48 | 0.02 |
| F:*sp*;H:*s*;bc:*sp* | [b] [h] | 5.663 546 | | 90.71 |
| F:*spd*;H:*sp*;bc:*spd* | [d] | 5.752 072 | -88.52 | 2.18 |
| F:*spdf*;H:*spd*;bc:*spdf* | [f] | 5.754 232 | -2.16 | 0.02 |
| [223 × 349; 65] | [i] | 5.754 252 7 | | |
| [223 × 379; 100] | [i] | 5.754 254 9 | | |
| [223 × 379; 100] | [j] | 5.77 | | |
| [223 × 379; 100] | [k] | 5.96 | | |
| [223 × 379; 100] | [l] | 5.754 255 0 | | |

[a] [F:30$s$15$p$; H:30$s$; bc:(0+30-18)$s$]. The energy tabulated should be compared with the energy of −100.056 662 2 Hartree obtained before this truncation of the basis set.
[b] [F:30$s$15$p$; H:30$s$; bc:(0+30-18)$s$(0+15-5)$p$]. The energy tabulated should be compared with the energy of −100.069 722 8 Hartree obtained before this truncation of the basis set.
[c] [F:30$s$15$p$15$d$; H:30$s$15$p$; bc:(-3+30-18)$s$(-3+15-5)$p$].
[d] [F:30$s$15$p$15$d$; H:30$s$15$p$; bc:(-3+30-18)$s$(-3+15-5)$p$(-3+15-5)$d$].
[e][F:30$s$15$p$15$d$15$f$; H:30$s$15$p$15$d$; bc:(-4+30-18)$s$(-4+15-5)$p$(-4+15-5)$d$].
[f] [F:30$s$15$p$15$d$15$f$; H:30$s$15$p$15$d$; bc:(-4+30-18)$s$(-4+15-5)$p$(-4+15-5)$d$(-4+15-5)$f$].
[g] The polarizability tabulated should be compared with that of 4.353 273 a.u. obtained before the truncation of the basis set described in Note [a].
[h] The polarizability tabulated tabulated should be compared with that of 5.663 544 a.u. obtained before the truncation of the basis set described in Note [b].
[i] Calculated from finite field dipole moment expectation values by 5 point central difference approximation according to Eq. (55)
[j] Calculated from finite field energy expectation values by 3 point central difference approximation according to Eq. (52).
[k] Calculated from finite field energy expectation values by 5 point central difference approximation according to Eq. (54).
[l] Calculated from finite field dipole moment expectation values by 3 point central difference approximation according to Eq. (53).

difference result.

Turning to the hyperpolarizability, Table 6, finite difference Hartree-Fock values have at most 3 or 4 significant figures because of errors introduced by the numerical differentiation required in the finite field approach. For the basis sets containing only atom-centred subsets, nearly all of the truncation error measured for the smallest basis set is recovered on adding the primary set of polarization functions to each centre. Adding sets of secondary polarization functions to each atomic centre leads to a hyperpolarizability component $\beta_{zzz}$ that is within $0.1 \times 10^{-3}$ of our reference value. The smallest basis set containing both atom- and bond-centred subsets supports an approximation to $\beta_{zzz}$ for which the error is some 67% of that for the smallest set containing only atom-centred functions. The error associated with the [F:$sp$;H:$s$;bc:$s$] set is reduced by some 71.5% by adding a set of function with $p$-symmetry to the bond centre. Over 98% of the error associated with the [F:$sp$;H:$s$;bc:$sp$] set is recovered on adding primary polarization sets to the atom centres and almost 97% of the error associated with the [F:$spd$;H:$sp$;bc:$spd$] set is recovered when it is supplemented by secondary, atom-centred polarization sets. For the largest basis set considered in Table 6 the difference between the matrix Hartree-Fock and the finite difference Hartree-Fock polarizabilities is $1.4 \times 10^{-3}$ a.u.

### 6.5. *Energy, dipole moment and polarizability curves for the* FH *ground state*

Matrix Hartree-Fock energy, dipole moment, dipole polarizability and hyperpolarizability values together with the corresponding finite difference results were calculated for the FH molecule as a function of the nuclear separation for $R_{\mathrm{FH}} = 1.5\,(0.1)\,2.2$ bohr and are displayed in Tables 7, 8, 9 and 10, respectively. The basis set employed in the matrix Hartree-Fock calculations is the largest defined in Table 1.

The points calculated on the energy curves are presented in Table 7. The difference between the matrix Hartree-Fock energies and the finite difference Hartree-Fock values, $\delta_E$, is of the order 1-3 $\mu$Hartree over the range of nuclear separations considered and increases from $\sim$1 $\mu$Hartree at $R_{\mathrm{FH}} = 1.5$ bohr to $\sim$2.5 $\mu$Hartree at $R_{\mathrm{FH}} = 2.2$ bohr. Basis set superposition errors are associated with basis set deficiencies which do not remain constant with changes in nuclear geometry. Such errors can affect both intermolecular and intramolecular interactions. Most often basis set superposition errors are estimated by introducing a correction. The function counterpoise correction, first suggested by Jansen and Ros [116] but usually known as the Boys-Bernardi [117] recipe, is the most commonly employed technique for obtaining this estimate. Using the present finite difference Hartree-Fock energies the basis set superposition error in the interaction energy can be determined to a sub-$\mu$Hartree level of accuracy.

Basis set superposition errors also arise in calculations of properties other than the energy. If the basis set truncation error for a property X and for geometry $R$ is defined as

$$\delta_X(R) = X^{(\mathrm{mHF})}(R) - X^{(\mathrm{fdHF})}(R), \tag{58}$$

then the corresponding basis set superposition error associated with the geometry change $R_1 \rightarrow R_2$ can be defined as

$$\Delta_X(R_1 \rightarrow R_2) = \delta_X(R_1) - \delta_X(R_2). \tag{59}$$

Tables 8, 9 and 10 contain data points for dipole moment, dipole polarizability and hyperpolarizability which can be used to examine the basis set truncation and superposition errors for these properties in the present matrix Hartree-Fock calculations for the FH molecule. The magnitude of the difference between the matrix Hartree-Fock dipole moments and the corresponding finite difference Hartree-Fock values increases from $2 \times 10^{-7}$ to $3.4 \times 10^{-6}$ over the range of nuclear separations considered which shows that for the basis set used the basis set superposition errors are very small. If finite difference Hartree-Fock values of polarizabilities have at most 7 or 8 significant figures the matrix Hartree-Fock values can be seen to deteriorate slightly with increasing $R_{\mathrm{FH}}$. In the case of hyperpolarizabilities (Table 10) we estimate that the finite difference Hartree-Fock values have at most 3 or 4 significant figures and thus the matrix Hartree-Fock and finite difference Hartree-Fock approximations seem to yield a roughly comparable level of accuracy and there is no sign of basis set superposition error.

Table 6

Comparison of approximations to the Hartree-Fock hyperpolarizability of the FH molecule at its equilibrium geometry. Hyperpolarizabilities are in a.u. Differences between calculated hyperpolarizabilities are in a.u. $\times 10^{-3}$. $\Delta_{\beta_{xxx}}$ and $\Delta_{\beta_{zzz}}$ record the changes in the dipole polarizability components $\beta_{xxx}$ and $\beta_{zzz}$ with respect to the previous row. $\delta_{\beta_{zzz}}$ is the difference between the dipole polarizability component $\beta_{zzz}$ supported by a given basis set and the finite difference value obtained with the [223 × 379; 100] grid using a 5 point central difference approximation applied to finite field dipole moment expectation values

| Basis set/Grid | Notes | $\beta_{zzz}$ | $\Delta_{\beta_{zzz}}$ | $\delta_{\beta_{zzz}}$ |
|---|---|---|---|---|
| F:$sp$;H:$s$ | | −18.707 6 | | −10260 |
| F:$sp$;H:$s$;bc:$s$ | [a] [g] | −15.338 8 | −3369 | −6891 |
| F:$sp$;H:$s$;bc:$sp$ | [b] [h] | −10.410 7 | -4928 | −1963 |
| F:$spd$;H:$sp$ | | -8.455 8 | | −8 |
| F:$spd$;H:$sp$;bc:$sp$ | [c] | −8.485 2 | 29 | −37 |
| F:$spd$;H:$sp$;bc:$spd$ | [d] | −8.490 0 | 5 | −42 |
| F:$spdf$;H:$spd$ | | −8.448 0 | | 0 |
| F:$spdf$;H:$spd$;bc:spd | [e] | −8.449 4 | 1.4 | −1 |
| F:$spdf$;H:$spd$;bc:$spdf$ | [f] | −8.449 5 | 0 | −1 |
| F:$sp$;H:$s$ | | −18.707 6 | | −10260 |
| F:$spd$;H:$sp$ | | −8.455 8 | −10252 | −8 |
| F:$spdf$;H:$spd$ | | −8.448 0 | −8 | 0 |
| F:$sp$;H:$s$;bc:$s$ | [a] [g] | −15.338 8 | | −6891 |
| F:$spd$;H:$sp$;bc:$sp$ | [c] | −8.485 2 | −6854 | −37 |
| F:$spdf$;H:$spd$;bc:$spd$ | [e] | −8.449 4 | −36 | −1 |
| F:$sp$;H:$s$;bc:$sp$ | [b] [h] | −10.410 7 | | −1963 |
| F:$spd$;H:$sp$;bc:$spd$ | [d] | −8.490 0 | −1921 | −42 |
| F:$spdf$;H:$spd$;bc:$spdf$ | [f] | −8.449 5 | −41 | −1 |
| [223 × 349; 65] | | −8.406 7 | | |
| [223 × 379; 100] | [i] | −8.448 1 | | |
| [223 × 379; 100] | [j] | −8.450 1 | | |

[a] [F:30$s$15$p$; H:30$s$; bc:(0+30-18)$s$]. The energy tabulated should be compared with the energy of −100.056 662 2 Hartree obtained before this truncation of the basis set.
[b] [F:30$s$15$p$; H:30$s$; bc:(0+30-18)$s$(0+15-5)$p$]. The energy tabulated should be compared with the energy of −100.069 722 8 Hartree obtained before this truncation of the basis set.
[c] [F:30$s$15$p$15$d$; H:30$s$15$p$; bc:(-3+30-18)$s$(-3+15-5)$p$].
[d] [F:30$s$15$p$15$d$; H:30$s$15$p$; bc:(-3+30-18)$s$(-3+15-5)$p$(-3+15-5)$d$].
[e] [F:30$s$15$p$15$d$15$f$; H:30$s$15$p$15$d$; bc:(-4+30-18)$s$(-4+15-5)$p$(-4+15-5)$d$].
[f] [F:30$s$15$p$15$d$15$f$; H:30$s$15$p$15$d$; bc:(-4+30-18)$s$(-4+15-5)$p$(-4+15-5)$d$(-4+15-5)$f$].
[g] The hyperpolarizability tabulated should be compared with that of −15.338 7 a.u. obtained before the truncation of the basis set described in Note [a].
[h] The hyperpolarizability tabulated should be compared with that of 10.410 6 a.u. obtained before the truncation of the basis set described in Note [b].
[i] Calculated from finite field dipole moment expectation values by 3 point central difference approximation according to Eq. (56).
[j] Calculated from finite field dipole moment expectation values by 5 point central difference approximation according to Eq. (57).

## 7. Conclusions

In this review, we have considered the determination of dipole, quadrupole, octopole and hexadecapole moments for the ground states of diatomic species within the Hartree-Fock approximation using both finite basis set expansions and the finite difference technique. For an appropriately chosen grid, the finite difference method provides exact expectation values of the total energies and multipole moments. Within the finite field approximation the method can yield highly accurate values of polarizabilities and hyperpolarizabilities.

The comparison of finite difference and finite basis set multipole moments has provided a more detailed

Table 7
Matrix Hartree-Fock energies, $E^{(\mathrm{mHF})}$, and finite difference Hartree-Fock energies, $E^{(\mathrm{fdHF})}$, in Hartree, obtained on [223 × 379; 100] grid, for the ground state of the FH molecule as a function of internuclear separation, $R_{\mathrm{FH}}$, in bohr. $\delta_{\mathrm{E}} = \mathrm{E}^{(\mathrm{mHF})} - E^{(\mathrm{fdHF})}$ in Hartree

| $R_{\mathrm{FH}}$ | $E^{(\mathrm{mHF})}$ | $E^{(\mathrm{fdHF})}$ | $\delta_{\mathrm{E}}$ |
|---|---|---|---|
| 1.5000 | −100.053 362 150 | −100.053 363 219 | $1.07 \times 10^{-6}$ |
| 1.6000 | −100.067 581 364 | −100.067 582 523 | $1.16 \times 10^{-6}$ |
| 1.7000 | −100.071 283 975 | −100.071 285 265 | $1.29 \times 10^{-6}$ |
| 1.7328 | −100.070 801 141 | −100.070 802 492 | $1.35 \times 10^{-6}$ |
| 1.8000 | −100.067 791 315 | −100.067 792 775 | $1.46 \times 10^{-6}$ |
| 1.9000 | −100.059 426 266 | −100.059 427 935 | $1.67 \times 10^{-6}$ |
| 2.0000 | −100.047 823 077 | −100.047 824 995 | $1.92 \times 10^{-6}$ |
| 2.1000 | −100.034 136 452 | −100.034 138 657 | $2.21 \times 10^{-6}$ |
| 2.2000 | −100.019 184 385 | −100.019 186 930 | $2.55 \times 10^{-6}$ |

Table 8
Total dipole moments, in a.u., from matrix Hartree-Fock calculations, $\mu_{\mathrm{z}}^{(\mathrm{mHF})}$, and from finite difference Hartree-Fock calculations obtained on [223 × 379; 100] grid, $\mu_{\mathrm{z}}^{(\mathrm{fdHF})}$, for the ground state of the FH molecule as a function of internuclear separation, $R_{\mathrm{FH}}$ (in bohr). In this table, $\delta_{\mu_{\mathrm{z}}} = \mu_{\mathrm{z}}^{(\mathrm{mHF})} - \mu_{\mathrm{z}}^{(\mathrm{fdHF})}$ in atomic units

| $R_{\mathrm{FH}}$ | $\mu_{\mathrm{z}}^{(\mathrm{mHF})}$ | $\mu_{\mathrm{z}}^{(\mathrm{fdHF})}$ | $\delta_{\mu_{\mathrm{z}}}$ |
|---|---|---|---|
| 1.5000 | −0.664 944 9 | −0.664 944 72 | $-1.8 \times 10^{-7}$ |
| 1.6000 | −0.703 241 | −0.703 241 44 | $-4.4 \times 10^{-7}$ |
| 1.7000 | −0.742 834 3 | −0.742 834 22 | $-8 \times 10^{-8}$ |
| 1.7328 | −0.756 075 9 | −0.756 075 86 | $-4 \times 10^{-8}$ |
| 1.8000 | −0.783 567 5 | −0.783 567 76 | $2.6 \times 10^{-7}$ |
| 1.9000 | −0.825 315 1 | −0.825 315 76 | $6.6 \times 10^{-7}$ |
| 2.0000 | −0.867 978 1 | −0.867 979 38 | $1.28 \times 10^{-6}$ |
| 2.1000 | −0.911 480 4 | −0.911 482 61 | $2.21 \times 10^{-6}$ |
| 2.2000 | −0.955 762 8 | −0.955 766 16 | $3.36 \times 10^{-6}$ |

measure of the quality of the finite basis set approximation than comparisons of total energy values and orbital energies for such systems [97,98], and also supplements our study of orbital amplitude difference functions [102].

The basis sets employed in this work are of the distributed universal even-tempered type [28]. They are not tailored to one particular property or indeed one particular molecular species or environment. They provide a systematic procedure for approaching the complete basis set limit for the chosen theoretical model. We note that the infinite basis set limit is not synonymous with the complete basis set limit. One can trivially generate a sequence of basis set of even-tempered functions of increasing size, $N$, which is demonstrably incomplete in the limit $N \to \infty$ merely by fixing the parameters $\alpha$ and $\beta$ defining the exponents. Energy values determined from such sequences can be easily shown to converge to an incorrect limit. For expectation values such as multipole moments, Klahn and Morgan [118] have shown that the situation is more complicated in that a sequence of basis sets which support convergence of the energy may not support convergence of other expectation values. We have provided numerical evidence that the $2^n$-pole moments, $n = 1, 2, 3, 4$, for diatomic molecules within the Hartree-Fock approximation can be obtained by systematic implementation of the algebraic approximation. Within the Hartree-Fock approximation, basis subsets located at the bond midpoint have been shown to be particularly effective

Table 9
Dipole polarizabilities, in a.u., from matrix Hartree-Fock calculations, $\alpha_{zz}^{(mHF)}$, and from finite difference Hartree-Fock calculations on [223 × 379; 100] grid, $\alpha_{zz}^{(fdHF)}$, for the ground state of the FH molecule as a function of internuclear separation. The polarizability is evaluated from finite difference Hartree-Fock calculations either by numerical differentiation of the energy, $\alpha_{zz}^{(fdHF)}(E)$, or by numerical differentiation of the dipole moment, $\alpha_{zz}^{(fdHF)}(\mu_z)$. In this table, $\delta_{\alpha_{zz}} = \alpha_{zz}^{(mHF)} - \alpha_{zz}^{(fdHF)}(\mu_z)$

| $R_{FH}$ | $\alpha_{zz}^{(mHF)}$ | $\alpha_{zz}^{(fdHF)}(E)$ | $\alpha_{zz}^{(fdHF)}(\mu_z)$ | $\delta_{\alpha_{zz}}$ |
|---|---|---|---|---|
| 1.5000 | 4.650 172 | 4.66 | 4.650 181 6 | $-9.5 \times 10^{-6}$ |
| 1.6000 | 5.084 807 | 5.08 | 5.084 822 5 | $-1.5 \times 10^{-5}$ |
| 1.7000 | 5.578 430 | 5.57 | 5.578 450 5 | $-2.0 \times 10^{-5}$ |
| 1.7328 | 5.754 232 | 5.77 | 5.754 255 0 | $-2.3 \times 10^{-5}$ |
| 1.8000 | 6.137 220 | 6.11 | 6.137 248 2 | $-2.8 \times 10^{-5}$ |
| 1.9000 | 6.767 220 | 6.79 | 6.767 259 3 | $-3.9 \times 10^{-5}$ |
| 2.0000 | 7.474 208 | 7.47 | 7.474 266 6 | $-5.9 \times 10^{-5}$ |
| 2.1000 | 8.263 591 | 8.24 | 8.263 677 2 | $-8.6 \times 10^{-5}$ |
| 2.2000 | 9.140 328 | 9.13 | 9.140 459 1 | $-1.3 \times 10^{-4}$ |

Table 10
Hyperpolarizabilities, in a.u., from matrix Hartree-Fock calculations, $\beta_{zzz}^{(mHF)}$, and from finite difference Hartree-Fock calculations on [223 × 379; 100] grid, $\beta_{zzz}^{(fdHF)}$, for the ground state of the FH molecule as a function of internuclear separation. The polarizability is evaluated from finite difference Hartree-Fock calculations by numerical differentiation of the dipole moment, $\beta_{zzz}^{(fdHF)}(\mu_z)$. In this table, $\delta_{\beta_{zzz}} = \beta_{zzz}^{(mHF)} - \beta_{zzz}^{(fdHF)}(\mu_z)$

| $R_{FH}$ | $\beta_{zzz}^{(mHF)}$ | $\beta_{zzz}^{(fdHF)}(\mu_z)$ | $\delta_{\beta_{zzz}}$ |
|---|---|---|---|
| 1.5000 | −3.950 7 | −3.950 3 | $-3.8 \times 10^{-4}$ |
| 1.6000 | −5.556 7 | −5.556 0 | $-7.0 \times 10^{-4}$ |
| 1.7000 | −7.642 9 | −7.641 8 | $-1.1 \times 10^{-3}$ |
| 1.7328 | −8.449 5 | −8.448 1 | $-1.4 \times 10^{-3}$ |
| 1.8000 | −10.316 0 | −10.310 5 | $-5.5 \times 10^{-3}$ |
| 1.9000 | −13.691 3 | −13.704 8 | $1.4 \times 10^{-2}$ |
| 2.0000 | −17.888 5 | −17.886 4 | $-2.1 \times 10^{-3}$ |
| 2.1000 | −23.027 2 | −23.035 2 | $8.0 \times 10^{-3}$ |
| 2.2000 | −29.221 6 | −29.218 9 | $-2.7 \times 10^{-3}$ |

in calculations of multipole moments. However, the absolute accuracy of the expectation values is seen to degrade somewhat as $n$ increases. The pattern of convergence of the multipole moment expectation values with increasing size of basis set has been shown to be very similar to that observed for the total Hartree-Fock energy.

We have also reviewed the determination of dipole polarizabilities and hyperpolarizabilities for diatomic species within the Hartree-Fock approximation using both finite basis set and finite difference techniques. For the FH molecule we have studied the dependence of the total energy, dipole moment and the polarizabilities on a sequences of distributed universal even-tempered basis sets and on the nuclear separation. For an appropriately chosen grid, the finite difference method provides total Hartree-Fock energies and dipole moment expectation values which are near exact for the purposes of the present study.

Polarizabilities and hyperpolarizabilities can be obtained in the finite difference Hartree-Fock technique by using the finite field approach but some care is required when calculating the polarizabilities via numerical differentiation. We have provided numerical evidence that by systematic implementation of the algebraic approximation these electrical properties can be obtained to an accuracy approaching that supported by the finite difference Hartree-Fock method. Basis subsets?? located at the bond midpoint have been shown to be particularly effective in reducing the differences between finite basis set and finite difference results. It has also been demonstrated that basis set truncation errors can be made negligible for these electric properties.

## Acknowledgments

JK acknowledges the support of the Komitet Badań Naukowych through Contract No. 8 T11F 028 18. DM acknowledges the support of the Office of Energy Research, Office of Basic Energy Sciences, Division of Chemical Sciences, US Department of Energy under grant DE-FG02-97ER-14758.

## References

[1] D.J. Williams, ed., *Properties of Organic and Polymeric Materials*, American Chemical Society, Washington D.C., 1983.
[2] Y.R. Shen, *The Principles of nonlinear Optics*, Academic Press, New York, 1984.
[3] A. Yariv, *Optical Electronics*, Holt Saunders, New York, 1985.
[4] D.S. Chemla and J. Zyss, eds, *Nonlinear Optical Properties of Organic Molecules and Crystals*, (Vols 1 and 2), Academic Press, Orlando, 1987.
[5] P.N. Butcher and D. Cotter, *The Elements of Nonlinear Optics*, Cambridge University Press, Cambridge, 1990.
[6] P.N. Prasad and D.J. Williams, *Introduction to Non-linear Optical Effects in Molecules*, Wiley, New York, 1991.
[7] R.J. Boyd, *Nonlinear Optics*, Academic Press, New York, 1992; 2nd Edition, 2003.
[8] J. Michl, ed., Optical Nonlinearities in Chemistry, *Chem. Rev.* **94** (1994).
[9] H.S. Singh and S. Miyata, eds, *Nonlinear Optics of Organic Molecules and Polymers*, CRC Press, Boca Raton, USA, 1997.
[10] D.M. Bishop, *Rev. Mod. Phys.* **62** (1990), 343.
[11] J. Kobus, *Chem. Phys. Lett.* **202** (1993), 7.
[12] J. Kobus, *Adv. Quantum Chem.* **28** (1997), 1.
[13] E.A. McCullough, in: *Encyclopedia of Computational Chemistry*, (Vol. 3), P.v.R. Schleyer, N.L. Allinger, T. Clark, J. Gasteiger, P.A. Kollamn, H.F. Schaefer III and P.R. Schreiner, eds, 1941, Wiley, Chichester, 1998.
[14] J. Kobus, in: *Handbook of Molecular Physics and Quantum Chemistry*, (Vol. 2), S. Wilson, P.F. Bernath and R. McWeeny, eds, Wiley, Chichester, 2003.
[15] D. Moncrieff and S. Wilson, *Chem. Phys. Lett.* **209** (1993), 423.
[16] D. Moncrieff and S. Wilson, *J. Phys. B: At. Mol. Opt. Phys.* **26** (1993), 1605.
[17] S. Wilson, in: *Handbook of Molecular Physics and Quantum Chemistry*, (Vol. 2), S. Wilson, P.F. Bernath and R. McWeeny, editors, Wiley, Chichester, 2003.
[18] S. Wilson, *Electron correlation in molecules*, Clarendon Press, Oxford, 1984.
[19] A. Szabo and N.S. Ostlund, *Modern Quantum Chemistry: Introduction to Advanced Electronic Structure Theory*, Revised 1st Edition, McMillan, New York, 1989.
[20] R. McWeeny, *Methods of Molecular Quantum Mechanics*, 2nd Edition, Academic Press, London, 1992.
[21] *Handbook of Molecular Physics and Quantum Chemistry*, (Vol. 2), *Molecular Electronic Structure*, S. Wilson, P.F. Bernath and R. McWeeny, eds, Wiley, Chichester, 2003.
[22] J. Kobus, D. Moncrieff and S. Wilson, *Phys. Rev. A* **62** (2000), 62503.
[23] J. Kobus, D. Moncrieff and S. Wilson, *J. Phys. B: At. Mol. Opt. Phys.* **34** (2001), 5127.
[24] J. Karwowski and I. Shavitt, in: *Handbook of Molecular Physics and Quantum Chemistry*, (Vol. 2), S. Wilson, P.F. Bernath and R. McWeeny, eds, Wiley, Chichester, 2003.

[25] S. Wilson, in: *Handbook of Molecular Physics and Quantum Chemistry*, (Vol. 2), S. Wilson, P.F. Bernath and R. McWeeny, eds, Wiley, Chichester, 2003.
[26] J. Paldus, in: *Handbook of Molecular Physics and Quantum Chemistry*, (Vol. 2), S. Wilson, P.F. Bernath and R. McWeeny, eds, Wiley, Chichester, 2003.
[27] S. Wilson, in: *Methods in Computational Molecular Physics*, (Vol. 113), G.H.F. Diercksen and S. Wilson, eds, NATO Advanced Study Institute Series C: Mathematical and Physical Sciences, Reidel, Dordrecht, 1983.
[28] S. Wilson, Ab initio methods in Quantum Chemistry, Part 1, K.P. Lawley, ed., *Adv. Chem. Phys.* **67** (1987), 439.
[29] E.R. Davidson and D. Feller, *Chem. Rev.* **86** (1986), 681.
[30] E.R. Davidson and D. Feller, *Rev. Comput. Chem.* **1** (1989), 1.
[31] S. Wilson, in: *Problem Solving in Computational Molecular Science: Molecules in Different Environments*, S. Wilson and G.H.F. Diercksen, eds, NATO ASI, Bad Windsheim, August 1996, Kluwer, Dordrecht, 1997.
[32] S. Wilson and D.M. Silver, *Phys. Rev. A* **14** (1976), 1949.
[33] H.M. Quiney, I.P. Grant and S. Wilson, *J. Phys. B: At. Mol. Phys.* **18** (1985), 2805.
[34] H.M. Quiney, I.P. Grant and S. Wilson, in Many-Body Methods in Quantum Chemistry, U. Kaldor, ed., *Lecture Notes in Chemistry* **52** (1989), 307.
[35] H.M. Quiney, in: *Handbook of Molecular Physics and Quantum Chemistry*, (Vol. 2), S. Wilson, P.F. Bernath and R. McWeeny, eds, Wiley, Chichester, 2003.
[36] J.A. Pople, *Rev. Mod. Phys.* **71** (1998), 1267.
[37] W. Heitler and F. London, *Z. Phys.* **44** (1927), 455.
[38] G.G. Hall, *Proc. Roy. Soc. (London) A* **205** (1951), 541.
[39] C.C.J. Roothaan, *Rev. Mod. Phys.* **23** (1951), 69.
[40] R. McWeeny, *Dissertation*, University of Oxford, 1948.
[41] S.F. Boys, *Proc. Roy. Soc. (London) A* **200** (1950), 542.
[42] C.C.J. Roothaan, *Rev. Mod. Phys.* **32** (1960), 179.
[43] C.C.J. Roothaan, L.M. Sachs and A.W. Weiss, *Rev. Mod. Phys.* **32** (1960), 186.
[44] W. Kolos and C.C.J. Roothaan, *Rev. Mod. Phys.* **32** (1960), 219.
[45] C.M. Schwartz, *Phys. Rev.* **126** (1962), 1015; see also *Methods in Computational Physics* **2** (1963), 241.
[46] C. Reeves, *J. Chem. Phys.* **39** (1963), 1.
[47] C. Reeves and M. Harrison, *J. Chem. Phys.* **39** (1963), 11.
[48] K. Ruedenberg, R.C. Raffenetti and R. Bardo, in: *Energy, Structure and Reactivity*, Proceedings of the 1972 Boulder Conference on Theoretical Chemistry, Wiley, New York, 1973.
[49] R.C. Raffenetti and K. Ruedenberg, *J. Chem. Phys.* **59** (1973), 5978.
[50] R.C. Raffenetti, *J. Chem. Phys.* **58** (1973), 4452.
[51] R.C. Raffenetti, *J. Chem. Phys.* **59** (1973), 5936.
[52] R.D. Bardo and K. Ruedenberg, *J. Chem. Phys.* **59** (1973), 5956.
[53] R.D. Bardo and K. Ruedenberg, *J. Chem. Phys.* **59** (1973), 5966.
[54] R.D. Bardo and K. Ruedenberg, *J. Chem. Phys.* **60** (1974), 918.
[55] D.M. Silver, S. Wilson and W. Nieupoort, *Intern. J. Quantum Chem.* **14** (1978), 635.
[56] D.M. Silver and W. Nieupoort, *Chem. Phys. Lett.* **57** (1978), 421.
[57] D.M. Silver and S. Wilson, *J. Chem. Phys.* **69** (1978), 3787.
[58] S. Wilson and D.M. Silver, *Chem. Phys. Lett.* **63** (1979), 367.
[59] S. Wilson and D.M. Silver, *J. Chem. Phys.* **72** (1980), 2159.
[60] S. Wilson and D.M. Silver, *J. Chem. Phys.* **77** (1982), 3674.
[61] P. Mezey, *Theoret. chim. Acta* **53** (1979), 187.
[62] D.L. Cooper and S. Wilson, *J. Chem. Phys.* **76** (1982), 6088; ibid, **77** (1982), 4551 ibid. **77** (1982), 5053; ibid. **78** (1983), 2456; *J. Physics B: At. Molec. Phys.* **15** (1982), 493.
[63] E. Clementi and G. Corongui, *Chem. Phys. Lett.* **90** (1982), 359; IBM Technical Report POK-11 (1982); E. Clementi, S.J. Chakravorty, G. Corongui and V. Sonnad, in MOTECC - Modern Techniques in Computational Chemistry, edited by E. Clementi, ESCOM, Leiden, 1990.
[64] D.L. Cooper and J. Gerratt, *J. Phys. B: At. Mol. Opt. Phys.* **16** (1983), 3703.
[65] D.L. Cooper, J. Gerratt and M. Raimondi, *Molec. Phys.* **56** (1985), 611.
[66] D.L. Cooper, J. Gerratt, M.J. Ford and M. Raimondi, *Phys. Rev. A* **34** (1986), 1752.
[67] S.A. Barnard, M.J. Ford, D.L. Cooper, J. Gerratt and M. Raimondi, *Molec. Phys.* **60** (1987), 1193.
[68] G. Doggett, *J. Chem. Soc. Faraday Trans. 2* **80** (1984), 1413.
[69] R. Custodio, J.D. Goddard, M. Giordan and N.H. Morgan, *Can. J. Chem.* **7** (1992), 580.
[70] G. Doggett and G.D. Fletcher, *J. Molec. Struct. (THEOCHEM)* **92** (1992), 313.
[71] G. Malli, A.B.F. da Silva and Y. Ishikawa, *Can. J. Chem.* **70** (1992), 1822.
[72] J.W. Thompson and S. Wilson, *J. Phys. B: At. Mol. Opt. Phys.* **25** (1992), 621.

[73] G. Malli, A.B.F. da Silva and Y. Ishikawa, *Chem. Phys. Lett.* **201** (1993), 37; *Phys. Rev.* **47** (1993), 143; A.B.F. da Silva, G. Malli and Y. Ishikawa, *Chem. Phys. Lett.* **203** (1993), 201; A.B.F. da Silva, G. Malli and Y. Ishikawa, *Can. J. Chem.* **71** (1993), 1713; G. Malli, A.B.F. da Silva and Y. Ishikawa, *J. Chem. Phys.* **101** (1994), 6829; G. Malli and J. Styszynski, *J. Chem. Phys.* **101** (1994), 10736; G. Malli, J. Styszynski and A.B.F. da Silva, *Intern. J. Quantum Chem.* **55** (1995), 213.
[74] D. Moncrieff and S. Wilson, *J. Phys. B: At. Mol. Opt. Phys.* **27** (1994), 1.
[75] F.A. Parpia and A.K. Mohanty, *Phys. Rev. A* **52** (1995), 962.
[76] S. Wilson, in: *Theoretical Chemistry*, (Vol. 4), Specialist Periodical Reports, Senior Reporter: C. Thomson, Royal Society of Chemistry, London, 1981.
[77] S. Huzinaga and M. Klobukowski, *Chem. Phys. Lett.* **120** (1985), 509.
[78] S. Huzinaga, M. Klobukowski and H. Tatewaki, *Can. J. Chem.* **63** (1985), 1812.
[79] S. Huzinaga and M. Miguel, *Chem. Phys. Lett.* **175** (1990), 289.
[80] M. Klobukowski, *Chem. Phys. Lett.* **214** (1993), 166.
[81] M. Klobukowski, *Can. J. Chem.* **72** (1994), 1741.
[82] G.A. Petersson, S. Zhong, J.A. Montgomery, Jr. and M.J. Frisch, *J. Chem. Phys.* **118** (2003), 1101.
[83] D. Feller and K. Ruedenberg, *Theor. chim. Acta* **52** (1979), 231.
[84] M.W. Schmidt and K. Ruedenberg, *J. Chem. Phys.* **71** (1979), 3951.
[85] S. Wilson, *Theor. Chem. Acta* **57** (1980), 53; ibid. **58**, 31; ibid. **59**, 71.
[86] H.M. Quiney, I.P. Grant and S. Wilson, *J. Phys. B: At. Mol. Opt. Phys.* **22** (1990), L15.
[87] E. Kryachko and S. Wilson, *Intern. J. Quantum Chem.* **93** (2003), 112.
[88] D.R. Hartree, *The Calculation of Atomic Structures*, Wiley, New York, 1957.
[89] C. Froese-Fischer, *The Hartree-Fock Method for Atoms. A Numerical Approach*, Wiley, New York, 1977.
[90] L. Laaksonen, P. Pyykkö and D. Sundholm, Intern, *J. Quantum Chem.* **23** (1983), 309, 319.
[91] J. Kobus, *Chem. Phys. Lett.* **202** (1993), 7.
[92] J. Kobus, *Comput. Phys. Commun.* **78** (1994), 247.
[93] L. Laaksonen, P. Pyykkö and D. Sundholm, *Comput. Phys. Rept.* **4** (1986), 313.
[94] J. Kobus, L. Laaksonen and D. Sundholm, *Comput. Phys. Commun.* **98** (1996), 346.
[95] J. Kobus, D. Moncrieff and S. Wilson, *J. Phys. B: At. Mol. Opt. Phys.* **27** (1994), 2867.
[96] J. Kobus, D. Moncrieff and S. Wilson, *J. Phys. B: At. Mol. Opt. Phys.* **27** (1994), 5139.
[97] D. Moncrieff, J. Kobus and S. Wilson, *J. Phys. B: At. Mol. Opt. Phys.* **28** (1995), 4555.
[98] J. Kobus, D. Moncrieff and S. Wilson, *Molec. Phys.* **86** (1995), 1315.
[99] J. Kobus, D. Moncrieff and S. Wilson, *Molec. Phys.* **93** (1998), 713.
[100] J. Kobus, D. Moncrieff and S. Wilson, *Molec. Phys.* **96** (1999), 1559.
[101] J. Kobus, D. Moncrieff and S. Wilson, *Molec. Phys.* **98** (2000), 401.
[102] J. Kobus, D. Moncrieff and S. Wilson, *Molec. Phys.* **92** (1997), 1015.
[103] J. Kobus, D. Moncrieff and S. Wilson, *Molec. Phys.* **99** (2001), 315.
[104] A.D. Buckingham, *J. Chem. Phys.* **30** (1959), 1580.
[105] A.D. Buckingham, *Quart. Rev.* **13** (1959), 183.
[106] A.D. Buckingham, *Adv. Chem. Phys.* **12** (1967), 107.
[107] A.D. McLean and M. Yoshimine, *J. Chem. Phys.* **47** (1967), 1927.
[108] A.J. Stone, *Chem. Phys. Lett.* **83** (1981), 233.
[109] J. Applequist, *Chem. Phys.* **85** (1984), 279.
[110] G. Maroulis, *J. Chem. Phys.* **108** (1998), 5432.
[111] P. Lazzeretti, in: *Handbook of Molecular Physics and Quantum Chemistry*, (Vol. 3), S. Wilson, P.F. Bernath and R. McWeeny, eds, Wiley, Chichester, 2003.
[112] D. Moncrieff and S. Wilson, *J. Phys. B: At. Mol. Opt. Phys.* **29** (1996), 2425.
[113] D. Moncrieff and S. Wilson, *J. Phys. B: At. Mol. Opt. Phys.* **31** (1998), 3819.
[114] M.J. Frisch, G.W. Trucks, H.B. Schlegel, P.M.W. Gill, B.G. Johnson, M.A. Robb, J.R. Cheeseman, T. Keith, G.A. Petersson, J.A. Montgomery, K. Raghavachari, M.A. Al-Laham, V.G. Zakrzewski, J.V. Ortiz, J.B. Foresman, J. Cioslowski, B.B. Stefanov, A. Nanayakkara, M. Challacombe, C.Y. Peng, P.Y. Ayala, W. Chen, M.W. Wong, J.L. Andres, E.S. Replogle, R. Gomperts, R.L. Martin, D.J. Fox, J.S. Binkley, D.J. Defrees, J. Baker, J.P. Stewart, M. Head-Gordon, C. Gonzalaz and J.A. Pople, GAUSSIAN94 Rev. C.3, Gaussian Inc., Pittsburgh PA, 1995; GAUSSIAN98 Rev. A.9, Gaussian Inc., Pittsburgh PA, 2000.
[115] K.P. Huber and G. Herzberg, *Molecular Spectra and Molecular Structure*, (Vol. 4), Constants of Diatomic Molecules, Van Nostrand Reinhold, 1979.
[116] H.B. Jansen and P. Ros, *Chem. Phys. Lett.* **3** (1969), 140.
[117] S.F. Boys and F. Bernardi, *Molec. Phys.* **19** (1970), 553.
[118] B.J. Klahn and J.D. Morgan III, *J. Chem. Phys.* **81** (1984), 410.

Journal of Computational Methods in Sciences and Engineering 4 (2004) 641–664
IOS Press


# A mean-field approach for the determination of the polarizabilities for the water molecule in liquid state

A.V. Gubskaya and P.G. Kusalik*
*Department of Chemistry, Dalhousie University, Halifax, Nova Scotia, B3H 4J3, Canada*

**Abstract.** A mean-field method is presented describing the electrostatic environment experienced by water molecule in liquid state, which is used to extract the corresponding hyper- and high-order polarizabilities. Within this approach, MD computer simulations of liquid water samples for two standard water potentials at several different temperatures are performed to characterize the distributions (specifically average values) of local fields and field gradients. The electric response properties (including non-linear contributions up to fourth-order) are then calculated using *ab initio* techniques in conjunction with a charge perturbation variant of a finite field method. Sets of fixed charges are used to generate the desired electric fields and electric field gradients. Calculations of dipole polarizability, hyper- and principal components of high-order polarizabilities of the water molecule in gas and liquid phase conditions are carried out at MP2 and MP4 levels of theory; the values obtained for three different liquid phase models are compared with those for gas phase. For a liquid phase water molecule the first hyperpolarizability ($\beta$) and first higher polarizability (A) increase markedly, actually changing sign. The second hyperpolarizability ($\gamma$) also increases but much less dramatically, and components of the second high-order polarizability tensor (B) demonstrate a rearrangement of contributions. We observe that a less symmetrical gradient model gives the most accurate representation of liquid-phase conditions. The excellent agreement of our gas-phase values with experimental results and the most accurate previous theoretical predictions is evident of the quality of our higher order polarizabilities and theoretical models.

Keywords: Water, mean-field methods, *ab initio*, molecular dynamics, electric field, field gradient, electric response properties, polarizabilities

PACS: 31.15.Ar; 31.15.Fx; 31.15.Qg; 31.25.Qm; 33.15.Kr

## 1. Introduction

The electrical response properties of the water molecule have been studied extensively both experimentally and theoretically during the last two decades. These are the multipole moments ($\mu$, $\Theta$, etc.), dipole polarizability ($\alpha$), and nonlinear optical polarizabilities (NOP), namely the first ($\beta$) and second ($\gamma$) hyperpolarizabilities and the high-order polarizabilities (dipole-quadrupole (A), dipole-dipole-quadrupole (B), etc.), which appear as coefficients in the energy expansion for a molecule in the presence of a static electric field. The importance of these quantities has long been recognized for understanding a wide variety of physical phenomena such as intermolecular interactions [1,2] and resonant or nonresonant optical phenomena. The latter are now of a particular interest because of very fast growing fields related with development of novel materials. Such investigations have very often focused on modeling of

*Corresponding author. Tel.: +1 902 494 3627; Fax: +1 902 494 1310; E-mail: Peter Kusalik@dal.ca.

the polarizabilities for large molecules as, for example, donor-acceptor complexes and oligomers [3]. However, the studies of small species like water, for which accurate gas phase polarizabilities exist, are very useful because they offer the possibility of monitoring the different contributions to molecular polarizabilities, including those in liquid phase, in greater detail.

A significant amount of data is currently available on the electrical response of an isolated water molecule. An accurate value for the dipole moment of a water molecule in gas phase has been known since 1973 from Stark effect measurements [4], and more recently data determined using molecular beam electric resonance spectroscopy (MBER) were reported by Shostak et al. [5]. The value of the static mean dipole polarizability is also well known from dipole oscillator strength distribution (DOSD) experiments [6]. The majority of gas-phase hyperpolarizability measurements of small molecules like water are based on the dc Kerr effect and the electric field induced second harmonic generation (ESHG or EFISH) technique. Unfortunately, a precise determination of higher-order terms appears to be considerably more difficult. To date the only source of the first ($\beta$) and second ($\gamma$) hyperpolarizabilities of $H_2O$ in the gas phase is the mean values reported by Ward and Miller [7] and Kaatz et al. [8]. Several theoretical studies of the nonlinear optical polarizabilities (NOP) of an isolate water molecule have been carried out using a variety of *ab initio* techniques (appropriate for gas-phase conditions). Significant success has been achieved in calculations where the molecule is perturbed by a weak homogeneous electric field. Starting with the paper of Purvis and Bartlett [9], where the importance of electron correlation was clearly shown, and ending with the work of Guan et al. [10] investigating the ability of a DFT approach to reproduce the molecular response to an applied electric field, a great deal of effort has been expended in obtaining accurate values for these properties. The most accurate values of $\beta$ and $\gamma$ determined to date appear to be those reported by Maroulis [11]. He has utilized Møller-Plesset (MP) and coupled cluster (CC) theories to examine the basis set limit and electron correlation effects.

Although experimental and theoretical data are available for the gas phase, this is not the case for liquid water. Until recently, there have been neither experimental multipole moments nor dipole polarizabilities reported in the literature for liquid water. The experimental values of $\overline{\beta}$ and $\overline{\gamma}$ for liquid water were determined by Levine and Bethea as related quantities in their liquid phase ESHG measurements of associating liquid mixtures [12]. These data remain the only source of experimental electrical properties of $H_2O$ in the liquid phase. The lack of experimental data for the electrical properties of water in condensed phase coupled with the importance of such information make their calculations a very attractive and challenging task.

There are only a few previous theoretical studies of the electrical response properties (specifically polarizabilities) for water in condensed state. Historically, ice was a natural starting point for systematic studies in this field because of its regular and predictable molecular structure (in contrast to that in liquid water). More than thirty years ago Coulson and Eisenberg [13] performed an ingenious calculation of the dipole moment in ice $I_h$ using an induction model. Recently Batista et al. [14] re-evaluated these results using a similar induction model but including several higher order terms in the multipolar expansion. This allowed the authors to study the local electric field in ice and to determine a dipole moment of 3.1 D for a water molecule in perfect ice $I_h$ crystal at 0 K. In principle, the induction model can be classified as a local-field method, which may be applicable to liquids, if a corresponding set of representative liquid structures is available. Unfortunately, a classical induction model can not be a source of values for molecular polarizabilities and one has to use quantum or a combination of quantum and classical calculations in order to determine them. Reis et al. [15] applied a local-field theory to calculate static and frequency-dependent effective in-crystal hyperpolarizabilities for several ice polymorphs. The local field experienced by a molecule in the crystal was calculated by summing over the field contributions from all surrounding molecules using a point dipole approximation to reproduce the induced charge density distributions. Average molecular electronic properties ($\mu$, $\alpha$, $\beta$, and $\gamma$) and their relative changes with respect to gas phase values were calculated in the presence of the electric dipole crystal fields for different ($I_h$, II, VIII and IX) ice polymorphs. The dipole moment and the first hyperpolarizability were found to be strongly affected by the fields, but dipole polarizability is only weakly changed compared to its gas phase value. The value of the first hyperpolarizability ($\beta$) for $I_h$, II and IX ices increases markedly (undergoing sign inversion analogously to that observed for liquid water, as discussed below), while for ice VIII it remains almost unchanged in comparison with the gas phase estimate. This can be explained by taking into account the geometric arrangement of the molecules in ice VIII, which leads to a nearly complete cancellation of the electric dipole field [15].

For liquid water, the work of Mikkelsen et al. was the first theoretical study where the same unusual feature, the sign change of the first hyperpolarizability upon liquefaction, has been shown [16]. They carried out calculations where the "solvent effect" due to neighboring water molecules was described using a continuum model, a supermolecular (cluster) model and a semicontinuum model. In the continuum model the solute molecule was surrounded by a dielectric medium that gives rise to a reaction-field contribution [17–21]. In the supermolecule approach calculations were carried out on a bare cluster containing the first solvation shell and in the semicontinuum model this cluster was again surrounded by a homogeneous dielectric medium. The continuum model gives only a small change in the first hyperplarizability ($\beta$) of liquid water compared to the vacuum results. In the bare cluster calculations the sign of hyperpolarizability is inverted; the combination of water imbedded in a cluster (as a first solvation shell) surrounded by a continuum solvent increased $\beta$ to a value that is in a good agreement with experiment [12]. Mikkelsen et al. suggested that the observed sign change of the first hyperpolarizability of liquid water can be rationalized by the presence of the strong hydrogen bonds and static dipolar interactions [16].

Tu and Laaksonen [22] developed a novel approach to calculate electrical response properties of single molecules in condensed state. This method was used to determine the first two multipole moments of a single water molecule in water clusters and in liquid water. The basic idea of this approach is to carry out the quantum calculation only on one constituent molecule at a time avoiding the time consuming calculation of the whole system. To achieve this these authors first simplified the interactions between the molecules into pure Coulombic form and then incorporated the corresponding terms into the Schrödinger equation for each molecule (see Ref. [22] for details). A calculation for each molecule of the system was performed using this equation and the whole system was then iterated until self-consistency was reached. Tu and Laaksonen described this approach as a compromise between classical induction models and quantum chemical supermolecular calculations. However, the results they obtained for the dipole and quadrupole moments, as well as for the components of dipole polarizability, appear somewhat lower in comparison with some other resent theoretical estimates [23–25]. This is perhaps not surprising because of the fact that other interactions such as exchange, charge transfer and dispersion interactions were not included in this theoretical formulation [22].

Although much theoretical effort have been expended to determine values for the total molecular dipole moment and the dipole polarizability of liquid water, we are not aware of any other theoretical work to calculate higher order nonlinear coefficients in the energy expansion and our recently published work [26] is the first such study. In the present paper we will provide further details on how a charge perturbation variant of the finite-field approach combined with results of MD simulations can be used to simulate liquid phase conditions for a water molecule in its ground state and then to extract its multipole moments and hyperpolarizabilities.

The representation of the liquid phase environment experienced by a water molecule is an important aspect of the current study. Reaction field approaches [16], where a "solute" molecule is surrounded by a continuum dielectric, have been commonly used to perform *ab initio* calculations for molecules in solution. Unfortunately, this approach has difficulties for hydrogen bonded solvents such as water [27], where a dielectric continuum may poorly represent the interactions with the surrounding hydration and the choice of cavity size and shape becomes crucial. Alternately, techniques have also been developed making it possible to carry out *ab initio* simulations of liquids, including water [24]. Such simulations are extremely computationally intensive, and moreover the fact that the "molecules" are not well define, together with the plane wave representation of the electronic wave function, makes it difficult to extract apparently simple molecular properties such as a dipole moment. Hybrid methods [28], where one or a few quantum molecules are immersed in a classical solvent, have also been examined, but sampling problems remain and the question of treatment of cross (quantum-classical) interactions now arises.

In the present study we introduce a somewhat different approach that can be viewed as an extension of mean-field methods [29]. In mean field theories the interactions of a central particle with all its neighbors are collapsed into the influence of a local field. Such an approach has been previously used to investigate successfully classical electrostatic models for water [30–32] and ice [13,14]. Thus, in the present *ab initio* calculations the influence of the surrounding water molecules will be modeled by a local electric field experienced by the (central) molecule of interest. The nature and strength of this local field are determined from average distributions obtained from classical simulations of liquid water using standard classical water potentials. It is sufficient here to note that one would expect the classical potential models employed to provide a reasonable representation of the local electrostatic environment in liquid water. In this study three different models of the local environment are examined, where both homogeneous and non-homogeneous local electric fields are considered. MD simulations were performed to determine the average values of the electric field and electric field gradient experienced by a water molecule in liquid state; these results are discussed in Section 2. *Ab initio* calculations were carried out for three liquid models and for gas-phase molecules at the MP2 and MP4 levels of theory. The finite field technique (and its charge perturbation variant) used to calculate molecular polarizabilities, our liquid phase models, and the scheme developed for energy partitioning, as well as other computational details, are described in Section 3. In Section 4 our gas-phase results are compared with those from previous work, and NOP for the three liquid phase models of water are presented and discussed. Our concluding remarks are given in Section 5: summarily, the present approach appears to work very well and we are able to determine the first *ab initio* values for the hyper- and high-order polarizabilities for liquid water.

## 2. Electric field and field gradient distributions

Molecular dynamics (MD) simulations have been performed with simple empirical water potential models in order to characterize the local environment (i.e. the local electric field and field gradient distributions) present in liquid water. In particularly, they provide average numerical values of these quantities which are further exploited in our *ab initio* calculations. The SPC/E [34] and TIP4P [33] models were chosen because of their success in reproducing liquid state properties. Their ability to reproduce the structure of liquid water indicates that they should provide a satisfactory description of the fluctuations in the local field and the relationship between the field and field gradient. Simulations were carried out at constant experimental densities at temperatures of 263, 298, 333 and 373K. We should remark that in the application of the present mean-field method only numerical values of the average field and field gradient measured at ambient temperature will be required. However, the temperature

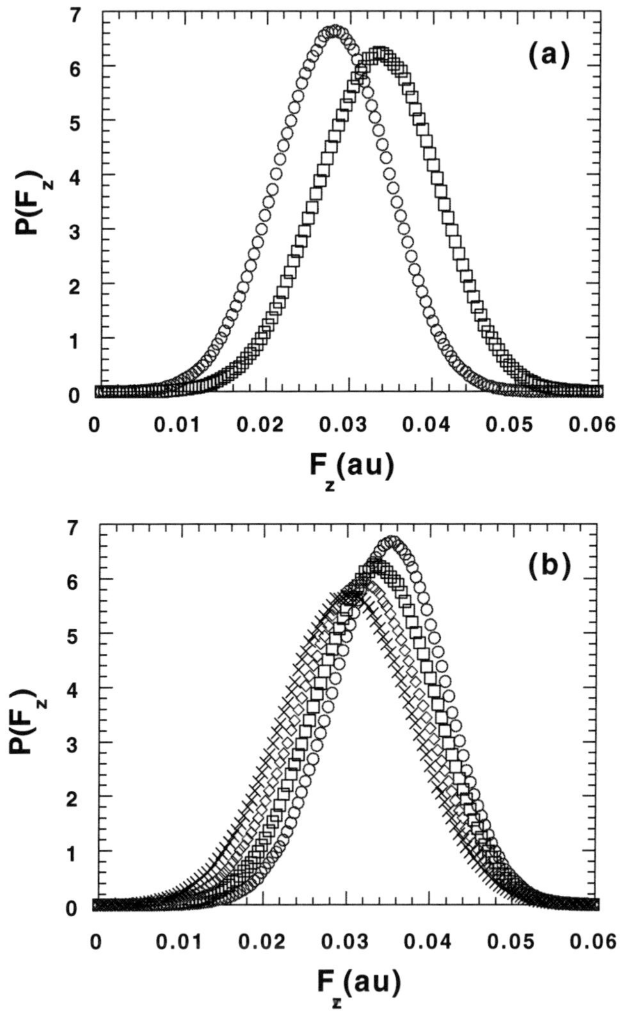


Fig. 1. (a) Probability distributions of the z-component of the local electric field, $F_z$, for SPC/E (squares) and TIP4P (circles) models at $T = 298$ K. (b) Temperature dependence of the $F_z$ probability distribution for SPC/E water. The open circles, squares, diamonds and crosses represent results at 263, 298, 333, and 373 K, respectively.

dependent distributions carefully analyzed and compared with previously reported results provide us essential confidence in their reliability.

Within our MD runs a truncated octahedral simulation cell containing 256 particles was utilized and all fields and field gradients were measured at the center of the oxygen atom. The Ewald summation technique [35] was used to calculate the Coulombic interactions and the electric fields. In the calculation of electric field gradients a spherical cutoff at $\sqrt{3}L/4$, the inscribed sphere radius, was employed.

For reasons of symmetry only the electric field component along the molecular dipole vector (the z-axis) possesses a non zero average value, and hence is sufficient to characterize the average total electric field within liquid water. The distributions for the z-component of the electric field for SPC/E and TIP4P water at a temperature of 298 K are shown in the Fig. 1(a), where some difference in the width and the position of maxima is evident. TIP4P water has a slightly narrower distribution, which is also shifted to lower values of the electric field in comparison with the SPC/E result. The temperature dependence of the z-component field distributions for TIP4P (not shown) and SPC/E water (see Fig. 1(b)) reveals the sharpest distributions at $T = 263$ K. The broadening of the distributions with increasing temperature reflects the presence of larger fluctuations and structural disorder in the liquid at higher temperatures. We note that, as expected, the electric field distributions in the x- and y directions have similar shapes and are centered about zero.

Distributions for all (diagonal and off-diagonal) components of the electric field gradient were examined. Only the diagonal components ($xx$, $yy$ and $zz$) have non zero average values; distributions for these field gradient components are shown in Fig. 2(a) and (b), respectively, for SPC/E and TIP4P water at 298 K. Since the $xy$, $xz$ and $yz$ (off-diagonal) components do not contribute in the average field gradient experienced by a water molecule, only the diagonal components will be considered further. Each of the field gradient distributions shown in Fig. 2(a) and (b) are essentially symmetric, with both water models providing similar results. The most notable difference between the SPC/E and TIP4P results evident in Figs 2(a) and 2(b) is the slight shift in the relative positions of the $yy$ and $zz$ distributions, where the latter model predicts field gradient values somewhat closer to those of an ideal tetrahedral structure [26]. The temperature dependence of the $zz$ field gradient distribution is given in Fig. 2(c) for the TIP4P model. In Fig. 2(c) we notice that the width of the distribution is virtually unchanged while there is a slight shift to a smaller average magnitude as the temperature increased.

Analysis of the distributions shown in Figs 1 and 2 reveals that they are well described by Gaussian forms,

$$P(x) = Ae^{-k(x-x_0)^2}, \tag{1}$$

where $x_0$ represents the average value, and $k = 1/2\sigma^2$ with $\sigma$ and A as the width and the normalizing factor, respectively, of the distribution. Consequently, field and field gradient distributions are easily characterized and determination of their average values becomes straightforward. In our analysis all distribution curves were fit to Gaussian functions; the average values of the nonvanishing components of the electric field and field-gradient have been determined and are presented in Table 1.

For the diagonal components of the field gradient one would expect that they be traceless (i.e. the sum of $\langle F_{xx}\rangle$, $\langle F_{yy}\rangle$ and $\langle F_{zz}\rangle$ to be equal to zero). The distributions of the $xx$-component for both water potentials are nearly Gaussian in form, while for the $yy$-component the field gradient exhibits a somewhat distorted Gaussian shape. This distortion becomes quite pronounced for the TIP4P model and increases with temperature. This distortion causes the sums of the average values (obtained directly from Eq. (1)) for the $xx$, $yy$ and $zz$ components of the field gradient tensor to deviate slightly ($\pm 1 \times 10^{-6}$ au for SPC/E and $\pm 3 \times 10^{-6}$ au for TIP4P) from zero.

Electric field and electric field gradient distributions for liquid water have been previously reported by Nymand et al. [36,37] for the polarizable NEMO intermolecular potential [38,39]. They have reported the temperature dependence of the electric field at the oxygen as well as at the hydrogen atoms. Comparing the present data with those given by Nymand et al. [37] for the electric field and field gradient at the oxygen atom, one can see that their numerical values for the electric field are significantly higher (approximately 1.5–2 times) than those recorded in Table 1. The NEMO potential [36] also exhibits a stronger temperature dependence in the electric field compared with the present results. Both these discrepancies are not surprising given that the SPC/E and TIP4P models both represent effective nonpolarizable interaction potentials. However, our probability distributions of the local electric field for these models are in good agreement with the results from the simulation study of Bursulaya et al. [25] obtained using *ab initio* potentials. The diagonal components of the field gradient from the present

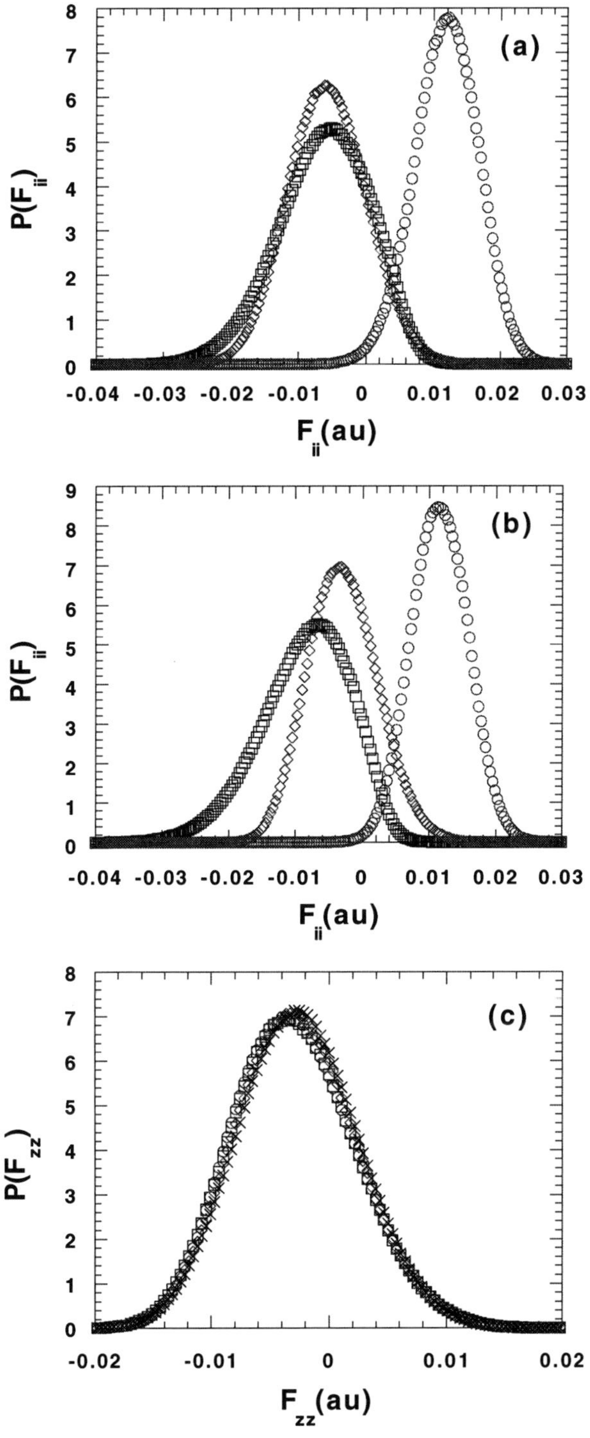


Fig. 2. Field gradient distributions for (a) SPC/E and (b) TIP4P water at $T = 298$ K. The open circles, squares and diamonds are results for P($F_{xx}$), P($F_{yy}$) and P($F_{zz}$), respectively. (c) Temperature dependence of the $F_{zz}$ probability distribution for TIP4P water. The symbols are defined as in Fig. 1(b).

Table 1
Electric field and field gradient components (in atomic units[a]) of SPC/E and TIP4P water

| | SPC/E | | | | TIP4P | | | |
|---|---|---|---|---|---|---|---|---|
| T(K) | $\langle F_z \rangle$ | $\langle F_{xx} \rangle$ | $\langle F_{yy} \rangle$ | $\langle F_{zz} \rangle$ | $\langle F_z \rangle$ | $\langle F_{xx} \rangle$ | $\langle F_{yy} \rangle$ | $\langle F_{zz} \rangle$ |
| 263 | 0.0351 | 0.0131 | −0.0072 | −0.0059 | 0.0296 | 0.0128 | −0.0100 | −0.0029 |
| 298 | 0.0334 | 0.0119 | −0.0062 | −0.0057 | 0.0279 | 0.0115 | −0.0086 | −0.0029 |
| 333 | 0.0317 | 0.0109 | −0.0057 | −0.0052 | 0.0266 | 0.0105 | −0.0078 | −0.0028 |
| 373 | 0.0301 | 0.0100 | −0.0052 | −0.0048 | 0.0252 | 0.0097 | −0.0071 | −0.0026 |

[a]Atomic unit conversion factors: 1 au = 51.42 V/Å; 1au = $5.142 \times 10^{-9}$V/Å$^2$.

work generally exhibit (with some differences in the magnitudes) the same trend reported by Nymand et al. [37].

The behavior of the field gradient distributions appears to be more sensitive to the details of the local structure (i.e. the molecular model or the temperature) than those for the electric field. This fact influenced our choice of numerical values of the average local electric field and electric field gradient utilized in the *ab initio* part of the present study. In building up the liquid phase models for water described in the Section 4.2 below, we have used values 0.03 au for the electric field and $\{0.01, -0.006, -0.0038\}$ for the field gradient, which lie somewhere in-between those found for SPC/E and TIP4P models at ambient temperature. We note that there were two reasons for this particular choice. Firstly, it was desirable to design a mean-field model that was consistent with the electrical properties for two standard water potentials. It was also advantageous for the values chosen to be "numerically convenient" for utilization within our modified scheme for the charge perturbation variant of the finite field method.

## 3. Ab initio calculations of the multipole polarizarilities and hyperpolarizabilities

### 3.1. Calculation in the static limit: the finite-field method

Most *ab initio* calculations determine the electronic contribution to the multipole polarizabilities and hyperpolarizabilities in the static limit. These properties usually are evaluated at the experimental molecular geometry or alternatively using a geometry optimized at a given level of theory. Vibrationally averaged electronic values have been estimated in a small number of cases and in these systems the difference between the vibrationally averaged value and the value determined at the equilibrium geometry is only a small percentage of the total electronic contribution [40]. Thus, on these grounds, vibrational averaging is generally neglected.

The static electronic contribution can be calculated by a variety of methods. Among them the finite-field method provides a straightforward computational technique applicable to almost any quantum chemical formalism. Traditionally in the finite-field method, either the energy or dipole moment is computed for several perturbing field strengths (the different conventions for the corresponding Taylor series expansions in field can be found in [1,40,41]). This technique is often the easiest way to extract the molecular polarizabilities since the finite field energies can be determined with trivial modification to a program that calculates the energy at a given level of theory. However, one has to remember that this method does require care to avoid numerically differentiating errors and is less efficient than analytical methods where they are available.

Guan et al. [10] slightly altered the conventional finite-field procedure in order to calculate polarizabilities and hyperpolarizabilities for several small molecules ($H_2$, $N_2$, CO, $CH_4$, $NH_3$, $H_2O$ and HF). In that study the components of $\alpha$ and $\beta$ were obtained from least squares fits of the dipole moments to a sixth-order polynomial. Fitting to a polynomial whose order is substantially higher than that of the highest order coefficients desired should reduce contamination from higher order terms. By using the dipole moment directly rather than the energies, one less parameter is needed in the fit. The field directions were chosen such that only one-dimensional fits were needed. The sets of field strength used consist of N (an odd number) equally spaced points ranging from $-F_{\max}$ to $F_{\max}$, except that when all the odd or all the even order coefficients were zero by symmetry. Most of the fits were done with $N = 9$ points because comparisons did not show significant differences in the values of $\alpha$ or $\beta$ between fits with $N = 9$ vs those when more points (up to 17) were included. For each fit, the precision of $\alpha$ and $\beta$ were assessed by looking not only at the standard deviation for each coefficient, but also at the convergence of the coefficients with respect to the order of polynomial fit. It has been suggested [10] that if the values of $\beta$ do not converge, the range of fields must be modified, i.e. the convergence of $\beta$ provides a convenient check on the fields used. There are two opposing requirements which govern the choice of field [10]. The fields must be large enough to make the imprecision in the calculated dipole moments small compared to the contribution in the dipole moment expansion of the term whose coefficient is desired. At the same time, the fields must be small enough so that the error caused by the truncation of the expansion is acceptably small. The convergence of $\beta$ with respect to the polynomial order indicates when fields are too large, or too small.

Bartlett and Purvis [42] and then Kurtz et al. [43] have proposed an approach based on expressions determined from the total energy and dipole moment expansions (which were truncated after the fourth-order term) to calculate the coefficients $\mu_i, \alpha_{ii}, \beta_{iii}$, and $\gamma_{iii}$. In this approach the energy must be evaluated for pairs of electric field strengths ($\pm F_i$ and $\pm 2F_i$) where the electric field is applied along an axis of the system. Following Kurtz et al. [43], Sim, Chin and Dupuis [41] made some further modifications and derived expressions for $\mu, \alpha, \beta$, and $\gamma$ in the terms of the field-perturbed energies and field-perturbed dipoles. Their formalism differs from that proposed by Kurtz et al. [43] in two ways. Firstly, their analytical expressions are based on the perturbation definition expansion [41] rather than the power series definition adopted in Ref. [43] (because of the different choices of convention their $\beta$ differs by a factor of 2 and $\gamma$ by a factor of 6). Secondly, their expression for $\beta_{iii}F_i^2$ eliminates the term in $F_i^4$ since $\mu^0$ is evaluated directly as $\mu_i(0)$ (see Ref. [41] for details). This so-called finite-difference approach [41–43] can be considered as an alternative to that based on fitting to a polynomial expansion [10].

A similar finite-difference approach has been utilized by Maroulis [44] for obtaining the first correlated result of the mean value of $\gamma$ for $H_2O$. His derivation starts from the classical representation of the Taylor series expansion of the energy proposed by Buckingham [1]

$$
\begin{aligned}
E = E^0 &- \mu_\alpha^0 F_\alpha - 1/2\alpha_{\alpha\beta}F_\alpha F_\beta - 1/6\beta_{\alpha\beta\gamma}F_\alpha F_\beta F_\gamma - 1/24\gamma_{\alpha\beta\gamma\delta}F_\alpha F_\beta F_\gamma F_\delta \\
&-1/3\Theta_{\alpha\beta}^0 F_{\alpha\beta} - 1/3A_{\alpha,\beta\gamma}F_\alpha F_{\beta\gamma} - 1/6B_{\alpha\beta,\gamma\delta}F_\alpha F_\beta F_{\gamma\delta} - 1/6C_{\alpha\beta,\gamma\delta}F_{\alpha\beta}F_{\gamma\delta} \\
&-1/15\Omega_{\alpha\beta\gamma}^0 F_{\alpha\beta\gamma} - 1/15E_{\alpha,\beta\gamma\delta}F_\alpha F_{\beta\gamma\delta} - \cdots,
\end{aligned}
\tag{2}
$$

where $E^0, \mu_\alpha^0, \Theta_{\alpha\beta}^0$ and $\Omega_{\alpha\beta\gamma}^0$ are the energy, permanent dipole, quadrupole and octopole moments of the free molecule, respectively, $\alpha_{\alpha\beta}, \beta_{\alpha\beta\gamma}, \gamma_{\alpha\beta\gamma\delta}$ and $A_{\alpha,\beta\gamma}, B_{\alpha\beta,\gamma\delta}, C_{\alpha\beta,\gamma\delta}, E_{\alpha,\beta\gamma\delta}$ are molecular polarizabilities describing the electronic distortion of the molecule by the external electric field ($F_\alpha$), field gradient ($F_{\alpha\beta}$) and gradient of field gradient. The Greek suffixes denote Cartesian components and a repeated suffix implies summation over $x, y, z$. Defining $E(F_x, F_y, F_z)$ as the energy of the molecule in the presence of a homogeneous field with $x, y$, and $z$ components $F_x, F_y, F_z$, Maroulis simplified the notation in Eq. (2) by dropping the suffix and introducing perturbing (scalar) fields of $\pm F$ and $\pm\sigma F$

where $\sigma$ (4/3 and 2) is a parameter of the calculations. The working equations used to compute the independent components of the dipole and hyperpolarizabilities can be found in Ref. [44].

To avoid the extra complexity of deriving equations to calculate each component of interest, a more general matrix formalism can be explored. For values of $E(F)$ determined at m values of the applied electric field, chosen to be $\pm F$ or $\pm\sigma_i F$, the coefficients of $F$ in Eq. (2) can be determined by solving a set of linear equations. These equations can be written in matrix notation as shown in [41]. All the variations of the finite-field approach described above, however, require great care be taken in the choice of appropriate field strengths and a sufficient number of terms must be included in Eq. (2) to ensure accurate values for higher order polarizabilities.

The charge perturbation method [45,46] is a variant of the conventional finite-field technique. In this approach a homogeneous external electric field, or a nonhomogeneous electric field (i.e. where electric field gradients etc. will be present), are generated by a charge (or charges) Q placed at a distance R from the molecule. This method can be used to determine the tensoral components of the electrical moments and higher polarizabilities. The Hamiltonian of such a system can be written as

$$H(Q,R) = H^{(0)} + H^{(1)} \tag{3}$$

with the perturbation $H^{(1)}$ given by

$$H^{(1)} = -\sum_{i=1}^{n_{el}} \frac{Q}{|r_i - R|} + \sum_{j=1}^{N_n} \frac{Z_j Q}{|r_j - R|} \tag{4}$$

where $n_{el}$ and $N_n$ represent the numbers of electrons and nuclei, respectively, $Q$ is a charge, $Z_j$ is an atomic number, vectors $r_i$ and $r_j$ denote positions of electrons and nuclei, respectively, and $R$ denotes the position of charge $Q$. Expanding the perturbed Hamiltonian about the unperturbed one the following expression in terms of molecular multipole moments can be obtained

$$H = H^0 - \mu_\alpha F_\alpha - 1/3\Theta_{\alpha\beta}F_{\alpha\beta} - 1/15\Omega_{\alpha\beta\gamma}F_{\alpha\beta\gamma} - \ldots \tag{5}$$

The corresponding energy of the system is described by Eq. (2) and multipole moments are given as

$$\begin{aligned} \mu_\alpha &= \mu_\alpha^0 + \alpha_{\alpha\beta}F_\beta + 1/3A_{\alpha,\beta\gamma}F_{\beta\gamma} + 1/3B_{\alpha\beta,\gamma\delta}F_\beta F_{\gamma\delta} + \ldots, \\ \Theta_{\alpha\beta} &= \Theta_{\alpha\beta}^0 + A_{\alpha,\beta\gamma}F_\alpha + C_{\alpha\beta,\gamma\delta}F_{\gamma\delta} + 1/2B_{\alpha\beta,\gamma\delta}F_\gamma F_\delta + \ldots, \\ \Omega_{\alpha\beta\gamma} &= \Omega_{\alpha\beta\gamma}^0 + E_{\delta,\alpha\beta\gamma}F_\delta + \ldots, \end{aligned} \tag{6}$$

where $\mu^0, \Theta^0, \Omega^0$ are again the permanent multipole moments of the free molecule and $\alpha, \beta, \gamma$, A, C, B, and E the static molecular polarizabilities. $F_\alpha, F_{\alpha\beta}$, etc. are the electric field, field gradient, etc. at the specified origin. The above relations along with the energy expansion Eq. (2) are amenable to analysis by means of the charge perturbation method [47].

This method has been used for the diatomic molecules LiH, BeH and BH and the ion $CH^+$. The corresponding set of equations for abstracting the polarizability tensors from the multipole moments has been given and described in detail elsewhere [48]. Bishop and Pipin [49] deduced the set of equations for molecules of $C_{2\nu}$ symmetry. In their theoretical formalism a charge $Q$ is placed at a particular distance, R, and orientation from the origin. Charges of $\pm e$ were used and values of the tensors found with $R = 30$, 32, and 34 au. On the basis of this approach all components of the polarizability tensors for $H_2O$ have been evaluated at the SCF level and, with the exception of $\beta$ and B, at the MCSCF level of approximation [49]. Following Bishop and Pipin [49], Pluta et al. reported results for the electric moments and polarizabilities for HF, $H_2O$ and the $F^-$ anion [47]. All multipole moments were computed as expectation values and were then used to determine the components of higher order electrical tensors. Electron correlation effects have been accounted for using the coupled cluster method with the inclusion of single, double and triple excitations (CCSDT-1) [47]. It has been shown that the correlated electric moments are in good agreement with experiment and that they may be used to obtain the polarizability tensors in an effective way. It was concluded that use of electron correlated methods in conjunction with the charge perturbation approach provides an effective and accurate tool for calculating electric tensors of small molecules.

### *3.2. Liquid phase models*

A charge perturbation variant of the finite-field method [47–49] was used in the present work to obtain multipole moments, hyper- and high-order (multipole) polarizabilities for a single water molecule in gas and in liquid phase surroundings. We have slightly modified a conventional finite-field procedure thereby allowing a water molecule to be simultaneously exposed by both an electric field and electric field gradient. Within this procedure the gas phase electric response properties for a water molecule were obtained by applications of small perturbing electric fields and field gradients with respect to a reference state condition (e.g. zero field and zero field gradient). To model condensed phase conditions the small perturbing fields and field gradients were applied in addition to a strong fixed electric field and field gradient (representing the influence of surrounding neighbors) to a single water molecule.

In order to achieve better control of the magnitudes and symmetry of the electric field and electric field gradient, a special scheme for the positioning of sets of charges has been developed. Considering the z direction (defined to be the direction of the molecular dipole moment), the magnitude of the applied electric field is given as

$$F_z = Q_z/R_z^2, \tag{7}$$

and electric field gradient by

$$F_{zz} = -2Q_z/R_z^3. \tag{8}$$

It follows from Eq. (7) that a charge of magnitude $Q_z = F_z R^2$ is required to generate the electric field $F_z$. However, by placing *a pair* of charges of magnitude $q_{1,2} = \pm Q_z/2$ at the distance of $\pm R$ from the origin, a greater degree of control is achieved since the net field gradient (at the origin) will be zero. This provides requisite flexibility needed in the modeling of the much more complex condensed phase conditions (see Fig. 3). In a similar way, to generate the field gradient $F_{zz}$ ($F_z$ is now zero), two identical charges $q_{3,4} = -F_{zz}R_z^3$, are placed at a distance $\pm R$ from the origin. If one wants to generate both a field $F_z$ and a field gradient $F_{zz}$, then two pairs of charges, one pair with opposite and one with like signs, are used.

In the present work the scheme described above has been implemented to produce small perturbing fields with magnitudes of 0.005 au and field gradients with magnitudes of 0.001 au with respect to a reference point (RP), which is zero for gas-phase conditions (neither fixed field nor field gradient applied). All perturbing charges were placed at a separation of R = ± 30 au from the centre of mass. Following the finite-field approach, single point energies were calculated for pairs of perturbing fields (± F and $\pm\sigma$F, $\sigma = 1.5$) and field gradients.

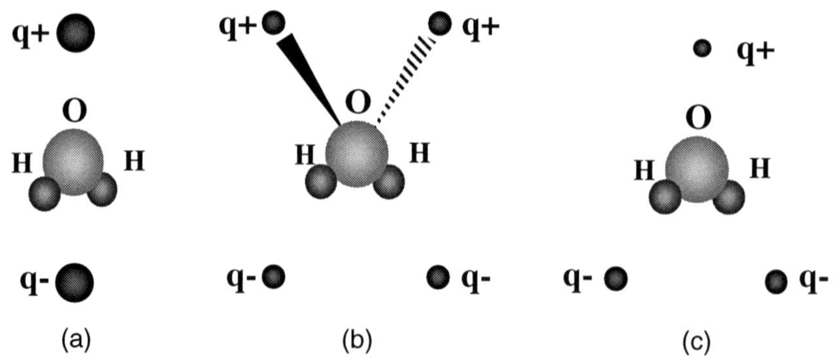


Fig. 3. (a) Model I: $q_{\pm} = \pm\, 13.5$, $r = 30$ au; (b) model II: $q_{\pm} = \pm\, 0.5$, $r = 6.204$ au; (c) model III: $q_{+} = 0.2$, $r_{+} = 4.7$ au, $q_{-} = -\, 0.5$, $r_{-} = 5.0$ au.

Here we do not consider how these perturbing fields might be produced in a real system. In the present context our perturbing fields would be the total (Maxwell) fields experience by a molecule which could, in principle, be written as a function of an externally (to the liquid system) applied field. Since the perturbing fields are small relative to the local (fluctuating) field found within the liquid [68], it can be reasonably assumed that they would have a minor impact on the characteristics of the local environment within the liquid.

With the aim of reproducing a local environment representative of the liquid state, three liquid-phase models were investigated in our calculations. In each case the model serves as our RP. Model I Fig. 3(a) assumes a strong electric field of 0.03 au generated by two charges with opposite signs, applied to the water molecule along its z-axis. Model II Fig. 3(b) is defined to have a strong electric field (0.03 au) and a highly symmetric field gradient ($\{0.008, -0.008, 0\}$) applied. For this model the water molecule has been placed in the xz plane and surrounded by two positive and two negative charges in an ideal tetrahedral arrangement (with an angle of 109.47°) at a separation of $r = 6.204$ au. Finally, in model III Fig. 3(c), an electric field of 0.032 au and a less symmetric field gradient of $\{0.01, -0.006, -0.0038\}$ were generated by three charges, in an arrangement reminiscent of a 3-site water potential model. The numerical values of field and field gradient for this model were chosen on the basis of the MD simulations results presented in the Section 2.

### *3.3. Determination of field and field-gradient dependent coefficients*

During the last decade several different approaches have been proposed to calculate the coefficients in Eq. (2). The general formalism used in the present investigation determines the coefficients of this energy expansion by solving a set of linear equations [41]. Equation (2) must first be truncated and we choose to do so at fourth order in field and at first order in field gradient. We are then assuming that higher order terms contribute negligibly to the total energy and that they can be safely ignored. Following the scheme proposed in [41], Eq. (2) has been further simplified (taking into account the presence of both field and field gradient dependent components) to yield

$$\begin{aligned} E(F,\nabla F) = E^0(F^0,\nabla F^0) - \mu F - 1/3Q\nabla F - 1/2\alpha F^2 - 1/3AF\nabla F \\ -1/6\beta F^3 - 1/6BF^2\nabla F - 1/24\gamma F^4, \end{aligned} \tag{9}$$

For the sake of convenience we are giving Eq. (9) in a general form using the symbols $F$ and $\nabla F$ to represent perturbations from the corresponding components of field and field gradient, respectively, and the symbols $F^0$ and $\nabla F^0$ to represent appropriate reference state conditions.

One further simplification can be made for separating terms dependent only upon field ($\mu, \alpha, \beta, \gamma$), only upon field gradient ($\Theta$), and upon both field and field gradient ($A$ and $B$). Retaining the terms in Eq. (9) that are only field dependent, we obtain

$$E_n(F) = E^0(F^0) - \alpha F - 1/2\mu F^2 - 1/6\beta F^3 - 1/24\gamma F^4, \quad (10)$$

where $E_n$ are single point energies when the perturbing fields $\pm F$ and $\pm\sigma F$ have been applied in the $x, y, z, xy, xz, yz$ directions. $E^0$ is the energy value when no perturbations are applied to the RP (see Section 3.2 for details). On the basis of Eq. (10) a set of linear equations can be derived to calculate the independent components of the $\mu, \alpha, \beta, \gamma$ tensors for the water molecule. They have been given in Appendix, Eq. (A1). Two different magnitudes of field perturbation were compared, $F = 0.005$ au with $\sigma = 1.5$ and $F = 0.0075$ au with $\sigma = 1.3$. It was found that for the gas phase conditions the results are essentially independent of the magnitude of the perturbing field (and $\sigma$). For condensed phase conditions with a reference field of 0.03 au, the smaller perturbation magnitude was found to be more appropriate. Mean values of the dipole polarizabilities, $\overline{\alpha}, \overline{\beta}, \overline{\gamma}$, were calculated from the definitions given in [44].

In order to calculate the field-gradient dependent components, Eq. (9) was rewritten as

$$E_n^*(F, \nabla F) = E_n(F) - 1/3Q\nabla F - 1/3AF\nabla F - 1/6BF^2\nabla F, \quad (11)$$

where $E_n^*(F, \nabla F)$ are single point energies when a perturbing field, $F$, and field gradient, $\nabla F$, are applied. The numerical value of the appropriate field dependent part of the total energy, $E_n(F)$, was utilized (as our fixed point energy) in each linear equation constructed. This allowed us to simplify greatly the solution of these equations, retaining exactly the same energy contributions from the field dependent part. For this case the working equations are also given in the Appendix, Eq. (A2). In addition, the determination of $\Theta$, A and B can take advantage of the extra sets of equations (see Eq. (A3)) reflecting the traceless requirements (e.g. $B_{yy,yy} + B_{yy,zz} + B_{yy,xx} = 0$) of these tensors.

### 3.4. Computational details

In this study many-body perturbation theory (MBPT) has been used to calculate the requisite perturbed energies and the resulting electrical response properties [11,44,50–52]. Calculations were carried out at the MP2 and MP4 levels of theory to account for electron correlations using GAUSSIAN 94 and GAUSSIAN 98 [53,54]. Inclusion of correlation is essential in determining accurate hyper- and high-order polarizabilities. MBPT wavefunctions do not satisfy the Hellmann-Feynman theorem [55–57], and so multipole moments and polarizabilities tensors have been determined as numerical derivatives (via finite-difference) of the total perturbed energy [41].

It is well known that accurate determination of higher-order polarizabilities and hyperpolarizabilities demands high precision in the total energies. Following the criteria developed in Ref. [58], a convergence of $10^{-12}$ was identified as providing a sensible compromise between the level of accuracy and computational time. In addition we have found that the single point energies required to determine the coefficients in our linear equations must be presented with at least 10 digits after the decimal point to maintain numerically stable results for the electrical properties of interest. Reducing the number of decimal digits in the single point energies to 9 slightly changes, and to 8 or less dramatically alters the values of the desirable quantities. In the present study values of the MP2 energies were always taken with 12 decimal places, while MP3 and MP4 energies were taken with 9 decimal digits, as they are given by default in the output provided by the GAUSSIAN programs.

The choice of basis set is known to be crucial for obtaining accurate electric response properties of small molecules like water [11,44,59], and particularly that the inclusion of diffuse polarization functions drastically alters the computed higher-order polarizabilities [42,44–46]. To the best of our knowledge there is only a single experimental study for the higher-rank polarizabilities of liquid water which is restricted only to values of $\overline{\beta}$ and $\overline{\gamma}$, and at present there are no other *ab initio* theoretical data for those tensors available for comparison. In this work the aim was to calculate the desired quantities using well balanced, reasonably sized basis sets which are already known to give "high quality" polarizabilities for the water molecule under gas-phase conditions. Six one-particle basis sets, namely two of GAUSSIAN's [53] built-in basis sets (6-311++G and AUG-CC-pVTZ), three large basis sets (W1, W3, W5) proposed by Maroulis [44], and Sadlej's [60] polarizability consistent (POL) basis set were chosen for preliminary testing. The 6-311++G, AUG-CC-pVTZ, W1, W3 and W5 basis sets were examined for their ability to give the field dependent (dipole) properties and the basis sets W1 and POL were used to examine field and field gradient dependent coefficients. We have repeated calculations of selected electrical properties under the same conditions used by previous authors to confirm the reproducibility of single point energies at the SCF, MP2 and MP4 levels (see [26] for a more detailed discussion of previous results obtained using the basis sets listed above).

It was found that at the SCF level the accuracy of the dipole properties depends on quality rather than size of the basis set; accuracy decreases with the basis set chosen in the order W5 = W3 > W1 > AUG-CC-pVTZ > 6-311++G. At higher levels of theory, the 6-311++G basis set could again be classified as the least satisfactory, particularly for $\beta$ and $\gamma$ while W1 proved to be superior for determining dipole properties. Calculations of the quadrupole moment were carried out to choose an appropriate basis set for the field-gradient dependent properties for $H_2O$. A comparison of our quadrupole moment data with the experimental values and those obtained with POL [61] and MPOL [47] leads to the conclusion that the W1 basis set is consistently superior for SCF calculations, while both the POL and W1 basis sets can, in principle, be utilized at higher levels of theory.

The experimental equilibrium gas-phase geometry for the water molecule [59] (OH bond length of 0.9572 Å and 104.52° HOH angle) was employed in all our calculations. The molecule is defined to be in the $xz$ plane with the z-axis along the symmetry axis and the centre of mass taken to be the origin.

Atomic units are used throughout this paper. The conversion factors to SI units are as follows: for distance, $r$, 1 au ($a_0$) $\cong$ 0.529177249 $\times$ 10 $^{-10}$ m; for dipole moment, $\mu$, 1 au ($ea_0$) $\cong$ 8.478358 $\times$ $10^{-30}$ Cm; for quadrupole moment, $\Theta$, 1 au ($ea_0^2$) $\cong$ 4.486554 $\times$ $10^{-40}$ Cm$^2$; for polarizability, $\alpha$, 1 au ($e^2a_0^2E_h^{-1}$) $\cong$ 1.648778 $\times$ $10^{-30}$ C$^2$m$^2$J$^{-1}$; for first hyperpolarizability, $\beta$, 1 au ($e^3a_0^3E_h^{-2}$) $\cong$ 3.206361 $\times$ $10^{-53}$ C$^3$m$^3$J$^{-2}$; for second hyperpolarizability, $\gamma$, 1 au ($e^4a_0^4E_h^{-3}$) $\cong$ 6.235378 $\times$ $10^{-65}$ C$^4$m$^4$J$^{-3}$; for dipole-quadrupole polarizability, A, 1 au ($e^2a_0^3E_h^{-1}$) $\cong$ 8.7250 $\times$ $10^{-52}$ C$^2$m$^3$J$^{-1}$; dipole -dipole-quadrupole polarizability, B, 1 au ($e^3a_0^4E_h^{-2}$) $\cong$ 1.6967 $\times$ $10^{-63}$ C$^3$m$^4$J$^{-2}$.

## 4. Results and discussion

### 4.1. Field dependent properties of water

For the first stage of the present study we have carried out gas phase calculations (i.e. no fixed electric field or field gradients applied) to confirm the reliability of our methodology and to identify an appropriate basis set by comparing our results with previous findings for polarizabilities of a single water molecule. Reliable experimental estimates of the low order field dependent quantities for water in the gas phase

Table 2
Previously published theoretical and experimental results for the polarizability and hyperpolarizability tensor components of the $H_2O$ molecule in the gas phase

| Level of theory | MP2 [65] | MP [44] | SCF + ECC [11] | Experiment |
|---|---|---|---|---|
| Basis set | LP | W1 | PA, KT1 | |
| $\mu$ | 0.7272 | 0.7219 | 0.7238 | 0.721[a], 0.7268[b], 0.7306[c] |
| $\alpha_{xx}$ | 10.03 | 10.13 | 9.93 | $10.31 \pm 0.08$[d] |
| $\alpha_{yy}$ | 9.54 | 9.76 | 9.34 | $9.55 \pm 0.08$[d] |
| $\alpha_{zz}$ | 9.8 | 9.93 | 9.59 | $9.91 \pm 0.08$[d] |
| $\overline{\alpha}$ | 9.79 | 9.94 | 9.62 | 9.92[e], 9.81[f] |
| $\beta_{zxx}$ | −9.4 | −10.1 | −9.8 | |
| $\beta_{zyy}$ | −5.9 | −6.5 | −5.5 | |
| $\beta_{zzz}$ | −13.7 | −15.8 | −13.8 | |
| $\overline{\beta}$ | −17.4 | −19.4 | −17.5 | $-22.0 \pm 0.9$[g], $-19.2 \pm 0.9$[h] |
| $\gamma_{xxxx}$ | 820 | 821 | 836 | |
| $\gamma_{yyyy}$ | 2800 | 2900 | 2650 | |
| $\gamma_{zzzz}$ | 1540 | 1575 | 1481 | |
| $\gamma_{xxyy}$ | 660 | 684 | 633 | |
| $\gamma_{zzyy}$ | 750 | 779 | 711 | |
| $\gamma_{zzxx}$ | 420 | 464 | 439 | |
| $\overline{\gamma}$ | 1764 | 1830 | 1706 | $2310 \pm 120$[g], $1800 \pm 150$[h] |

[a] Werner and Meyer [62]. [b] Clough et al [4]. [c] Shostak et al [5].
[d] Murphy [54] (dynamic results obtained at 514.5 nm).
[e] Murphy [64] (averaged dynamic results obtained at 514.5 nm).
[f] Spackman [63] (cummulative result of several experiments).
[g] Ward and Miller [7] (rescaled by Shelton and Rice). [h] Kaatz et al [8].

are available from the literature [4,7,62–64], and recently some previous results for $\mu, \beta$, and $\gamma$ were updated [5]. These results, as well as the best theoretical estimates for field-dependent properties, are given in Table 2. As mentioned above the importance of electron correlation effects in calculations of nonlinear dipole properties has already been confirmed in the previous literature [9,44]. In this study calculations were performed at the MP2, MP3 and MP4 (accounting for contributions from single (S), double (D), triple (T) and quadruple (Q) substitutions from the zeroth order wave function) levels. The results obtained at the highest levels of theory are discussed in detail elsewhere [26] and the present MP2 results are given in Table 3. We only remark here that a move from the MP2 to MP4 level results in only a minor change in all dipole properties.

Comparing data from Tables 2 and 3 one can see that our MP2 value for the dipole moment of 0.7312 au is in excellent agreement with the ground state value of 0.7298 au reported by Shostak et al. [5] (they have determined precise dipole moments of $H_2O$ in its ground state and in each of its singly exited vibrational states). In Table 3 the results for the polarizabilities and hyperpolarizabilities (especially for $\overline{\alpha}$ and $\overline{\beta}$) obtained with the W1 basis set are in very good agreement with those determined by Sekino and Bartlett using an unconventional LP basis set [65] (see Table 2). Our MP2 value for $\overline{\alpha}$ (9.85 au) is in good agreement with the MP4 $\overline{\alpha}$ (9.82 au) reported by Diercksen et al. [61], while our MP4 value of $\overline{\alpha}$ (9.94 au) is virtually identical to that reported by Maroulis [44] and very close to the experimental value.

The experimental values for $\overline{\beta}$ and $\overline{\gamma}$, $-22.0 \pm 0.9$ au and $2310 \pm 120$ au, respectively, were deduced from gas-phase EFISH measurements by Ward and Miller at a frequency of 694.3 nm [7]. More recently Kaatz et al. published new data for the hyperpolarizabilities of $H_2O$ ($-19.2 \pm 0.9$ au for $\overline{\beta}$ and $1800 \pm 150$ au for $\overline{\gamma}$) obtained at 1064 nm using an analogous experimental technique [8]. These two results clearly indicate the sensitivity of both hyperpolarizabilities to the experimental details, including the

Table 3
Dipole properties of the water molecule in the gas and the liquid phase. All calculations were carried out with the W1 basis set at the MP2 level with a perturbing field of 0.005 au and $\sigma = 1.5$ used throughout

| | Gas phase | Gas phase* | Condensed phase | | | | Experiment |
|---|---|---|---|---|---|---|---|
| | | | (model I) | (model II) | (model III) | (model III)* | |
| $\mu$ | 0.7312 | 0.7205 | 1.029 | 1.047 | 1.065 | 1.0575 | – |
| $\alpha_{xx}$ | 10.02 | 10.11 | 9.94 | 9.59 | 9.55 | 9.62 | |
| $\alpha_{yy}$ | 9.70 | 9.78 | 9.86 | 11.30 | 10.69 | 10.80 | |
| $\alpha_{zz}$ | 9.84 | 9.93 | 10.10 | 10.24 | 10.39 | 10.49 | |
| $E_n$ | 9.85 | 9.94 | 9.97 | 10.37 | 10.21 | 10.30 | – |
| $\beta_{zxx}$ | −8.8 | −9.8 | 4.1 | 4.3 | 5.3 | 5.7 | |
| $\beta_{zyy}$ | −5.1 | −6.1 | 18.8 | 31.7 | 10.2 | 10.9 | |
| $\beta_{zzz}$ | −14.1 | −16.3 | 35.0 | 31.2 | 29.6 | 31.6 | |
| $\overline{\beta}$ | −16.8 | −19.3 | 34.7 | 40.3 | 27.1 | 28.9 | 15.96[a] |
| $\gamma_{xxxx}$ | 756 | 840 | 756 | 622 | 710 | 768 | |
| $\gamma_{yyyy}$ | 2705 | 2990 | 3124 | 5486 | 4434 | 4936 | |
| $\gamma_{zzzz}$ | 1459 | 1587 | 2343 | 1997 | 2273 | 2504 | |
| $\gamma_{xxyy}$ | 598 | 666 | 695 | 729 | 814 | 902 | |
| $\gamma_{zzyy}$ | 678 | 749 | 1128 | 1481 | 1074 | 1197 | |
| $\gamma_{zzxx}$ | 399 | 438 | 551 | 392 | 446 | 490 | |
| $\overline{\gamma}$ | 1654 | 1825 | 2234 | 2662 | 2417 | 2677 | 2858[a] |

*data obtained at MP4 level.
[a]Revised value comparing with [26]. The original experimental value reported by Levine and Bethea [12] was rescaled following the scheme proposed by Mikkelsen et al. [16].

frequency of the EFISH measurements, that compound the difficulties faced in comparing them with theoretical data obtained at the static limit.

The need of at least a crude estimation of the vibrational contributions to polarizabilities and hyperpolarizabilities has given rise to discussions in the literature with regard of their significance. For example, Shelton and Rice [40] argued that the vibrational contributions to EFISH hyperpolarizabilities are only about 2% and 1% of $\overline{\beta}$ and $\overline{\gamma}$, respectively. This agrees well with the estimates of the pure vibrational contributions at optical frequencies by Bishop et al. [66], while Luo et al. [67] deduced that zero point vibrational averaging contributions are about 10% of the value of $\overline{\beta}$. Kaatz et al. [8] suggest that the difference between experimental values of the hyperpolarizabilities for $H_2O$ and $D_2O$ gives an approximate indication of the size of the vibrational contribution; the difference is 8% for both $\beta$ and $\gamma$ and so is consistent with the theoretical estimate by Luo et al. Relying on the result of Kaatz et al., one can deduce a static $\overline{\beta}$ of $-17.3 \pm 0.8$ au and a static $\overline{\gamma}$ of $1620 \pm 135$ au, which are in excellent agreement with our MP2 values. However, taking into account the apparent sensitivity of hyperpolarizabilities to the experimental frequencies, as well as the discrepancy in the assessments of the size of vibrational contributions and the necessity of a dispersion correction, the accurate comparison of theoretically determined static values with the experimental results available still remains problematic.

Comparing estimates of the hyperpolarizabilities obtained in this work with theoretical data previously reported in the literature one can see that our MP2 value of $\overline{\beta}$ ($-16.8$ au) compares favorably with that of Sekino and Bartlett [65], and the present MP4 result ($-19.3$ au) is also in excellent agreement with Maroulis' static value of $-19.4$ au [44] as well as the $-19.6$ au result reported by Guan et al. [10]. Our MP2 value of $\overline{\gamma}$ is slightly lower in comparison with the value obtained in [65] at the same level of theory.

Three sets of data, corresponding to our three models for liquid phase conditions, were determined and presented in Table 3. We emphasize here that model I was constructed to assess the influence of the field on numerical values of the polarizabilities of the water molecule, while models II and III were developed to assess the importance of the symmetry and magnitude of the field gradient experienced by a water molecule.

Comparison of the only experimental result for hyperpolarizabilities obtained from liquid-phase EFISH measurements (at 1064 nm) with those reported earlier for gas phase shows that the value of $\overline{\beta}$ increases dramatically and actually undergoes a sign inversion. The change in $\overline{\gamma}$ is significantly smaller, although its increase from 1800 to 2858 au is still a 60% increase in magnitude. Unfortunately, for such limited data, any estimates of the static limit of the electronic hyperpolarizability of water in the liquid state are not possible and a vibrational contribution and dispersion correction could not be taken into account in the following analysis. However, for liquid-phase conditions, with the lack of both theoretical and experimental data, even qualitative or semi-quantitative comparison with experimental results will be informative.

The analysis of the results from model I together with those from models II and III (see Table 3) reveals the "response" of the dipole properties to the presence and the characteristics of a field gradient in the liquid phase model. One can see that $\mu_z$ steadily increases from model I (when there is no gradient applied) to model III. The dipole polarizability, $\overline{\alpha}$, exhibits only slight variation across the three models. At the same time, $\overline{\beta}$ and $\overline{\gamma}$ increase from model I to model II and then both drop for model III. In particularly, for $\overline{\beta}$ this change is quite significant and the value of $\overline{\beta}$ for model III becomes considerably closer to the experimental estimate. Relying on results of MD simulations [68] to characterize the local environment in liquid water, and comparing the dipole properties for models II and III with each other and with experimental ($\overline{\beta}$ and $\overline{\gamma}$) results, we conclude that the less symmetrical gradient (model III) described in Section 3.2 gives the most accurate representation of liquid-phase conditions. Comparison of this most successful model (III) with gas phase values in Table 4 reveals that $\mu_z$ increases considerably from 0.7312 to 1.065 au (MP2 values), $\overline{\alpha}$ increases only marginally, while $\overline{\beta}$ increases significantly (changing sign) from $-16.8$ to 27.1 au and $\overline{\gamma}$ grows from 1654 to 2417 au (approaching the experimental value). In addition, the value of the dipole moment itself can serve as an independent criterion for the quality of our model representations. Recent theoretical estimates for the dipole moment of $H_2O$ in condensed state [14,24,69] range from 2.4 to 3.1 D. Our dipole moment values of 1.065 au (2.71 D) and 1.057 (2.69 D) for model III at the MP2 and MP4 levels, respectively, fall nicely within this range. Moreover, they are in an excellent agreement with the very recent result of $2.64 \pm 0.17$ D reported by Tu and Laaksonen [22]. Taken together all these results suggest that model III is a reasonable representation of the local electrical environment within liquid water.

To the best of our knowledge only one other theoretical study confirming the dramatic increase in numerical value (the changing sign) of the first hyperpolarizability of liquid water has been published [16]. This previous work investigates the solvent effects on the hyperpolarizabilities and their frequency dispersions for liquid water using continuum, semicontinuum and supermolecular models. The semicontinuum calculations at $\omega = 0$ produced a $\overline{\beta}$ of 15.52 au which is in good agreement with that reported by Levine and Bethea [12]. Although the $\overline{\beta}$ reported in the present study is somewhat larger than the only experimental value available, our results could be classified as the first *ab initio* estimates of the first hyperpolarizability for $H_2O$ in a liquid state.

### *4.2. Field-gradient dependent properties of water*

The field-gradient dependent properties of water ($\Theta$, A and B tensors) were calculated under *nonhomogeneous* field conditions with perturbing fields of 0.005 au and perturbing field gradients of 0.001 au applied either to gas phase or to liquid-phase reference state conditions. The quadrupole moment is

Table 4
Independent components of the A (2 of 4) and the B (6 of 9) tensors for water

| Conditions | Basis set | $A_{z,xx}$ | $A_{z,zz}$ | $B_{zz,zz}$ | $B_{xx,xx}$ | $B_{xx,zz}$ | $B_{yy,zz}$ | $B_{zz,xx}$ | $B_{yy,xx}$ |
|---|---|---|---|---|---|---|---|---|---|
| Gas phase | | | | | | | | | |
| MCSCF [a] | [c] | 2.54 | 2.44 | – | – | – | – | – | – |
| CCSDT-1 [b] | MPOL | 2.31 | 2.28 | −133.0 | −94.9 | 50.2 | 109.8 | 82.8 | 144.7 |
| Present work | | | | | | | | | |
| | POL* | 1.58 | 2.02 | −129.5 | −91.6 | 44.2 | 92.8 | 75.0 | 118.4 |
| | W1* | 0.89 | 2.73 | −154.2 | −104.2 | 58.3 | 110.9 | 90.9 | 126.1 |
| | POL** | 1.66 | 2.09 | −133.5 | −93.6 | 44.6 | 94.2 | 79.1 | 121.8 |
| | W1** | 0.91 | 2.83 | −159.6 | −107.2 | 60.0 | 114.1 | 94.7 | 130.9 |
| Cond. phase | | | | | | | | | |
| model I | POL* | 4.05 | −2.00 | −153.9 | −83.5 | 43.2 | 95.8 | 96.7 | 129.1 |
| | W1* | 3.87 | −2.12 | −197.8 | − 96.1 | 55.1 | 113.6 | 124.0 | 147.5 |
| model II | POL* | 4.07 | −1.25 | −137.8 | −66.0 | 42.5 | 138.2 | 105.0 | 185.0 |
| | W1* | 3.95 | −1.20 | −169.8 | −76.1 | 58.2 | 175.7 | 139.7 | 232.6 |
| model III | POL* | 3.54 | −2.26 | −165.9 | −67.4 | 41.7 | 124.3 | 110.2 | 159.1 |
| | W1* | 3.00 | −2.12 | −206.2 | −79.9 | 58.4 | 158.3 | 138.4 | 180.8 |
| | POL** | 3.67 | −2.29 | −170.9 | −68.7 | 42.5 | 127.4 | 114.7 | 165.7 |
| | W1** | 3.11 | −2.16 | −214.0 | −81.4 | 60.0 | 164.6 | 145.4 | 190.6 |

*MP2 level of theory, **MP4 level of theory.
a – Bishop and Pipin [49].
b – Pluta et al. [47].
c – [5s, 8p, 3d / 2s, 2p].

the only coefficient in the truncated total energy expansion dependent upon field gradient only, while the A and B tensors (high-order polarizabilities) couple to both field and field gradients, appearing as coefficients of $F\nabla F$ and $F^2\nabla F$, respectively, in the energy expansion Eq. (11). We have calculated the values of all the components of quadrupole moment for water in gas and in liquid phase; results were reported in [26]. Comparison of our gas phase values for $\Theta$ with other theoretical determinations and with those from experiment guided our choice of appropriate basis sets for all field gradient dependent properties. The numerical values of $\Theta$ for a liquid phase of water molecule display approximately 10% elevation in comparison with an isolated molecule [11,47,49]. We also would like to direct readers to [26] for MP3 and MP4 results for the A and B tensors.

There has been no experimental evaluation of the high-order tensors A and B for $H_2O$ in gas phase and the number of theoretical studies is rather limited. For the A tensor two sources of theoretical data at a correlated level are available, these are multiconfigurational self consistent field (MCSCF) [49] and coupled cluster density matrix (CCSDT-1) [47] calculations for an isolated water molecule. The components of A calculated by means of these methods are compared with the results of the present work in Table 4. The values obtained in the present investigation are generally lower than those previously reported for the components $A_{z,xx}$ and $A_{z,zz}$ with both the POL and W1 basis sets. The results obtained with the POL basis set can be considered to be in acceptable agreement (our value of $A_{z,zz}$ is 2.02 au vs 2.28 au [47] and 2.44 [49]), while the values calculated with W1, and particularly $A_{z,xx}$, are not as close to those of other authors. It is also not surprising that our results agree better with those of [47] than [49]. Similar basis sets (POL and MPOL) were used in the calculations, and moreover, the CCSDT-1 and MBPT approaches capture more electron correlation effects than does MCSCF [49].

In the present work, 6 of the 9 independent components of the B tensor were calculated; they are also given in Table 4. There were no MCSCF values for the B tensor determined in [49] with which to compare. The values of $B_{zz,zz}$, $B_{xx,xx}$, and $B_{xx,zz}$ determined with POL agree well with those calculated with the similar basis set by Pluta et al. [47], while $B_{yy,zz}$, $B_{zz,xx}$ and $B_{yy,xx}$ are slightly lowered. In general, the values of B reported in [47] lie approximately in between our estimates calculated with POL and W1. This numerical behavior is indicative of the high sensitivity of the B tensor to the quality rather than to the size of a basis set. On the basis of careful analysis and our experience in related calculations of the temperature dependence for the total molecular dipole moment of liquid water (performed with the values for the B tensor given in Table 4), we would recommend the POL basis set to be preferable for the calculations of these quantities.

To our knowledge, there are no other theoretical or experimental data reported for the A or B tensors for a condensed phase water molecule. These properties calculated with the POL and W1 basis sets for a liquid phase molecule are given in Table 4. The components of the A tensor, like $\beta$, display a much stronger dependence on the local environment. The values of $A_{z,xx}$ were found to be more than two times higher for the liquid phase than for the gas phase, and the $A_{z,zz}$ component changes sign. Although for models I and II there is no significant discrepancy in the values of $\Theta$ and A obtained with the two different basis sets, for model III the discrepancy becomes more noticeable.

Analysis of the available components of the B tensor in liquid phase conditions reveals that they are of similar magnitudes as those for gas phase, with the signs of all the components remaining unchanged (two negative and four positive). There are no general overall trends in the behavior of these coefficients for all three models. For example, $B_{zz,zz}$ and $B_{xx,xx}$ increase as model II < model I < model III. $B_{xx,zz}$ remains almost unchanged, while $B_{yy,zz}$ is smaller for model I than for model II and III, with model III < model II. $B_{zz,xx}$ is enhanced slightly in the sequence model I < model II < model III. Sometimes the rearrangement of contributions for each component appears to be quite significant as evidenced by the values for $B_{yy,xx}$. The numerical values of the B's calculated with the POL basis set are approximately 25–35% lower than those calculated with W1. It would seem reasonable to extend to liquid phase condition the conclusion from our analysis of the gas-phase results, namely that while both of the basis sets used can be considered appropriate for determination of the quadrupole moment for the water molecule, the POL basis set is preferable for field and field gradient dependent properties.

## 5. Conclusions

In the present work a new method has been outlined which allows the determination of the molecular polarization response properties for small (e.g. water) molecules in the liquid state. The principal results of this work include both methodological development and the results of its application to the water molecule. A mean-field approach has been utilized to represent the environment experienced by a water molecule in liquid phase. Our methodology consists of two critical components. In the first component, liquid phase MD simulations for two popular water models were performed to determine the average local electric field and field gradient experienced by each water molecule due to its surrounding. In the second component, *ab initio* calculations were carried out to determined molecular polarization response properties (in particular polarizabilities) of the water molecule when conditions reproducing a liquid phase environment (the strong field electric field and field gradient determined above) were applied to the single water molecule. To accomplish this second component we have developed a scheme whereby the strong electric fields and electric field gradients required are produced by sets (arrangements) of fixed charges. We then adapted a special energy partitioning method to extract nonlinear optical coefficients in the presence of these fixed fields and field gradients, where we have found that the method requires very accurate values of the single point energies. A careful choice (or construction) of basis set and inclusion of electron correlation effects were both basic requirements necessary to determine accurate values of the NOP, especially for high-order coefficients.

There may be some restriction, however, on the applicability of this mean-field approach. Fowler and co-workers [70–74] performed *ab initio* studies of multipole polarizabilities of anions and ionic crystals and identified large short-range contributions to the induced multipoles in addition to the regular (Coulombic) contributions arising from electric field and field gradient effects. The origin of these short-range contributions is associated with the overlap of the cation and anion electron densities and results in a sharp drop in the polarizability and related hyperpolarizabilities of the anion [70,71]. Nevertheless, on the basis of the previous results obtained for the variety of molecular species, it is reasonable to expect these short-range contributions to be negligible for small water-like molecules in which the local environment in liquid state differs considerably from that of ionic crystals (such as, for example, LiF).

To determine the numerical values representing the electric fields and electric field gradients experienced by a water molecule in the liquid, basic characteristics (averages, width and proportionality constants) of the appropriate distributions are first required. These parameters were obtained in the first part of the present study from MD simulations employing the SPC/E [34] and TIP4P [33] water models for liquid systems from 263 to 373 K. The values obtained for the average local electric fields and field gradients demonstrate some sensitivity to the choice of potential model, but this sensitivity becomes less pronounced when considering the temperature-dependent behavior of these distributions.

In the second part of this work, our *ab initio* study, five basis sets were examined for their ability to reproduce the gas phase NOP for water [26]. Two of them (W1 [44] and POL [60]) give very good, correlated results. Further detailed analysis showed that for field dependent properties, the basis set W1 is somewhat superior. For field-gradient dependent properties both basis sets could be successfully explored. However, careful analysis led us to conclude that while both the POL and W1 basis sets give excellent results for the quadrupole moment, for both the A and B tensors POL appears preferable.

We have explored three different models (I, II and III) in our *ab initio* calculations to mimic the electrostatic environment in liquid water. For all models the required electric fields and field gradients were generated with sets of fixed charges. Because of the fact that model III attempts to reproduce most closely the field and field gradient conditions observed in classical simulations of water, one could expect this model to be the most "realistic". For model III, calculations were performed at both the MP2 and MP4 levels (see also [26]), although no significant difference was noted between the MP2 and MP4 coefficients. In comparison with gas phase results the value of the dipole moment, $\mu_z$, for liquid water increases substantially, by approximately 45%. The dipole polarizability, $\overline{\alpha}$, remains almost unchanged, while the first hyperpolarizability, $\overline{\beta}$, increases dramatically and changes sign, in accord with experiment. The second-order hyperpolarizability, $\overline{\gamma}$, also increases (by about 50%) and demonstrates sensitivity to modeling conditions. The magnitudes of the major components ($xx$ and $yy$) of the quadrupole moment are observed to increase (by about 15%) while the changes in the components of the A tensor are again more spectacular. Some values of the B tensor for the liquid phase are higher (typically by 30%) than those for gas phase conditions, however this trend is not universal. In general, the NOP we have determined should reasonably represent those of liquid water.

The excellent agreement of our gas-phase results with experiment and with the most accurate previous theoretical predictions provides confirmation of the soundness of our methodology for extracting NOP for water or other small molecules (e.g. ammonia). It is also worth noting that the MP2 values for hyper- and high-order polarizabilities reported here were used as a crucial part of a theoretical formalism developed to predict the temperature dependence of the total molecular dipole moment for liquid water [68]. The consistent trend observed in these results as well as excellent agreement with other studies for the absolute value of the total dipole moment at ambient temperature provides further evidence of the quality of our higher order polarizabilities and the reliability of our theoretical models. In conclusion we suggest that improvements in the characterization of the required properties of the field and field gradient distributions (e.g. a better choice of potential model), a superior choice of reference conditions together with utilization of a larger, "quality" basis set would allow for some further refinements of the results reported here.


## Acknowledgment

We are thankful to the Natural Science and Engineering Research Council of Canada for the financial support of this project.


## Appendix

The linear equations to calculate the field dependent properties ($\mu, \alpha, \beta, \gamma$) for the water molecule are as follows:

$$
\begin{aligned}
E_1 &= E^0 - \mu_z(F_z) - 1/2\alpha_{zz}(F_z)^2 - 1/6\beta_{zzz}(F_z)^3 - 1/24\gamma_{zzzz}(F_z)^4, \\
E_2 &= E^0 - \mu_z(\sigma F_z) - 1/2\alpha_{zz}(\sigma F_z)^2 - 1/6\beta_{zzz}(\sigma F_z)^3 - 1/24\gamma_{zzzz}(\sigma F_z)^4, \\
E_3 &= E^0 - \mu_z(-F_z) - 1/2\alpha_{zz}(-F_z)^2 - 1/6\beta_{zzz}(-F_z)^3 - 1/24\gamma_{zzzz}(-F_z)^4, \\
E_4 &= E^0 - \mu_z(-\sigma F_z) - 1/2\alpha_{zz}(-\sigma F_z)^2 - 1/6\beta_{zzz}(-\sigma F_z)^3 - 1/24\gamma_{zzzz}(-\sigma F_z)^4, \\
E_5 &= E^0 - 1/2\alpha_{xx}(F_x)^2 - 1/24\gamma_{xxxx}(F_x)^4, \\
E_6 &= E^0 - 1/2\alpha_{xx}(\sigma F_x)^2 - 1/24\gamma_{xxxx}(\sigma F_x)^4, \\
E_7 &= E^0 - 1/2\alpha_{yy}(F_y)^2 - 1/24\gamma_{yyyy}(F_y)^4, \\
E_8 &= E^0 - 1/2\alpha_{yy}(\sigma F_y)^2 - 1/24\gamma_{yyyy}(\sigma F_y)^4, \qquad \text{(A1)} \\
E_9 &= E^0 - \mu_z(F_z) - 1/2\alpha_{zz}(F_z)^2 - 1/2\alpha_{xx}(F_x)^2 - 1/2\beta_{zxx}(F_z)(F_x)^2 - 1/6\beta_{zzz}(F_z)^3 \\
&\quad -1/4\gamma_{zzxx}(F_z)^2(F_x)^2 - 1/24\gamma_{zzzz}(F_z)^4 - 1/24\gamma_{xxxx}(F_x)^4, \\
E_{10} &= E^0 - \mu_z(-F_z) - 1/2\alpha_{zz}(-F_z)^2 - 1/2\alpha_{xx}(F_x)^2 - 1/2\beta_{zxx}(-F_z)(F_x)^2 - 1/6\beta_{zzz}(-F_z)^3 \\
&\quad -1/4\gamma_{zzxx}(-F_z)^2(F_x)^2 - 1/24\gamma_{zzzz}(-F_z)^4 - 1/24\gamma_{xxxx}(F_x)^4, \\
E_{11} &= E^0 - \mu_z(F_z) - 1/2\alpha_{zz}(F_z)^2 - 1/2\alpha_{yy}(F_y)^2 - 1/2\beta_{zyy}(F_z)(F_y)^2 - 1/6\beta_{zzz}(F_z)^3 \\
&\quad -1/4\gamma_{zzyy}(F_z)^2(F_y)^2 - 1/24\gamma_{zzzz}(F_z)^4 - 1/24\gamma_{yyyy}(F_y)^4, \\
E_{12} &= E^0 - \mu_z(-F_z) - 1/2\alpha_{zz}(-F_z)^2 - 1/2\alpha_{yy}(F_y)^2 - 1/2\beta_{zyy}(-F_z)(F_y)^2 - 1/6\beta_{zzz}(-F_z)^3 \\
&\quad 1/4\gamma_{zzyy}(-F_z)^2(F_y)^2 - 1/24\gamma_{zzzz}(-F_z)^4 - 1/24\gamma_{yyyy}(F_y)^4, \\
E_{13} &= E^0 - 1/2\alpha_{xx}(F_x)^2 - 1/2\alpha_{yy}(F_y)^2 - 1/4\gamma_{xxyy}(F_x)^2(F_y)^2 \\
&\quad -1/24\gamma_{xxxx}(F_x)^4 - 1/24\gamma_{yyyy}(F_y)^4.
\end{aligned}
$$

The set of linear equations to calculate field-gradient terms (Θ, A and B) in the total energy expansion are as follows:

$$
\begin{aligned}
E_1^* &= E_1 - 1/3\Theta_{zz}F_{zz} - 1/3\Theta_{yy}F_{yy} - 1/3\Theta_{xx}F_{xx} - 1/3A_{z,zz}(F_z)F_{zz} - 1/3A_{z,xx}(F_z)F_{xx} \\
&\quad -1/3A_{z,yy}(F_z)F_{yy} - 1/6B_{zz,zz}(F_z)^2F_{zz} - 1/6B_{zz,yy}(F_z)^2F_{yy} - 1/6B_{zz,xx}(F_z)^2F_{xx}, \\
E_2^* &= E_1 - 1/3\Theta_{xx}F_{zz} - 1/3\Theta_{zz}F_{xx} - 1/3\Theta_{yy}F_{yy} - 1/3A_{z,xx}(F_z)F_{zz} - 1/3A_{z,zz}(F_z)F_{xx} \\
&\quad -1/3A_{z,yy}(F_z)F_{yy} - 1/6B_{zz,xx}(F_z)^2F_{zz} - 1/6B_{zz,zz}(F_z)^2F_{xx} - 1/6B_{zz,yy}(F_z)^2F_{yy}, \\
E_3^* &= E_3 - 1/3\Theta_{zz}F_{zz} - 1/3\Theta_{yy}F_{yy} - 1/3\Theta_{xx}F_{xx} - 1/3A_{z,zz}(-F_z)F_{zz} - 1/3A_{z,xx}(-F_z)F_{xx} \\
&\quad -1/3A_{z,yy}(-F_z)F_{yy} - 1/6B_{zz,zz}(-F_z)^2F_{zz} - 1/6B_{zz,yy}(-F_z)^2F_{yy} - 1/6B_{zz,xx}(-F_z)^2F_{xx}, \\
E_4^* &= E_3 - 1/3\Theta_{xx}F_{zz} - 1/3\Theta_{zz}F_{xx} - 1/3\Theta_{yy}F_{yy} - 1/3A_{z,xx}(-F_z)F_{zz} - 1/3A_{z,zz}(-F_z)F_{xx} \\
&\quad -1/3A_{z,yy}(-F_z)F_{yy} - 1/6B_{zz,xx}(-F_z)^2F_{zz} - 1/6B_{zz,zz}(-F_z)^2F_{xx} - 1/6B_{zz,yy}(-F_z)^2F_{yy}, \\
E_5^* &= E_5 - 1/3\Theta_{zz}F_{zz} - 1/3\Theta_{yy}F_{yy} - 1/3\Theta_{xx}F_{xx} - 1/6B_{xx,xx}(F_x)^2F_{zz} - 1/6B_{xx,yy}(F_x)^2F_{yy} \\
&\quad -1/6B_{xx,zz}(F_x)^2F_{xx} \\
E_6^* &= E_5 - 1/3\Theta_{zz}F_{zz} - 1/3\Theta_{yy}F_{yy} - 1/3\Theta_{xx}F_{xx} - 1/6B_{xx,zz}(-F_x)^2F_{zz} \\
&\quad -1/6B_{xx,xx}(-F_x)^2F_{xx} - 1/6B_{xx,yy}(-F_x)^2F_{yy}, \\
E_7^* &= E_7 - 1/3\Theta_{xx}F_{zz} - 1/3\Theta_{zz}F_{xx} - 1/3\Theta_{yy}F_{yy} - 1/6B_{yy,zz}(F_y)^2F_{zz} \\
&\quad -1/6B_{yy,yy}(F_y)^2F_{yy} - 1/6B_{yy,xx}F_{xx}, \\
E_8^* &= E_7 - 1/3\Theta_{xx}F_{zz} - 1/3\Theta_{zz}F_{xx} - 1/3\Theta_{yy}F_{yy} - 1/6B_{yy,xx}(-F_y)^2F_{zz} \\
&\quad -1/6B_{yy,yy}(-F_y)^2F_{yy} - 1/6B_{yy,zz}(-F_y)^2F_{xx}, \\
E_9^* &= E^0 - 1/3\Theta_{zz}F_{zz} - 1/3\Theta_{yy}F_{yy} - 1/3\Theta_{xx}F_{xx} \\
E_{10}^* &= E^0 - 1/3\Theta_{xx}F_{zz} - 1/3\Theta_{zz}F_{xx} - 1/3\Theta_{yy}F_{yy}
\end{aligned}
\tag{A2}
$$

where $E^0$ again is the energy value when no perturbing field and perturbing gradient are applied, $E_1$ to $E_{13}$ are single point energies when the appropriate perturbing field is applied, $E_1^*$ to $E_8^*$ are the energies when both perturbing field and field gradient are applied, and $E_9^*$ to $E_{10}^*$ are the energies when only the appropriate perturbing gradient is applied. In addition, the following extra set of equations must be solved together with Eq. (A2) to account for the traceless requirements of the Θ, A and B tensors:

$$
\begin{aligned}
\Theta_{yy} &= -\Theta_{xx} - \Theta_{zz}, \\
A_{z,yy} &= -A_{z,zz} - A_{z,xx}, \\
B_{zz,yy} &= -B_{zz,zz} - B_{zz,xx}, \\
B_{yy,yy} &= -B_{yy,zz} - B_{yy,xx}.
\end{aligned}
\tag{A3}
$$

## References

[1] A.D. Buckingham, *Adv. Chem. Phys.* **12** (1967), 107.
[2] A.J. Stone, *The Theory of Intermoleculr Forces*, Clarendon Press: Oxford, 1996.

[3] A.D. McAulay, *Optical Computer Architectures*, Wiley: New York, 1991.
[4] S.A. Clough, Y. Beers, G.P. Klein and S.L. Rothman, *J. Chem. Phys.* **59** (1973), 2254.
[5] S.L. Shostak, W.L. Ebenstein and J.S. Muenter, *J. Chem.Phys.* **94** (1991), 5875.
[6] G.D. Zeiss and W.J. Meath, *Mol. Phys.* **33** (1977), 1155.
[7] J.F. Ward and C.K. Miller, *Phys. Rev. A* **19** (1979), 826.
[8] P. Kaatz, E.A. Donley and D.P. Shelton, *J. Chem. Phys.* **108** (1998), 849.
[9] G.D. Purvis, III. and R.J. Bartlett, *Phys. Rev. A* **23** (1981), 1594.
[10] J. Guan, P. Duffy, J.T. Carter, D.P. Chong, K.C. Casida, M.E. Casida and M. Wrinn, *J. Chem. Phys.* **98** (1993), 4753.
[11] G. Maroulis, *Chem. Phys. Lett.* **289** (1998), 403.
[12] B.F. Levine and C.G. Bethea, *J. Chem. Phys.* **65** (1976), 2429.
[13] C.A. Coulson and D. Eisenberg *Proc. R. Soc. London, A* **291** (1966), 445.
[14] E.R. Batista, S.S. Xantheas and H. Jonsson, *J. Chem. Phys.* **109** (1998), 4546.
[15] H. Reis, S.G. Raptis and M.G. Papadopoulos, *Chem. Phys.* **263** (2001), 301.
[16] K.V. Mikkelsen, Y. Luo, H. Agren and P. Jorgensen, *J. Chem. Phys.* **102** (1995), 9362.
[17] J.G. Kirkwood and F. Westheimer, *J. Chem. Phys.* **6** (1936), 506.
[18] L. Onsager, *J. Am. Chem. Soc.* **58** (1936), 1486.
[19] K.V. Mikkelsen, H. Agren, H.J.A. Jensen and T. Helgaker, *J. Chem. Phys.* **89** (1988), 3086.
[20] G. Karlström, *J. Phys. Chem.* **93** (1989), 4952.
[21] H. Agren and K.V. Mikkelsen, *Theochem.* **234** (1991), 425.
[22] Y. Tu and A. Laaksonen, *Chem. Phys. Lett.* **329** (2000), 283.
[23] S.W. Rick, *J. Chem. Phys.* **114** (2001), 2276.
[24] P.L. Silvestrelli and M. Parrinello, *Phys. Rev. Lett.* **82** (1999), 3308.
[25] B.D. Bursulaya, J. Jeon, D.A. Zichi and H.J. Kim, *J. Chem. Phys.* **108** (1998), 3286.
[26] A.V. Gubskaya and P.G. Kusalik, *Mol. Phys.* **99** (2001), 1107.
[27] S. Sirois, E.L. Proynov, D.T. Nguyen and D.R. Salahub, *J. Chem. Phys.* **107** (1997), 6770.
[28] D.Q. Wei and D.R. Salahub, *Chem. Phys. Lett.* **224** (1994), 291.
[29] G. Stell, G.N. Patey and J.S. Høye, *Adv. Chem. Phys.* **38** (1981), 183.
[30] S.L. Carnie and G.N. Patey, *Mol. Phys.* **47** (1982), 1129.
[31] J.M. Caillol, D. Levesque, J.J. Weis, P.G. Kusalik and G.N. Patey, *Mol. Phys.* **55** (1985), 65.
[32] P.G. Kusalik and G.N. Patey, *Mol. Phys.* **65** (1988), 1105.
[33] W.L. Jorgensen, J. Chandrasekhar, J.D. Madura, R.W. Impley and M.L. Klein, *J. Chem. Phys.* **79** (1983), 926.
[34] H.G.C. Berendsen, J.R. Grigera and T.P. Straatsma, *J. Phys. Chem.* **91** (1987), 6269.
[35] M.P. Allen and D.J. Tildesley, *Computer Simulation of Liquids*, Clarendon Press: Oxford, 1987.
[36] T.M. Nymand and P.-O Astrand, *J. Chem. Phys.* **106** (1997), 8332.
[37] T.M. Nymand, P.-O. Astrand and K.V. Mikkelsen, *J. Phys. Chem. B* **101** (1997), 4105.
[38] A. Wallqvist and G. Karlström, *Chem. Scr.* **29 A** (1989), 131.
[39] P.-O. Astrand, P. Linse and G. Karlström, *Chem. Phys.* **191** (1995), 195.
[40] D.P. Shelton and J.E. Rice, *Chem. Rev.* **94** (1994), 3.
[41] F. Sim, S. Chin, M. Dupuis and J.E. Rice, *J. Phys. Chem.* **97** (1993), 1158.
[42] R.J. Bartlett and G.D. Purvis, *Phys. Rev. A* **20** (1979), 1313.
[43] H.A. Kurtz, J.J.P. Stewart and K.M. Dieter, *Comput. Chem.* **11** (1990), 82.
[44] G. Maroulis, *J. Chem. Phys.* **94** (1991), 1182.
[45] A.D. McLean and J. Yoshimine, *J. Chem. Phys.* **46** (1967), 3682.
[46] D.G. Truhlar, R.A. Dixon and J. Eades, *Phys. B* **12** (1979), 1913.
[47] T. Pluta, J. Noga and R.J. Bartlett, *International Journal of Quantum Chemistry: Quantum Chemistry Symposium*, **28** (1994), 379.
[48] M. Bishop and G.Maroulis, *J. Chem. Phys.* **82** (1985), 2380.
[49] D.M. Bishop and J. Pipin, *Theor. Chim. Acta.* **71** (1987), 247.
[50] R.J. Bartlett, *Annu. Rev. Phys. Chem.* **32** (1981), 359.
[51] A. Szabo and N.S. Ostlund, *Modern Quantum Chemistry*, (Chap. 6), MacMillan: New York, 1982.
[52] M. Urban, I. Cernusak, V. Kellö and J. Noga, *Method Comput. Chem.* **1** (1987), 117.
[53] M.J. Frisch, G.W. Trucks, H.B. Schlegel, P.M.W. Gill, B.G. Johnson, M.A. Robb, J.R. Cheeseman, T.A. Keith, G.A. Petersson, J.A. Montgomery, K. Raghavachari, M.A. Al-Laham, V.G. Zakrewski, J.V. Ortiz, J.B. Foresman, J. Cioslowski, B.B. Stefanov, A. Nanayakkara, M. Challacombe, C.Y. Peng, P.Y. Ayala, W. Chen, M.W. Wong, J.L. Andres, E.S. Replogle, R. Gomperts, R.L. Martin, D.J. Fox, J.S. Binkley, D.J. Defrees, J. Baker, J.P. Stewart, M. Head-Gordon, C. Gonzalez and J.A. Pople, *Gaussian 94 (Revision B.2)* , Gaussian, Inc., Pittsburgh PA, 1995.
[54] M.J. Frisch, G.W. Trucks, H.B. Schlegel, G.E. Scuseria, M.A. Robb, J.R. Cheeseman, V.G. Zakrzewski, J.A. Montgomery, R.E. Stratmann, J.C. Burant, S. Dapprich, J.M. Millam, A.D. Daniels, K.N. Kudin, M.C. Strain, O. Farkas, J. Tomasi,

V. Barone, M. Cossi, R. Cammi, B. Mennucci, C. Pomelli, C. Adamo, S. Clifford, J. Ochterski, G.A. Petersson, P.Y. Ayala, Q. Cui, K. Morokuma, D.K. Malick, A.D. Rabuck, K. Raghavachari, J.B. Foresman, J. Cioslowski, J.V. Ortiz, B.B. Stefanov, G. Liu, A. Liashenko, P. Piskorz, I. Komaromi, R. Gomperts, R.L. Martin, D.J. Fox, T.A. Keith, M.A. Al-Laham, C.Y. Peng, A. Nanayakkara, C. Gonzalez, M. Challacombe, P.M.W. Gill, B.G. Johnson, W. Chen, M.W. Wong, J.L. Andres, M. Head-Gordon, E.S. Replogle and J.A. Pople, *Gaussian 98*, Gaussian, Inc., Pittsburgh, PA, 1998.
[55] R.P. Feynman, *Phys. Rev.* **56** (1939), 340.
[56] H. Hellmann, *Die Entfuhrung in die Quantenchchemie*, Franz Deuticke: Leipzig, 1937.
[57] P. Chopra, L. Carlacci, H.F. King and P.N. Passard, *J. Chem. Phys.* **93** (1993), 7120.
[58] D.P. Chong and S.R. Langhoff, *J. Chem. Phys.* **93**(1990), 570.
[59] D. Feller, C.M. Boyle and E.R. Davidson, *J. Chem. Phys.* **86** (1987), 3424.
[60] A.J. Sadlej, *Coll. Czech. Chem. Comm.* **53** (1988), 1995.
[61] G.H.F. Diercksen, V. Kellö and A.J. Sadlej, *J. Chem. Phys.* **79** (1983), 2918.
[62] H.-J. Werner and W. Meyer, *Mol. Phys.* **31** (1976), 855.
[63] M.A. Spackman, *J. Phys Chem.* **93** (1989), 7594.
[64] W.F. Murphy, *J. Chem.Phys.* **67** (1977), 5877.
[65] H. Sekino and R.J. Bartlett, *J. Chem. Phys.* **98** (1993), 3022.
[66] D.M. Bishop, B. Kirtman, H.A. Kurtz and J.E. Rice, *J. Chem. Phys.* **98** (1993), 8024.
[67] Y. Luo, H. Agren, O. Vahtras, P. Jorgensen, V. Spirko and H. Hettema, *J. Chem. Phys.* **98** (1993), 7159.
[68] A.V. Gubskaya and P.G. Kusalik, *J. Chem. Phys.* **117** (2002), 5290.
[69] L. Delle Site, A. Alavi and R.M. Lynden-Bell, *Mol. Phys.* **96** (1999), 1683.
[70] P.W. Fowler and P.A. Madden, *Phys. Rev.* **30** (1984), 6131.
[71] P.W. Fowler, F. Ding and R.W. Munn, *Mol. Phys.* **84** (1995), 787.
[72] P. Jemmer, P.W. Fowler, M. Wilson and P.A. Madden, *J. Phys Chem.* **102** (1998), 8377.
[73] P. Jemmer, M. Wilson, P.A. Madden and P.W. Fowler, *J. Chem. Phys.* **111** (1999), 2038.
[74] P.W. Fowler, H.M. Kelly and N. Vaidehi, *Mol. Phys.* **82** (1994), 211.

Journal of Computational Methods in Sciences and Engineering 4 (2004) 665–676
IOS Press


# Can aromaticity be connected with molecular polarizability? A theoretical study of benzene isomers and five-membered heterocyclic molecules

R. Soriano Jartín[a], I. García Cuesta[b], A. Sánchez de Merás[b] and P. Lazzeretti[a,*]
[a]*Dipartimento di Chimica dell'Università degli Studi di Modena e Reggio Emilia, Via Campi 183, 41100 Modena, Italy*
[b]*Departament de Química Física, Institut de Ciència Molecular, Universitat de València, Carrer Dr. Moliner 50, 46100 Burjassot, València, Spain*

**Abstract.** Extended calculations of molecular electric dipole polarizability tensor, at Hartree-Fock and correlated level of accuracy (MP2, CCS, CC2, CCSD, and CCSD(T)) have been carried out to investigate whether aromaticity could be related to the electric dipole polarizability of planar ring systems. The calculations prove the exaltation of the average property of conjugated molecules, which is possibly due to their easily polarizable $\pi$-electron cloud. On the other hand, theoretical out-of-plane polarizability components are smaller in benzene than in any other $C_6H_6$ isomer. The aromatic stabilization energies of monosubstituted five-membered conjugated cyclic molecules increase in the same direction as theoretical out-of-plane polarizabilities.

Keywords: Molecular electric polarizabilites, aromaticity, MP2 and coupled cluster correlation contributions, pentatomic heterocyclic molecules

## 1. Introduction

"Aromatic" molecules are conventionally regarded as possessing some peculiar physico-chemical properties, for instance, remarkable stability, related to high resonance energy, and ability to undergo ionic substitution rather than addition. They are characterized by typical structural and geometrical parameters, thermodynamic stability data, etc. At any rate, a universally accepted, nondisputable definition of aromaticity has not yet been proposed.

A special role is attributed to the delocalized $\pi$ electrons in determining the properties of aromatic molecules. However, some recent theoretical work puts in questions well established concepts on the source of stability of aromatics and suggests that "a different story of benzene" can be told [1] on the basis of the distortive propensity of $\pi$ electrons [2–4].

As a matter of fact, quantitative indicators are needed to specify the vague notion of aromaticity. The idea that the enhanced magnetic anisotropies and the down-field proton chemical shift of planar

*Corresponding author. E-mail: lazzeret@unimo.it.

conjugated molecules can be related to aromatic character has been proposed several times in the literature. It is commonly held that the delocalization of $\pi$ electron cloud in these systems, to which the properties of aromatic molecules are usually ascribed, also implies high intrinsic mobility in the presence of external magnetic field. According to the Ring-Current Model (RCM) of Pauling [5], Lonsdale [6], and London [7,8], a magnetic field perpendicular to the molecular plane of benzene induces intense circulation of $\pi$ electrons [9–16], which is responsible for the large out-of-plane component of the magnetic susceptibility tensor and for deshielding observed in proton magnetic spectroscopy. Quantitative measures of aromaticity can then be provided in terms of out-of-plane magnetic susceptibility, or magnetic anisotropy [17], or by using NMR [12].

The natural question arises whether there is any connection between $\pi$ electron delocalization and other response properties of benzene which may be used to clarify the concept of aromaticity. Now, the deviation from simple Pascal additivity rules for magnetic susceptibility of planar conjugated hydrocarbons is customarily interpreted as a fingerprint of aromaticity, and the concept of diamagnetic susceptibility exaltation [18–20] provides a semiquantitative measure of it. Then the existence of a possible connection between electric polarizability and aromatic character may be guessed at [21–26] for the fact that also total average polarizabilities of aromatic molecules are unusually large and difficult to rationalize in terms of transferable atomic contributions *à* la Denbigh [27–33], whereas theoretical additivity schemes are easy to define for the alkane series [34,35]. Katrizky et al. have claimed that a polarizability-based indicator captures a mixture of "magnetic" and "classical" aromaticity [36].

An exaltation of electric dipole polarizability of aromatics could possibly be defined [25,26]. This paper sets out to examine electric dipole polarizability as a possible indicator of aromatic character. To this purpose a systematic approach was applied. Firstly, a series of planar benzene isomers, whose magnetic properties have been recently discussed [37], was taken into account in the present study. The main concern was that of reporting a consistent set of accurate theoretical polarizabilities including electron correlation. Details on the procedures employed are described in Section 2 and the results are discussed in Sections 3 and 4.

Two related cyclic systems having six $\pi$ electrons, $C_5H_5^-$ and $C_7H_7^+$, have been investigated in connection with benzene. Then the analysis has been extended to a number of variously substituted five-membered conjugated molecules, out of a large set which has recently been scrutinized in a recent paper on aromaticity [38]. An attempt was made to achieve some statistical information on the reliability of the electric polarizability as measure of aromaticity in Section 5 in connection with other indicators.

## 2. Computational details

Five planar $C_6H_6$ isomers, that is, benzene (**1**), cyclohexatriene (**2**), fulvene (**3**), 3-4-dimethylenecyclobutene (**4**), and trimethylenecyclopropane (**5**) are shown in Fig. 1. The molecular geometries adopted in the calculations are the same as in Ref. [37]. Although cyclohexatriene (**2**) is hypothetical, it has been adopted as a useful reference molecule in discussing aromatic character [37,39].

Earlier calculations on benzene and fulvene, within finite field Hartree-Fock and correlated MBPT2 methods, retaining Sadlej's basis sets [40] and slightly different molecular geometries, have been reported [41]. Benzene polarizability has also been calculated by other authors [42–46]. Fulvene polarizability was also recently computed by Hinchliffe et al. [47] via the density functional theory (DFT).

In the present investigation, electric dipole polarizability tensors have been evaluated theoretically at the Hartree-Fock level of accuracy and by other methods accounting for electron correlation effects. Thus, MP2 polarizabilites were determined within the framework of the second-order polarization propagator

Table 1
Polarizabilities of benzene[a]

| | $\alpha_{xx}$ | $\alpha_{zz}$ | $\alpha_{iso}$ | $\Delta\alpha$ |
|---|---|---|---|---|
| HF[b] | 79.58 | 45.61 | 68.26 | 33.97 |
| HF[c] | 79.60 | 45.63 | 68.28 | 33.97 |
| MP2[b] | 84.13 | 46.14 | 71.47 | 37.99 |
| MP2[c] | 81.71 | 45.57 | 69.66 | 36.14 |
| CCS[b] | 85.90 | 49.31 | 73.71 | 36.59 |
| CC2[b] | 85.24 | 46.23 | 72.24 | 39.01 |
| CCSD[b] | 81.40 | 45.29 | 69.36 | 36.11 |
| CCSD[c] | 79.62 | 44.70 | 67.98 | 34.92 |
| CCSD(T)[c] | 80.19 | 45.01 | 68.46 | 35.18 |
| EXP[d] | 83.54 | 43.1 | 70.05 | 40.4 |
| EXP[e] | 79.16 | 44.1 | 67.48 | 35.0 |
| EXP[f] | 79.2 | 47.7 | 68.7 | 31.5 |

[a]in au for all the Tables, see text for conversion factors to CGS and SI units. $\alpha_{xx} = \alpha_{yy}$ and $\Delta\alpha \equiv \alpha_{xx} - \alpha_{zz}$ in Tables 1,2, and 5. $\Delta\alpha \equiv (1/2)(\alpha_{xx} + \alpha_{yy}) - \alpha_{zz}$ in Tables 3 and 4.
[b]analytic result, [c]from finite field method, [d]Ref. [64], [e]Ref. [65], [f]Ref. [66].

Table 2
Polarizabilities of cyclohexatriene[a]

| | $\alpha_{xx}$ | $\alpha_{zz}$ | $\alpha_{iso}$ | $\Delta\alpha$ |
|---|---|---|---|---|
| HF[b] | 79.11 | 45.57 | 67.93 | 33.54 |
| HF[c] | 79.13 | 45.59 | 67.95 | 33.54 |
| MP2[b] | 83.91 | 46.38 | 71.40 | 37.53 |
| MP2[c] | 81.46 | 45.96 | 69.63 | 35.50 |
| CCS[b] | 85.16 | 49.28 | 73.20 | 35.88 |
| CC2[b] | 85.32 | 46.54 | 72.40 | 38.78 |
| CCSD[b] | 80.90 | 45.40 | 69.07 | 35.50 |
| CCSD[c] | 79.19 | 44.80 | 67.73 | 34.39 |
| CCSD(T)[c] | 80.03 | 45.21 | 68.42 | 34.82 |

[a]See footnotes to Table 1, [b]analytic result, [c]from finite field method.

(SOPPA) scheme [48]. We have also used a hierarchy of different coupled cluster methods implemented in a direct algorithm [49,50] that, in addition to the traditional CCS and CCSD methods, includes CC2 [51], an iterative method in which some approximations are made in the CCSD equations. The corresponding polarizabilites have been calculated within the coupled cluster linear response (CCLR) approach [52], which in the frequency-independent limit coincides with the so-called orbital-unrelaxed energy derivatives computational scheme (thus giving the orbital-unrelaxed finite field results).

In this context, it should be mentioned that orbital relaxation effects cannot explicitly be included in the calculation of response functions, but they are indirectly gotten through single excitations, as otherwise artificial poles would appear, with the subsequent unphysical behavior of the response function. In addition, when referring to triples corrections to polarizability, it is also well known [53] that the extremely popular CCSD(T) method is not adequate to compute unrelaxed properties. We have then estimated the effect of triples correction via the orbital-relaxed finite field method. Concerning the systems examined in this study, these estimates have then been obtained through a second-degree polynomial fit of nine calculated values for each molecule, with external static fields of strengths $\pm 0.004$, $\pm 0.003$, $\pm 0.002$,

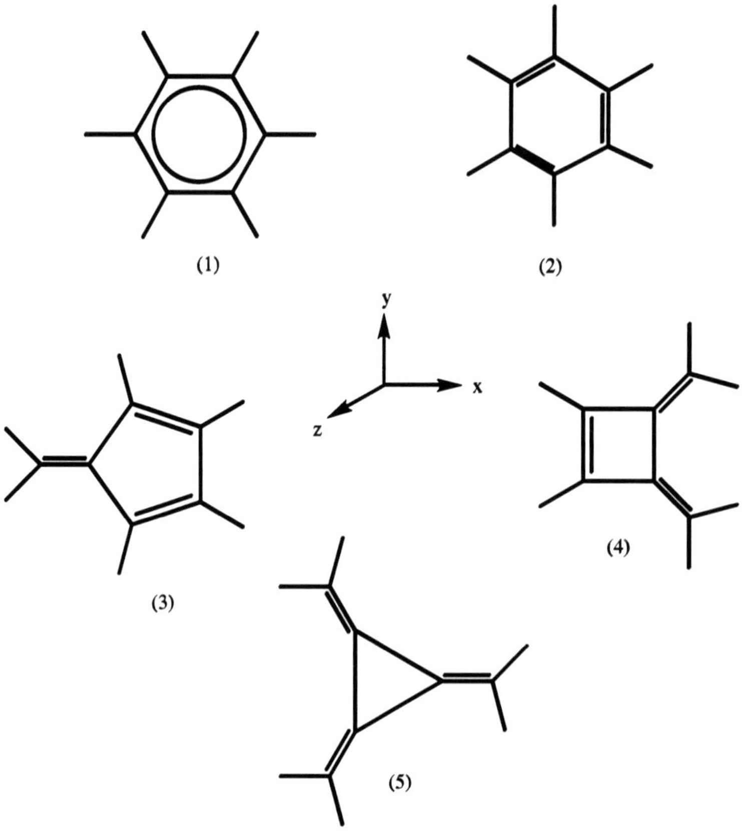


Fig. 1.

±0.001, and 0.000 au These values were chosen to minimize the numerical errors introduced in the least-square procedure.

The calculations were performed using the polarized basis sets of Sadlej [40], which were specifically designed to compute molecular polarizabilities. In order to comply with previous work [37], we have used a Cartesian representation of 3d orbitals, which introduces small differences with respect to the values calculated with the standard spherical harmonic basis. Additional test calculations were also made using Dunning's aug-cc-pVTZ basis [54]. The differences observed in the two sets of numerical results were not negligible. In any event, the same general trends were obtained, which constituted the matter of major concern in this study. Accordingly, we believe that the conclusions arrived at by using Sadlej's basis sets are quite reliable.

All the calculations were carried out using the DALTON program [55], in which the above-mentioned SOPPA and coupled cluster algorithms have been implemented.

## 3. Discussion of results for benzene isomers from different correlated approaches

Theoretical results are reported in Tables 1 to 5. Values are given in atomic units (au). An overall agreement was found between coupled-cluster estimates for benzene in Table 1 and earlier results [42, 46]. Other ab initio determinations are found in Refs. [43–45]. Allowing for the 1998 Adjustment of the

Table 3
Polarizabilities of fulvene[a]

| | $\alpha_{xx}$ | $\alpha_{yy}$ | $\alpha_{zz}$ | $\alpha_{iso}$ | $\Delta\alpha$ |
|---|---|---|---|---|---|
| HF[b] | 67.58 | 109.53 | 47.01 | 74.70 | 41.55 |
| HF[c] | 67.54 | 109.54 | 47.03 | 74.70 | 41.51 |
| MP2[b] | 72.23 | 107.67 | 47.75 | 75.89 | 42.20 |
| MP2[c] | 70.65 | 100.50 | 47.28 | 72.81 | 38.30 |
| CCS[b] | 72.33 | 116.08 | 50.79 | 79.73 | 43.42 |
| CC2[b] | 73.49 | 108.89 | 47.79 | 76.72 | 43.40 |
| CCSD[b] | 69.44 | 101.96 | 46.40 | 72.60 | 39.30 |
| CCSD[c] | 68.29 | 100.25 | 45.81 | 71.45 | 38.46 |
| CCSD(T)[c] | 69.40 | 100.20 | 46.23 | 71.94 | 38.57 |

[a]See footnotes to Table 1, [b]analytic result, [c]from finite field method.

Table 4
Polarizabilities of 3,4-dimethylenecyclobutene[a]

| | $\alpha_{xx}$ | $\alpha_{yy}$ | $\alpha_{zz}$ | $\alpha_{iso}$ | $\Delta\alpha$ |
|---|---|---|---|---|---|
| HF[b] | 85.13 | 101.80 | 47.60 | 78.18 | 45.87 |
| HF[c] | 85.15 | 101.82 | 47.62 | 78.20 | 45.87 |
| MP2[b] | 89.61 | 98.94 | 48.27 | 78.94 | 46.01 |
| MP2[c] | 86.45 | 92.05 | 48.05 | 75.52 | 41.20 |
| CCS[b] | 90.73 | 106.74 | 51.38 | 82.95 | 47.36 |
| CC2[b] | 91.27 | 100.45 | 48.45 | 80.06 | 47.41 |
| CCSD[b] | 84.83 | 94.17 | 46.97 | 75.32 | 42.53 |
| CCSD[c] | 83.09 | 92.86 | 46.40 | 74.12 | 41.58 |
| CCSD(T)[c] | 84.21 | 93.01 | 46.86 | 74.69 | 41.75 |

[a]See footnotes to Table 1, [b]analytic result, [c]from finite field method.

Fundamental Physical Constants [56], the conversion factor to cgs units is $1.481\,846\,093 \times 10^{-25}$ cm$^3$. The conversion factor to SI units is $1.648\,777\,251 \times 10^{-41}$ $\mathrm{C^2m^2J^{-1}}$.

In order to rationalize the effect of correlation we have compared the analytic CCLR results at the CCS, CC2 and CCSD levels of approximation. It can be observed in the tables that CCSD estimates are always smaller than corresponding CCS. The latter are much bigger than SCF, as CCS, an essentially non-correlated approach, does not account for orbital relaxation.

The difference between SCF and CCS predictions should not be explained in terms of correlation, as it depends on the different parameterization of the complement of states. As a matter of fact, orbital relaxation terms were not included in the CCS calculations in order to make a direct comparison with the CC2 and CCSD results (that are also unrelaxed).

Interestingly enough, electron correlation gives a negative contribution to the three components of polarizability in all cases, which constitutes a first interesting conclusion of this work. The effect is more pronounced for in-plane components, which provide dominant contribution to the average polarizability.

This pattern is similar to that observed on comparing electric dipole polarizabilities from Tamm-Dancoff approximation (TDA) [57], equivalent to single-excitation configuration interaction [58], and from random-phase approximation (RPA) [57]. The latter is the same as the coupled SCF retained here. In fact, diagonal components from TDA are invariably bigger than RPA [59–62].

A less accurate treatment of correlation, e.g., within the CC2 method, yields an erratic behavior of the correlation contribution to polarizability. For instance, theoretical $\alpha_{yy}$ of 3-4-dimethylenecyclobutene

Table 5
Polarizabilities of trimethylenecyclopropane[a]

| | $\alpha_{xx}$ | $\alpha_{zz}$ | $\alpha_{iso}$ | $\Delta\alpha$ |
|---|---|---|---|---|
| HF[b] | 107.79 | 47.92 | 87.83 | 59.87 |
| HF[c] | 107.81 | 47.94 | 87.85 | 59.87 |
| MP2[b] | 109.15 | 48.59 | 88.97 | 60.56 |
| MP2[c] | 101.96 | 48.33 | 84.08 | 53.63 |
| CCS[b] | 113.59 | 51.68 | 92.95 | 61.91 |
| CC2[b] | 111.00 | 48.76 | 90.25 | 62.24 |
| CCSD[b] | 102.02 | 47.29 | 83.77 | 54.73 |
| CCSD[c] | 99.83 | 46.74 | 82.13 | 53.09 |
| CCSD(T)[c] | 100.86 | 47.20 | 82.97 | 53.66 |

[a] See footnotes to Table 1, [b] analytic result, [c] from finite field method.

are 106.74 au and 100.45 au for CCS and CC2, respectively, whereas the opposite trend is observed for the $\alpha_{xx}$ component of the same isomer, compare 90.73 a.u and 91.27 au for CCS and CC2, respectively.

Two main trends can also be observed: (i) first, the CCS perpendicular component of polarizability always diminishes on introducing correlation (ii) the $\alpha_{xx}$ component increases on passing from CCS to CC2 in cyclohexatriene, fulvene, and 3-4-dimethylenecyclobutene, but the average in-plane component becomes smaller in all the benzene isomers except for the hypothetical cyclohexatriene.

In fact, as can be seen by comparison of analytic results with the corresponding finite field estimates, orbital relaxation plays an important role. Clearly this effect is related to the strain affecting molecular structure, and therefore it is negligible for the out-of-plane component of polarizability, whereas its role cannot be understated in the case of (**5**).

In this context let us recall that a better treatment of the $t_1$ amplitudes in CCSD significantly corrects the extreme deviations between CCS and CC2 for (**4**) and (**5**). Thus, in (**4**), $\alpha_{xx}$ decreases from the CC2 value 91.27 au to 84.83 au at the CCSD level. Similarly, in (**5**) a diminution of almost 9 au is observed. In both cases, as well as in the other benzene isomers considered, the analytic CCLR/CCSD result is rather close to the finite field CCSD(T) result.

Eventually, and in accordance with the conclusions of Ref. [41], it can be reasonably argued that correlation plays a minor role for both in-plane and out-of-plane polarizabilities of planar $C_6H_6$ isomers. However, the need for a correlated treatment should be emphasized for the very strained trimethylenecyclopropane.

It can be observed that MP2, especially in the analytic calculations, seems to overestimate the value of the polarizability tensor components. In addition, as previously pointed out by Bartlett and coworkers [63], the inclusion of linked triples corrections increases the magnitude of the polarizability with respect to the CCSD value. However, electron correlation accounted for in the CCSD(T) formalism, systematically decreases the HF value of the $\alpha_{zz}$ polarizability component in all the five benzene isomers studied here. On the other hand, the correlated parallel component for (**1**) and (**2**) from CCSD(T) calculations is slightly bigger, see, for instance, the $\alpha_{xx}$ component of fulvene.

Experimental polarizability tensors of benzene have been reported in 1975 by Proutière and Camail [64], and by Alms, Burnham, and Flygare [65]. More recently, Okruss, Müller, and Hese determined ground and excited state polarizabilities from static Stark effect [66]. The importance of the vibrational contributions has been discussed in Ref. [65], reporting experimental polarizabilities from measurements of the dispersion of the gas phase depolarization ratio of Rayleigh scattering. The extrapolation to zero frequency to obtain static polarizabilities does not include the vibrational part [65]. The results from the Kerr effect [64] should be considered with this dependence included [23,66]. The values from the

Table 6
Aromatic stabilization energies (ASE), (in kcal/mol), and CCSD polarizabilities of pentatomic heterocyclic systems (in au)*

| compound | ASE | $\alpha_{zz}$ | $\Delta\alpha$ | $\alpha_{Av}$ |
|---|---|---|---|---|
| furan | 14.77 | 35.02 | 21.20 | 48.98 |
| thiophene | 18.57 | 45.02 | 30.31 | 64.83 |
| pyrrole | 20.57 | 39.64 | 23.80 | 55.48 |
| phosphole | 3.20 | 74.53 | 31.76 | 71.62 |
| $CH^-$ | 22.05 | 73.86 | 20.60 | 87.60 |
| $CH_2$ | 0.00 | 42.24 | 22.85 | 57.30 |
| $CF_2$ | −11.88 | 43.25 | 20.60 | 56.36 |
| $N^-$ | 19.56 | 61.02 | 23.07 | 76.34 |
| $NH_2^+$ | −2.05 | 32.28 | 20.47 | 45.87 |
| $SiH^+$ | −26.58 | 38.44 | 32.20 | 58.47 |
| $SiH_2$ | −4.61 | 53.58 | 27.88 | 71.03 |
| $P^-$ | 23.12 | 80.47 | 37.84 | 102.15 |
| $PH_2^+$ | −8.31 | 41.73 | 27.30 | 58.82 |
| $GeH_+$ | −23.92 | 56.69 | 35.77 | 61.78 |
| $GeH_2$ | −2.97 | 39.83 | 29.54 | 74.80 |
| $As^-$ | 22.21 | 89.06 | 42.60 | 111.49 |
| AsH | 1.71 | 78.86 | 32.46 | 76.47 |
| $AsH_2^+$ | −6.55 | 45.57 | 29.10 | 63.43 |
| Se | 16.74 | 50.47 | 34.15 | 72.08 |
| C = O | −14.65 | 38.18 | 33.95 | 58.61 |
| C = O | −11.96 | 50.27 | 61.26 | 81.50 |
| C = Se | −11.44 | 55.73 | 72.14 | 91.23 |
| 3-azafuran | 12.37 | 30.60 | 19.85 | 43.63 |
| 3-azathiophene | 17.43 | 40.87 | 28.60 | 59.63 |
| 3-azapyrrole | 18.78 | 34.64 | 22.75 | 49.77 |
| 2,5-diazafuran | 20.19 | 27.23 | 18.76 | 39.73 |
| 2,5-diazathiophene | 22.67 | 36.38 | 29.18 | 54.53 |
| 2,5-diazapyrrole | 26.66 | 30.59 | 21.15 | 44.60 |
| 2,5-diazaphosphole | 3.14 | 59.56 | 30.75 | 60.48 |

*The nomenclature is the same as in Ref. [38]. The perpendicular component is $\alpha_{zz}$, the anisotropy is defined via $(\Delta\alpha)^2 = 1/2[(\alpha_{xx}-\alpha_{yy})^2+(\alpha_{xx}-\alpha_{zz})^2+(\alpha_{yy}-\alpha_{zz})^2+3(\alpha_{xy}^2+\alpha_{yz}^2+\alpha_{xx}^2+\alpha_{zx}^2+\alpha_{yz}^2+\alpha_{zy}^2)]$, and the average polarizability is $\alpha_{Av} = 1/3(\alpha_{xx}+\alpha_{yy}+\alpha_{zz})$.

Stark effect [66] involve vibronic contributions. In Ref. [67] the experimentally estimated effect on the average polarizability amounts to 1.94 au.

In any event, comparison with experimental values for the electric dipole polarizability of benzene [65, 66] seems to confirm that correlation contribution are small. The SCF-HF predictions in Table 1 for in-plane and out-of-plane components, and trace of the polarizability tensor, are closer to measured properties than those from any other approximation.

## 4. Out-of-plane polarizability and aromaticity of $C_5H_5^-$, $C_6H_6$ isomers, and $C_7H_7^+$

Theoretical values for gross atomic average polarizability of carbon and hydrogen, transferable within the alkane series, have been estimated at the Hartree-Fock level of accuracy. Assuming that $\alpha_{Av}^{C} \approx 5.7$ and $\alpha_{Av}^{H} \approx 2.7$ au [34,35], the average polarizability for benzene isomers should be $\approx 50$ au. Then, allowing for the results of Tables 1–5, the polarizability exaltation of these molecules, estimated via the Hartree-Fock approach, should range from $\approx 20$ to $\approx 40$ au.

A smaller exaltation, $\approx$ 6 au, i.e., 10%, would be found allowing for the results of Ref. [30], in which an average polarizability of benzene as large as $\approx$ 61–62 au was estimated by using a reviewed set of Thole's atomic parameters determined from experimental average polarizabilities [29]. In any event, the polarizability of alkenes and aromatics is systematically bigger than that predicted via atomic terms optimized from a large learning set, see, for instance, the average experimental polarizabilities of benzene, chlorobenzene, nitrobenzene, *o*- and *m*-dichlorobenzene in Table 11 of Ref. [30]. Then the anomalous value of polarizability in molecules containing unsaturated carbon bonds is a fact.

*Could the exaltation of the average polarizability be imputed to a peculiar behavior of the delocalized* $\pi$*-electrons in the presence of an external electric field?* This is the first fundamental question which should be answered to investigate a possible connection between aromatic character and electric polarizability.

To begin with, exaltation of electric polarizability also occurs for ethylene and acetylene, compare the experimental polarizabilites of these molecules with those evaluated from atomic terms in Table 11 of Ref. [30]. Thus exaltation seems to affect any unsaturated molecule.

In addition, the exaltation estimated for the benzene isomers allowing for a given set of atomic polarizabilities [34–36] can by no means be interpreted as a measure of aromaticity, as it characterizes all of them, although to a different extent, see Tables 1 to 5. In any event, it can be observed that a given component of the polarizability tensor increases with the molecular extension in the corresponding direction. For instance, the largest component in fulvene corresponds to the two-fold symmetry axis, that is the $y$ axis in Fig. 1. Analogously, the largest component of 3-4-dimethylenecyclobutene is $\alpha_{yy}$. Therefore theoretical in-plane components are not useful to make any assessment of relative aromaticity of planar benzene isomers. On the same ground, neither the average polarizability nor its anisotropy seem to be useful indicators.

On the other hand, the out-of-plane component of the polarizability tensor would seem to provide some information in the $C_6H_6$ isomers. It is almost the same for the molecules of benzene and cyclohexatriene, and increases steadily on passing from these systems to fulvene, 3-4-dimethylenecyclobutene, and trimethylenecyclopropane. The difference between theoretical predictions for (**5**) and (**1**), that is, $\approx$ 47.9 and $\approx$ 45.6 au respectively, at the Hartree-Fock level, is sizeable. The same trend is observed for MP2, CCS, CC2, CCSD, and CCSD(T) estimates: for a given approximation the perpendicular component of the electric polarizability always increases in the order

$$(\mathbf{1}) \approx (\mathbf{2}) < (\mathbf{3}) < (\mathbf{4}) < \mathbf{5}), \quad (1)$$

with the numbering of Fig. 1.

Remarkably enough, the aromaticity scale obtained by a magnetic criterion in Ref. [37], that is, by ordering the molecules according to absolute value of the out-of-plane component of the magnetic susceptibility, is the same. Therefore, one can test two hypotheses. The first one is that the aromatic character of benzene isomers is related to the magnitude of the polarizability component normal to the molecular plane. More aromatic molecules would seem to have smaller out-of-plane polarizability components, which may be related to the fact that polarization of the electron cloud in the presence of electric field at right angles to the molecular plane takes place to a lesser extent, compared with conjugated cyclic polyenes (**4**) and (**5**).

The second possibility is that also the out-of-plane component is mainly biased by molecular size, so that Eq. (1) merely accounts for this fact. In order to evaluate the second possibility, it is meaningful to compare theoretical polarizabilities of the cyclopentadienyl anion $C_5H_5^-$, benzene, and the cycloheptatrienylium (tropylium) cation $C_7H_7^+$. $C_5H_5^-$, $C_7H_7^+$, and $C_6H_6$ are ascribed to the aromatic series by the Hückel $4n + 2$ $\pi$ electrons rule with the same $n = 1$. Denoting by $\alpha_{\parallel}$ ($\alpha_{\perp}$) the component along (perpendicular to) the highest symmetry axis, calculated CCSD polarizabilities are $\alpha_{\parallel} = 73.87$ and $\alpha_{\perp} = 94.47$ au for $C_5H_5^-$, $\alpha_{\parallel} = 42.17$ and $\alpha_{\perp} = 94.35$ au for $C_7H_7^+$. The out-of-plane polarizability increases in the order

$$C_7H_7^+ < \text{benzene} < C_5H_5^-, \tag{2}$$

that is, in the direction opposite to molecular size and to total electron number, that is, 48, 42, and 38, respectively, for these cyclic molecules. However, such a result is not a proof that the first hypothesis is correct, although it discounts the second.

Therefore, we decided to take into account a series of monosubstituted pentatomic ring compounds, related by geometrical affinities, from the large set recently examined in an attempt at analyzing the phenomenon of aromaticity via statistical criteria [38].

## 5. Five-membered conjugated cyclic molecules

The electric dipole polarizabilities of 29 five-membered cyclic systems were predicted at the CCSD level of accuracy. They are reported in Table 6. Previous theoretical results for $C_4H_4X$, X = BH, AlH, $CH_2$, NH, PH, O and S, are reported in Ref. [68]. Calculated polarizabilities for azoles [24] and for aza-substituted thiophene derivatives [69] are available. Many previous experiments and other computations have been reported, but are not discussed, because the focus is on aromaticity.

The existence of a connection between the perpendicular component and aromatic character was tested by analyzing its dependence on the aromatic stabilization energy (ASE) evaluated in Ref. [38]. The used molecular geometries were those optimized by Cyrański et al. in Ref. [38], and which are available from the ACS web page.

If all the $C_4H_4X$ molecules with any X in Table 11 are considered, no apparent dependence is observed. As a matter of fact, linear fitting has a very poor correlation coefficient $\approx 0.421$.

The most striking feature is that different quasi linear relations are observed for different groups of related substituents. For instance, for the eight monosubstituted ring molecules $C_4H_4X$, with

$$X = O,\ NH,\ S,\ Se,\ N^-,\ CH^-,\ P^-,\ As^- \tag{3}$$

there exists a fairly linear relation between the out-of-plane component (hereafter denoted by $\alpha$ for simplicity) and ASE, with correlation coefficient $\approx 0.803$. This set of molecules could be regarded as "aromatic", with both ASE and out-of-plane polarizability increasing in the order given for the $X$ list Eq. (3). The pentatomic cyclic systems with X = PH and AsH are not planar, and their perpendicular components, defined with respect to the molecular plane, do not fit in the correlation.

The four molecules with

$$X = CF_2,\ C=O,\ C=S,\ C=Se \tag{4}$$

seem to belong to a separate family, for which the linear dependence between the out-of-plane component and ASE has correlation coefficient 0.817. Relying on ASE, these molecular systems, containing a double bond external to the pentatomic ring, should be regarded as "antiaromatic". Accordingly, the antiaromaticity would seem to decrease on increasing $\alpha$.

The slope of the two straight lines for aromatic Eqs (3) and antiaromatic systems (4) is roughly the same, and the proportionality coefficients between $\alpha$ and ASE are quite close to one another, i.e., 5.56 and 4.31, respectively, for the two series, in the units of Table 11.

Quite scatter plots are observed for other set of molecules. Trends analogous to that of Eq. (2), i.e.,

$$C_4H_4YH_2^+ < C_4H_4YH < C_4H_4Y^-, \quad (5)$$

are found for Y = N, P, S, Si, and As, with the exception of Y = Ge.

Only a rough increase of $\alpha$ versus ASE was found for the series 2,5-diazafuran, 2,5-diazathiophene, 2,5-diazapyrrole, and 2,5-diazaphosphole (the optimized geometry of this molecule is not planar [38]), and for the series 3-azafuran, 3-azathiophene, 3-azapyrrole. These molecules belong to two different series.

## 6. Conclusions

Attempts have been made to relate the concept of aromaticity to a measurable molecular property, the electric dipole polarizability tensor. It is argued that, possibly, only the out-of-plane component could serve as descriptor.

Three sets of molecules were taken into account, (i) five planar benzene isomers, (ii) three 6 $\pi$-electron cyclic molecules, i.e., cyclopentadyenil anion, benzene, and tropylium cation, (iii) twenty-nine substituted five-membered cyclic molecules.

Accurate calculations for (i), (ii), and (iii) confirm the exaltation of electric polarizability of aromatic molecules in comparison with saturated systems. Electron correlation effects, estimated via a series of approaches, do not seem to be very important for the molecule of set (i).

The existence of a connection between aromaticity and polarizability for planar benzene isomers is difficult to assess. Either the former increases as the latter decreases, or there is no connection at all.

Separate linear relationships were observed for aromatic and antiaromatic $C_4H_4X$ ring compounds on investigating plots of the out-of-plane polarizability versus aromatic stabilization energies. According to the findings of this paper, aromaticity would increase (and antiaromaticity would decrease) on increasing $\alpha$.

Therefore, the out-of-plane polarizability could function as an indicator of *relative aromaticity and antiaromaticity*, but only within sets of related systems.

Nevertheless, although other investigations are necessary to enlarge the statistics, it appears that electric dipole polarizability is not so reliable a descriptor of aromaticity as the out-of-plane magnetic susceptibility. This can be understood by recalling that perpendicular components of magnetic susceptibility and electric dipole polarizability tensors are properties accounting for response in different parts of the molecular domain. The first one depends on quantum mechanical operators which weigh the charge distribution, and provide some in-the mean information on its induced stationary motion, in the intermediate regions, whereas the second one samples the outer reaches of a molecule.

Whereas the out-of-plane magnetic susceptibility seems to define diatropicity as a typical feature of aromatics, the out-of-plane polarizability can limitedly concur to shed light on the concept of aromaticity. In other words, aromaticity, whatever it is, does not seem to depend on the electron distribution in the tail regions.

## Acknowledgments

Financial support to the present research from the European research and training network "Molecular Properties and Materials (MOLPROP)", contract No. HPRN-CT-2000-00013, from DGI (Dirección

General de Investigación) of the Spanish MCyT (Ministerio de Ciencia y Tecnología) under project BQU2001-2935-C02-01, and from the Italian MURST (Ministero dell'Università e della Ricerca Scientifica e Tecnologica), via 60% and 40% funds, is gratefully acknowledged.

## References

[1] S. Shaik, A. Shurki, D. Danovich and P.C. Hiberty, *J. Mol. Struct.* (*THEOCHEM*) **398–399** (1997), 155.
[2] S.S. Shaik, P.C. Hiberty, J.-M. Lefour and G. Ohanessian, *J. Am. Chem. Soc.* **109** (1987), 363.
[3] P.C. Hiberty, D. Danovich, A. Shurki and S. Shaik, *J. Am. Chem. Soc.* **117** (1995), 7760.
[4] S.S. Shaik, P.C. Hiberty, G. Ohanessian and J.-M. Lefour, *J. Phys. Chem.* **92** (1988), 5086.
[5] L. Pauling, *J. Chem. Phys.* **4** (1936), 673.
[6] K. Lonsdale, *Proc. Roy. Soc.* (*London*) **A159** (1937), 149.
[7] F. London, *J. Phys. Radium* **8** (1937), 397, 7ème Série.
[8] F. London, *J. Chem. Phys.* **5** (1937), 837.
[9] P. Lazzeretti, Ring currents, in: *Prog. Nucl. Magn. Res. Spectr.*, (Vol. 36), Elsevier, 2000, pp. 1–88.
[10] J.A.N.F. Gomes and R.B. Mallion, *Chem. Rev.* **101** (2001), 1349.
[11] P. von Ragué Schleyer, *Chem. Rev.* **101** (2001), 1115.
[12] R.H. Mitchell, *Chem. Rev.* **101** (2001), 1301.
[13] E. Steiner and P.W. Fowler, *J. Phys. Chem. A* **105** (2001), 9553.
[14] E. Steiner and P.W. Fowler, *J. Chem. Soc. Chem. Comm.* **2220** (2001).
[15] P.W. Fowler, E. Steiner and R.W.A. Havenith, *Chem. Phys. Lett.* **40** (2001), 85.
[16] A. Ligabue, A.Soncini and P. Lazzeretti, *J. Am. Chem. Soc.* **124** (2002), 2008.
[17] W.H. Flygare, *Chem. Rev.* **74** (1974), 653.
[18] J. Hoarau, *Ann. Chim.* (*Paris*) **1** (1956), 544, 13e Série.
[19] H.J. Dauben, J.D. Wilson and J.L. Laity, *J. Am. Chem. Soc.* **90** (1968), 811.
[20] H.J. Dauben, J.D. Wilson and J.L. Laity, *J. Am. Chem. Soc.* **91** (1969), 1991.
[21] S.B. Bulgarevich, T.A. Yusman and O.A. Osipov, *J. Gen. Chem. USSR* **54** (1984), 1427.
[22] P. Lazzeretti and J.A. Tossell, *J. Mol. Struct.* (*Theochem*) **234** (1991), 403.
[23] E.F. Archibong and A.J. Thakkar, *Mol. Phys.* **81** (1994), 557.
[24] N. El-Bakali Kassimi, R.J. Doerksen and A.J. Thakkar, *J. Phys. Chem.* **99** (1995), 12790.
[25] S. Millefiori and A. Alparone, *J. Mol. Struct.* (*Theochem*) **431** (1998), 59.
[26] F.D. Proft and P. Geerlings, *Chem. Rev.* **101** (2001), 1451.
[27] K.G. Denbigh, *J. Chem. Soc. Farafay Trans.* **36** (1940), 936.
[28] J. Applequist, *Acc. Chem. Res.* **10** (1977), 79.
[29] B.T. Thole, *Chem. Phys.* **59** (1981), 341.
[30] P.T van Duijnen and M. Swart, *J. Phys. Chem. A* **102** (1998), 2399.
[31] N. Celebi, J.G. Ángyán, F. Dehez, C. Millot and C. Chipot, *J. Chem. Phys.* **112** (2000), 2709.
[32] L. Jensen, P.-O. Åstrand and K. Mikkelsen, *Int. J. Quantum Chem.* **84** (2001), 513.
[33] J.R. Maple and C.S. Ewig, *J. Chem. Phys.* **115** (2001), 4981.
[34] M.B. Ferraro, M. Caputo and P. Lazzeretti, *J. Chem. Phys.* **109** (1998), 2987.
[35] M.E. Zitto, M.C. Caputo, M.B. Ferraro and P. Lazzeretti, *J. Chem. Phys.* **114** (2001), 4053.
[36] A.R. Katrizky, P. Barczynski, G. Musumarra, D. Pisano and M. Szafran, *J. Am. Chem. Soc.* **111** (1989), 7.
[37] A. Ligabue and P. Lazzeretti, *J. Chem. Phys.* **116** (2002), 964.
[38] M. Cyrañski, T.M. Krygowski, A.R. Katrizky and P. von Ragué Schleyer, *J. Org. Chem.* **67** (2002), 1333.
[39] U. Fleischer, W. Kutzelnigg, P. Lazzeretti and V. Mühlenkamp, *J. Am. Chem. Soc.* **116** (1994), 5298.
[40] A.J. Sadlej, *Collect. Czech. Chem. Commun.* **53** (1988), 1995.
[41] J.J. Perez and A.J. Sadlej, *J. Mol. Struct.* (*Theochem*) **371** (1996), 31.
[42] J.F. Stanton and R.J. Bartlett, *J. Chem. Phys.* **99** (1993), 5178.
[43] C. Adant, J.L. Brédas and M. Dupuis, *J. Phys. Chem. A* **101** (1997), 3025.
[44] S. Millefiori and A. Alparone, *J. Mol. Struct.* (*Theochem*) **422** (1997), 179.
[45] P. Norman, D. Jonsson and H. Ågren, *Chem. Phys. Lett.* **268** (1997), 337.
[46] O. Christiansen, C. Hättig and P. Jørgensen, *Spectr. Acta* **55** (1999), 509.
[47] A. Hinchliffe and H.S. Machado, *Int. J. Mol. Sci.* **1** (2000), 39.
[48] E.S. Nielsen, P. Jørgensen and J. Oddershede, *J. Chem. Phys.* **73** (1980), 6238.
[49] H. Koch, A. Sánchez de Merás, T. Helgaker and O. Christiansen, *J. Chem. Phys.* **104** (1996), 4157.
[50] H. Koch, O. Christiansen, P. Jørgensen, A.Sánchez de Merás and T. Helgaker, *J. Chem. Phys.* **106** (1997), 1808.

[51] O. Christiansen, H. Koch and P. Jørgensen, *Chem. Phys. Lett.* **243** (1995), 409.
[52] O. Christiansen, A. Halkier, H. Koch, P. Jøgensen and T. Helgaker, *J. Chem. Phys.* **108** (1998), 2801.
[53] J. Gauss, Molecular properties, in: *Modern Methods and Algorithms of Quantum Chemistry*, (Vol. 3 of NIC Series), J. Grotendorst, ed., Jülich, 2000, John von Neumann Institute for Computing, pp. 541–592.
[54] T.H. Dunning, Jr., *J. Chem. Phys.* **90** (1989), 1007.
[55] T. Helgaker, H.J. Jensen, P. Jørgensen, J. Olsen, K. Ruud, H. Ågren, A.A. Auer, T. Andersen, K.L. Bak, V. Bakken, O. Christiansen, S. Coriani, P. Dahle, E.K. Dalskov, T. Enevoldsen, B. Fernandez, C. Hättig, K. Hald, A. Halkier, H. Heiberg, H. Hettema, D. Jonsson, S. Kirpekar, R. Kobayashi, H. Koch, K.V. Mikkelsen, P. Norman, M.J. Packer, T.B. Pedersen, T.A. Ruden, A. Sanchez, T. Saue, S.P.A. Sauer, B. Schimmelpfenning, K.O. Sylvester-Hvid, P.R. Taylor and O. Vahtras, *Dalton, An electronic structure program, Release 1.2*, Dalton, 2001.
[56] P.J. Mohr and B.N. Taylor, *Rev. Mod. Phys.* **72** (2000), 351, Table XXVIII.
[57] P. Jørgensen, *Annu. Rev. Phys. Chem.* **26** (1975), 359.
[58] T.K. Rebane, *Opt. Spectros.* **8** (1960), 242.
[59] T.C. Caves and M. Karplus, *J. Chem. Phys.* **50** (1969), 3649.
[60] P. Lazzeretti, B. Cadioli and U. Pincelli, *Int. J. Quantum Chem.* **10** (1976), 771.
[61] P. Lazzeretti, E. Rossi and R. Zanasi, *J. Chem. Phys.* **79** (1983), 889.
[62] P. Lazzeretti and R. Zanasi, *J. Chem. Phys.* **83** (1985), 1218.
[63] P. Rozyczko, A. Perera, M. Nooijen and R. Bartlett, *J. Chem. Phys.* **107** (1997), 6736.
[64] A. Proutière and M. Camail, *Mol. Phys.* **29** (1975), 1473.
[65] R.J. Alms, A.K. Burnham and W.H. Flygare, *J. Chem. Phys.* **63** (1975), 3321.
[66] M. Okruss, R. Müller and A. Hese, *J. Chem. Phys.* **110** (1999), 10393.
[67] D.M. Bishop and L.M. Cheung, *J. Phys. Chem. Ref. Data* **11** (1982), 119.
[68] A. Hinchliffe and H.J. Soscún, *J. Mol. Struct.* (*Theochem*) **331** (1995), 109.
[69] N. El-Bakali Kassimi and Z. Lin, *J. Phys. Chem. A* **102** (1998), 9906.

Journal of Computational Methods in Sciences and Engineering 4 (2004) 677–701
IOS Press


# Third-order nonlinear optical responses of molecules in the intermediate and strong correlation regime: Variation of second hyperpolarizability in the bond dissociation

Masayoshi Nakano[a,*], Satoru Yamada[b] and Kizashi Yamaguchi[b]
[a]*Division of Chemical Engineering, Department of Materials Engineering Science, Graduate School of Engineering Science, Osaka University, Toyonaka, Osaka 560-8531, Japan*
[b]*Department of Chemistry, Graduate School of Science, Osaka University, Toyonaka, Osaka 560-0043, Japan*

**Abstract.** The variations in the longitudinal second hyperpolarizabilities ($\gamma$) in the bond-dissociation processes of molecules ($H_2$ and LiH) are investigated by the finite-field (FF) approach using several *ab initio* molecular orbital (MO) methods, i.e., spin-restricted methods (RHF, RMP$n$ ($n$ = 2,4), RHF coupled cluster (RCC)) and spin-unrestricted methods (UHF, UMP$n$ ($n$ = 2,4), UHF coupled cluster (UCC)) as well as approximate spin-projected methods (APUHF and APUMP$n$). The variation of $\gamma$ in the bond dissociation process is expected to elucidate the feature of third-order nonlinear optical (NLO) properties of molecules, e.g., open-shell organic and transition metal compounds, which belong to the intermediate and strong correlation regime. The calculation methods except for CC exhibit a cusp in the variation of $\gamma$ at the triplet instability threshold, and remarkable electron-correlation dependency of the variation in $\gamma$ is observed in the intermediate and large bond-distance region, corresponding to intermediate and strong correlation regime, respectively. From the results of the most accurate methods (UCCSD(T) or RCCSD(T)) employed in this study, not only significant increase and decrease behavior of the magnitude of $\gamma$ value is observed in the intermediate dissociation region of these molecules but also even the change in the sign of $\gamma$ value appears in the dissociation process of LiH.

Keywords: Nonlinear optics, second hyperpolarizability, open-shell, electron-correlation, *ab initio* MO methods

Mathematics Subject Classification: 6.5L12, 6.5L15

## 1. Introduction

Theoretical and experimental research of nonlinear optical (NLO) properties of molecular systems has been actively done over the past 20 years or more [1–7] because organic and organo-metallic conjugated compounds exhibit large nonlinear optical susceptibilities (hyperpolarizabilities at the microscopic level) with fast response times and the feasibility of controlling hyperpolarizabilities by the modification of molecular electronic structures. Most of molecules for NLO materials have been thermally stable

*Corresponding author. E-mail: mnaka@cheng.es.osaka-u.ac.jp.

closed-shell neutral systems. In our previous studies [5,8–23], however, several charged and/or open-shell systems are expected to have remarkable enhancement or reduction of the magnitude of the second hyperpolarizabilities, $\gamma$, and further drastic change of their signs by slightly modifying the structure and charged/spin states of such molecules. In particular, several compounds involving transition metals are predicted to have extraordinary large magnitude of $\gamma$ and to show unique contributions of various orbital interactions concerning $p$ and $d$ electrons, the feature of which have not been observed in conventional hydrocarbon systems [5,11,20]. Recently, various molecular based nano-structures involving transition metals have been synthesized and have shown remarkably large hyperpoalrizabilities [5,24].

Such charged and/or open-shell systems also tend to exhibit significant electron-correlation effects as compared to conventional closed-shell neutral systems [5,8–23] in addition to the large basis set dependency, which usually appears in the calculation of hyperpolarizability. This feature is considered to be a source of high sensitivity of hyperpoalrizabilities with respect to slight changes of structures and charged and/or spin states of molecules. Therefore, the study on the NLO responses of molecules in the intermediate and strong correlation regime is now one of the current topics in chemical physics and is also important from the viewpoint of engineering science. In quantum chemistry, the electron-correlation dependences of physical and chemical properties such as response properties have been often investigated using the behavior of those properties in the bond-dissociation processes of relatively small-size molecules/atoms, which can be treated by the *ab initio* molecular orbital (MO) methods involving various electron-correlation effects with extended basis sets. Actually, there are lots of sophisticated *ab initio* MO studies on the hyperpoalrizabilities of small-size molecules/atoms [25–32]. For example, the variations in polarizability ($\alpha$) and $\gamma H_2$ in the bond-dissociation process have been investigated and it has been found that they exhibit enormous enhancements of $\alpha$ and $\gamma$ in the intermediate dissociation region [33]. Inversely, the behaviors of such properties in the bond-dissociation process are useful for clarifying the features of various *ab initio* electron-correlation approaches because such properties exhibit distinct electron-correlation dependence as compared with the total energy. In this study, therefore, we investigate the behavior of longitudinal static $\gamma$ in the dissociation processes of relatively small-size systems, i.e., hydrogen molecule ($H_2$) and lithium hydrite (LiH) by the finite-field (FF) approach [24] using spin-restricted and spin-unrestricted Hatree-Fock (HF) and post-HF methods. In the former approach, RHF, restricted $n$th-order Møller-Plesset (RMP$n$), restricted coupled cluster (RCC) methods are employed, whereas the latter approach includes UHF, UMP$n$, and UCC methods. We also examine the applicability of approximate spin-projected UHF and UMP$n$ (APUHF and APUMP$n$) methods, which approximately but effectively remove spin contamination errors and have greatly succeeded in reproducing the shape of potential energy curves calculated by the large-scale CASSCF and CASPT2 levels [35]. In addition to the variation in $\gamma$ values, the spatial contributions of electrons to $\gamma$ are examined using the "second hyperpoalrizability ($\gamma$) density" [10,36] in the bond-dissociation process. For the dissociation process of $H_2$ molecule, we further investigate "two-electron $\gamma$ density" distributions of a one-dimensional model of $H_2$ calculated by the many-electron wavepacket (MEWP) approach [33, 37], which can provide numerically exact solutions of multi-electron systems. Since this density plot can provide a quasi-classical picture of hyperpolarization, i.e., the deviation of electron pair from its average distribution, we can analyze the real-space configuration of electron pair in the virtual excitation process describing $\gamma$ in the bond-dissociation process. The results obtained in this study, though they are concerned with simple small-size molecules, are expected to be useful for considering the applicability of various *ab initio* MO methods to the calculation of $\gamma$ values for interesting intermediate- and strong-correlated molecular systems such as transition-metal and charged radical compounds.

This paper is organized as follows. In Section 2, we first explain our classification rule of $\gamma$ as well as the condition providing static negative $\gamma$, which is rare in general, and its model systems, most of which belong to intermediate- and strong-correlation regime. The calculated models and calculation procedures are presented in Section 3 as well as a brief explanation of the FF-MEWP method. The hyperpolarizability density analysis and its extension to multi-electron case are explained in Section 4. The variations in $\gamma$ values and their density plots in the dissociation processes for $H_2$ and LiH calculated by various *ab initio* MO methods are discussed in Section 5, including the classical picture of $\gamma$ for the dissociation process of a one-dimensional $H_2$ model calculated by the MEWP method. This is followed by a conclusion in Section 6.

## 2. Classification rule of second hyperpolarizability ($\gamma$)

In this section, we explain our classification rule, i.e., structure-property relation, of $\gamma$ based on the perturbation theory and a condition providing negative $\gamma$ in order to clarify the importance of the intermediate- and strong- electron-correlated molecules as candidates of novel third-order NLO systems.

The perturbative formula of $\gamma$ can be partitioned into three types of contributions (I, II and III) [5, 8]. For convenience, we here consider the partition in the static $\gamma$, which is considered to be a good approximation to the off-resonant $\gamma$. The analytical expression of $\gamma$ is given by

$$\gamma = \gamma^I + \gamma^{II} + \gamma^{III} \tag{1}$$

where

$$\gamma^I = \sum_{n=1} \frac{(\mu_{n0})^2(\mu_{nn})^2}{E_{n0}^3}, \tag{2}$$

$$\gamma^{II} = -\sum_{n=1} \frac{(\mu_{n0})^4}{E_{n0}^3}, \tag{3}$$

and

$$\gamma^{III} = \sum_{\substack{m,n=1 \\ (m \neq n)}} \frac{(\mu_{n0})^2(\mu_{mn})^2}{E_{n0}^2 E_{m0}}. \tag{4}$$

Here, $\mu_{n0}$ is the transition moment between the ground and the $n$th excited states, $\mu_{mn}$ is the transition moment between the $m$th and the $n$th excited states, $\mu_{nn}$ is the difference of dipole moments between the ground and the $n$th excited states and $E_{n0}$ is the transition energy given by $(E_n - E_0)$. From these equations, apparently, the contributions of types I and III are positive in sign, whereas the contribution of type II is negative. For conventional molecular compounds with large positive $\gamma$, there are two characteristic cases: i $|\gamma^I|, |\gamma^{III}| > |\gamma^{II}| (\gamma > 0)$ and ii $|\gamma^I| = 0, |\gamma^{II}| < |\gamma^{III}| (\gamma > 0)$. In the case i, the compounds have large asymmetric charge distributions which are responsible for large $\mu_{nn}$, whereas in the case ii, the compounds are centrosymmetric systems in which the contributions of type I disappear. The third case, i.e., iii $|\gamma^I| = 0$, $|\gamma^{II}| > |\gamma^{III}| (\gamma < 0)$, is interesting because the systems with negative static $\gamma$ is rare in general and the sign of real part of third-order nonlinear optical susceptibility $\chi^{(3)}_{real}$ is important for the application in nonlinear optics [38]: positive and negative $\chi^{(3)}_{real}$ exhibit self-focusing and self-defocusing effects, respectively. Case iii systems are symmetric ($\mu_{nn} = 0$) and mostly exhibit strong virtual excitation between the ground and the first excited states ($|\mu_{n0}| > |\mu_{mn}|$). This indicates

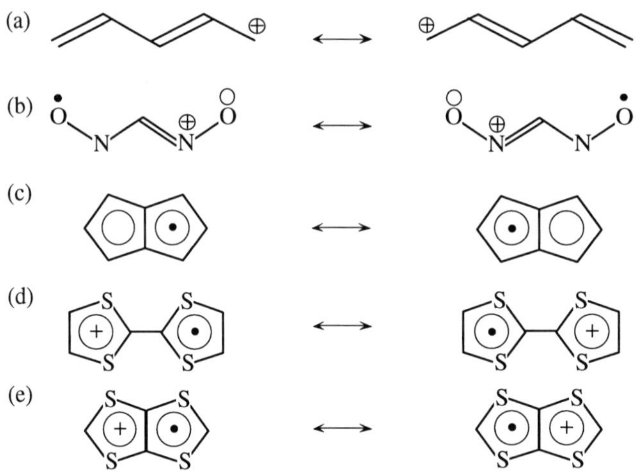

Fig. 1. Symmetric resonance structures with invertible polarization (SRIP) for various $\pi$-conjugated systems, i.e., (a) charged soliton-like oligomer, (b) nitronyl nitroxide radical, (c) anion radical state of pentalene, (d) cation radical state of tetrathiafulvalene ($TTF^+$) and (e) cation radical state of tetrathiapentalene ($TTP^+$).

that the symmetric systems with large ground-state polarizability ($\alpha_0$) tend to exhibit negative $\gamma$. Figure 1 shows resonance structures mainly contributing to the ground state of $\pi$-conjugated systems with large $\alpha_0$, i.e., (a) charged soliton-like oligomer [10,16,39], (b) nitronyl nitroxide radical [14,15,17], (c) anion radical state of pentalene [8,13] and (d) cation radical state of TTF ($TTF^+$) and (e) cation radical state of TTP ($TTP^+$) [18,19,22]. These indicate the resonances between polarized structures with mutually opposite directions, and can significantly contribute to the stability of ground-state electronic structures. The large contributions of these resonance structures also correspond to the fact that the magnitude of transition moment ($\mu_{n0}$) between the ground and the first excited states are large. As seen from the previous results for these systems [8,10,13–19,22,39], a system with resonance structures contributing to the stability of the ground state and inducing the polarization in the mutually opposite direction tends to exhibit negative $\gamma$ and remarkable electronic structure dependences of the $\gamma$. The large contributions of the stable resonance structures with large dipole moments correspond to an enhancement of the magnitude of the transition moment ($\mu_{n0}$) between the ground and the allowed first excited states. This contribution also leads to a reduction of the transition energy ($E_{n0}$) between the ground and the allowed first excited states with large contributions from the resonance structures. As a result, a system with large contribution of symmetric resonance structures with invertible polarization, which is referred to as SRIP, satisfies our criteria for the system to have a negative $\gamma$ [5,13].

Since most of systems with large contributions of SRIP are found to be in charged radical and/or highly polarized neutral states, the electron-correlation effects on their $\gamma$ values are expected to be significant. Namely, the magnitude and sign of $\gamma$ for such systems are expected to be drastically changed by slight chemical and physical perturbations. This feature suggests a possibility that systems with large SRIP contributions are candidates for novel third-order NLO molecular systems, e.g., controllable NLO systems. In this study, therefore, we investigate the relation between $\gamma$ and the degree of correlations using the bond-dissociation (to neutral species) models, in which the diradical character [40–42] increases as the bond dissociation proceeds from the equilibrium bond-length region (stable bonding region) to strong correlation (magnetic or diradical) region through the intermediate correlation region.

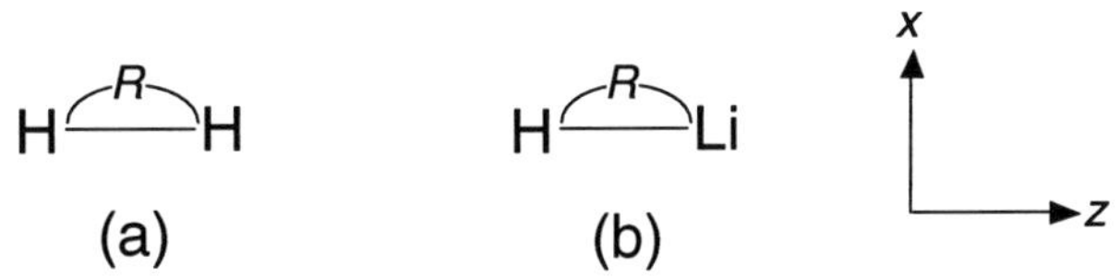


Fig. 2. Schematic illustrations of calculated models: (a) $H_2$ and (b) LiH. $R$ indicates the bond distance.

## 3. Calculated models and calculation procedures

### 3.1. Calculated simple molecules and various ab initio MO methods

Figure 2 shows calculated models: (a) hydrogen molecule ($H_2$) and (b) lithium hydrite (LiH). The longitudinal components of $\gamma$ ($\gamma_{zzzz}$) for these models are examined in their bond-dissociation processes in order to investigate the variation of $\gamma$ in the weak-, intermediate- and strong-correlation regime and to elucidate the characteristics of several computational methods. To this end, we employ the spin-restricted (RHF) and spin-unrestricted (UHF) type post-HF methods, i.e., R(U)MP$n$ ($n$ = 2–4), R(U)CCSD, R(U)CCSD(T). The RHF-based post-HF methods are known to break down in the dissociation (strong correlation) regime since the starting RHF solution is heavily triplet-unstable [44–46]. Higher excitation operators such as SD(T) are necessary for removing the deficient behavior for systems with electrons more than three in the case of RCC scheme [35]. On the other hand, the UHF-based methods provide smooth dissociation energy curves in the whole region. However, it is well-known that the UHF based methods including relatively low-order electron correlation effects such as UMP$n$ suffer from spin contamination effects and give a small hump in the intermediate bond-dissociation region of the potential energy curve [35]. In order to remove such spin contamination, an approximate spin-projection (AP) scheme has been presented by Yamaguchi et al. [40,47]. In this scheme, the lowest spin (LS) UHF-based solutions are projected only by the corresponding highest spin (HS) solutions. The APUHF X energy is given by [35]

$$^{\mathrm{LS}}E_{\mathrm{APUHF\ X}} = ^{\mathrm{LS}} E_{\mathrm{UHF\ X}} + f_{\mathrm{SC}}[^{\mathrm{LS}}E_{\mathrm{UHF\ X}} - ^{\mathrm{HS}} E_{\mathrm{UHF\ X}}], \tag{5}$$

where $^{Y}E_{\mathrm{UHF\ X}}$ denotes the total energy of the spin state Y by UHF X: X denotes various post-HF methods, e.g., MP$n$, CCSD and quadratic CI (QCI) SD methods. The factor $f_{\mathrm{SC}}$ for a spin multiplet ($2s+1$) is the fraction of spin contamination given by

$$f_{\mathrm{SC}} = \frac{^{\mathrm{LS}}\langle S^2\rangle_{\mathrm{UHF\ X}} - s(s+1)}{^{\mathrm{HS}}\langle S^2\rangle_{\mathrm{UHF\ X}} - {}^{\mathrm{LS}}\langle S^2\rangle_{\mathrm{UHF\ X}}}. \tag{6}$$

The AP UHF X methods (X = MP$n$, CC) have been successfully applied to calculations of potential energy surfaces [35] and effective exchange integrals in the Heisenberg models for open-shell clusters [41, 48–50]. In this study, $\langle S^2\rangle$ value of the singlet and triplet UHF based correlated methods (X) beyond UMP2 is estimated by extrapolation on the basis of the UMP2 approximation:

$$^{W}\langle S^2\rangle = ^{W} \langle S^2\rangle_{(\mathrm{UMP2})} + g\left[^{W}\langle S^2\rangle_{(\mathrm{UMP2})} - {}^{W}\langle S^2\rangle_{(\mathrm{UHF})}\right], \tag{7}$$

where

$$g = \frac{^{W}E(\mathrm{UHF\ X}) - ^{W} E(\mathrm{UMP2})}{^{W}E(\mathrm{UMP2}) - ^{W} E(\mathrm{UHF})}. \tag{8}$$

It is well-known that the multi-reference (MR) post-HF method is inevitable for quantitative descriptions of unstable molecules in the intermediate correlation regime if the many-body perturbation (PT$n$) and coupled cluster expansions are truncated in the low orders [40,41,51]. The UNO CASSCF can provide a relevant non-dynamical correlation corrections arising from the triplet instability of the RHF solution [51]. Although the UNO CASSCF with large CAS space and/or the second-order perturbation (PT2) method [52,53] based on the UNO CASSCF give reliable potential curves in the whole region [35], the use of small CAS space, which is sufficient for reproducing reliable potential energy curves, is predicted to fail in giving converged values of $\gamma$ for systems in the intermediate correlation regime [54]. The UNO CASSCF based MRCCSD(T) calculations will be necessary for providing reliable and sufficiently converged values of $\gamma$ for such systems though such calculations would be hard to apply to larger systems.

### 3.2. *Finite-field* (*FF*) *method*

Various theoretical approaches for calculating $\gamma$ have been presented in the literature. The sum-over-state (SOS) approach [1–8,55] based on the time-dependent perturbation theory (TDPT) is useful for elucidating the contribution of transitions between electronic excited states to $\gamma$. However, the application of this approach at the *ab initio* MO level is limited to the relatively small-size molecules since this needs transition properties (transition moments and transition energies) over many excited states, which are hard to describe precisely at the *ab initio* MO level at the present time. Alternatively, the approaches using numerical or analytical derivatives of total energy with respect to the applied electric field are widely employed for calculating static $\gamma$. Particularly, the FF approach using the CHF method at the semiempirical [56] and *ab initio* [57] MO levels, various electron-correlated *ab initio* MO methods [25–32,58] and the DF methods [12,21,59,60] has been widely employed for calculating static (hyper)polarizabilities. We here briefly explain the FF approach to the calculation of static $\gamma$ for the following discussion. The total Hamiltonian in the presence of a uniform electric field $\boldsymbol{F}$ is expressed as

$$H = H_0 + \sum_i \boldsymbol{F} \cdot \boldsymbol{r}_i - \sum_i Z_I \boldsymbol{F} \cdot \boldsymbol{R}_I, \tag{9}$$

where indices $i$ and $I$ signify electrons and nuclei, respectively. $Z_I$ is the atomic number of the $I$th nucleus and $H_0$ is the field-free Hamiltonian. The total energy $E$ can be obtained as the expectation values $\langle \Psi | H | \Psi \rangle$ for the wavefunctions $|\Psi\rangle$ in the presence of the electric field. Similarly, the dipole moment $\boldsymbol{\mu}$ is expressed as

$$\boldsymbol{\mu} = \langle \Psi | \sum_i Z_I \boldsymbol{R}_I - \sum_i \boldsymbol{r}_i | \Psi \rangle . \tag{10}$$

The differentiation of total energy $E$ with respect to $F^i$ gives

$$\frac{\partial E}{\partial F^i} = \left\langle \frac{\partial \Psi}{\partial F^i} \right| H | \Psi \rangle + \langle \Psi | \frac{\partial H}{\partial F^i} | \Psi \rangle + \langle \Psi | H \left| \frac{\partial \Psi}{\partial F^i} \right\rangle . \tag{11}$$

If $|\Psi\rangle$ is the true wavefunction, the first and third terms on the right-hand side of Eq. (11) is equal to zero by the Hellmann-Feynman theorem [61]. The variational methods such as CHF satisfy the theorem. If the Hellmann-Feynman theorem is satisfied, the dipole moment defined by Eq. (10) can be expressed as

$$\mu^i = -\frac{\partial E}{\partial F^i} . \tag{12}$$

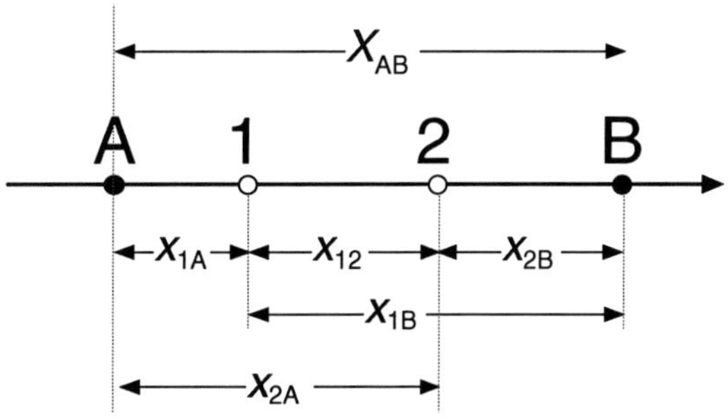


Fig. 3. One-dimensional $H_2$ composed of two nuclei ( **A** and **B**) and two electrons ( **1** and **2**).

The total energy and dipole moment can be expanded as the power series of the applied field:

$$E = E_0 - \sum_i \mu_0^i F^i - \frac{1}{2}\sum_{i,j} \alpha_{ij} F^i F^j - \frac{1}{3}\sum_{i,j.k} \beta_{ijk} F^i F^j F^k - \frac{1}{4}\sum_{i,j.k} \gamma_{ijk} F^i F^j F^k F^l - \ldots, \quad (13)$$

and

$$\mu^i = \mu_0^i + \sum_j \alpha_{ij} F^j + \sum_{jk} \beta_{ijk} F^j F^k + \sum_{jkl} \gamma_{ijk} F^j F^k F^l - \ldots, \quad (14)$$

where $\mu_0^i$ is the permanent dipole moment. The Hellmann-Feynman theorem asserts that Eqs (13) and (14) are compatible. In this study, we use the definition of hyperpolarizability based on Eq. (13). Since we focus on the longitudinal components of $\gamma$ ( $\gamma_{zzzz}$) for small molecules shown in Fig. 2, the $\gamma_{zzzz}$ values are calculated by the numerical differentiation of the total energy $E$ with respect to the applied electric field:

$$\begin{aligned}\gamma_{iiii} = \{&E(3F^z) - 12E(2F^z) + 39E(F^z) - 56E(0) + 39E(-F^z) - 12E(-2F^z) \\ &+E(-3F^z)\}/\{36(F^z)^4\},\end{aligned} \quad (15)$$

where $E(F^i)$ represents the total energy in the presence of the electric field $F$ applied in the $i$direction. This method is referred to as the FF method. The total energies are calculated by various *ab initio* MO methods explained above. In order to avoid numerical errors, we use several minimum field strengths (0.0002 a.u – 0.01 a.u.). After numerical differentiations using these fields, we adopt a numerically stable $\gamma$.

### 3.3. *Finite-field many-electron wave-packet* (*FF-MEWP*) *method*

We here explain the FF-MEWP method for $H_2$ molecule [33], in which the wave-packet dynamics is performed in the spatial coordinates for a two-electron system at a fixed spin state. A one-dimensional $H_2$ is examined here as shown in Fig. 3. In the Born-Oppenheimer approximation, the electronic Hamiltonian involving the interaction with static electric field $E(F^i)$ is written (in atomic units) by

$$H_e = -\frac{1}{2}\frac{\partial^2}{\partial x_1^2} - \frac{1}{2}\frac{\partial^2}{\partial x_2^2} - \frac{1}{x_{1A}} - \frac{1}{x_{1B}} - \frac{1}{x_{2A}} - \frac{1}{x_{2B}} + \frac{1}{x_{12}} + Fx_1 + Fx_2, \quad (16)$$

where $x_1$ and $x_2$ denote the coordinates for electrons **1** and **2**, respectively, and $x_{ij}$ indicates the distance between particles $\boldsymbol{i}$ and $\boldsymbol{j}$. The solution to a time-dependent Schrödinger equation involving the electronic Hamiltonian,

$$i\frac{\partial}{\partial t}\Phi(x_1, x_2, t) = H_e\Phi(x_1, x_2, t), \tag{17}$$

is the electronic wavefunction,

$$\Phi(x_1, x_2, t) = \psi(x_1, x_2, t)\varphi(\omega_1, \omega_2, t), \tag{18}$$

where $\psi(x_1, x_2, t)$ and $\varphi(\omega_1, \omega_2, t)$ represent the spatial and spin wave-functions, respectively, in which $\omega_1$ and $\omega_2$ indicate each spin coordinate. The singlet state of the two-electron system is considered here since we focus on the ground-state (hyper)polarizability of the $H_2$. For the singlet-state two-electron system, the spatial wave-function, which must be symmetric with respect to the interchange of the spatial coordinates of two electrons, is generally represented as

$$\psi_s(x_1, x_2, t) = \phi_1(x_1, t)\phi_2(x_2, t) + \phi_2(x_1 t)\phi_1(x_2, t), \tag{19}$$

where $\phi_i(x, t)$ is the $i$th single-particle wave-function. Therefore, the time-dependent Schrödinger equation to solve is

$$i\frac{\partial}{\partial t}\psi_x(x_1, x_2, t) = H_e\psi_s(x_1, x_2, t). \tag{20}$$

The singlet spatial wave-function $\psi_s(x_1, x_2, t)$ is simply written as $\psi(x_1, x_2, t)$ for convenience hereafter. In the FF-MEWP method, a Gaussian wave-packet is considered as the initial $i$th single-particle wavefunction expressed by

$$\phi_i(x_j, t = 0) = C\exp\left[ip(x_j - x_{i0}) - \frac{|x_j - x_{i0}|^2}{2\sigma^2}\right]. \tag{21}$$

Here, $P$, $x_{i0}$, and $\sigma$ indicate momentum, coordinate of the center and width of the $i$th initial wavepacket, respectively. The symbol $C$ denotes a normalization constant. Using Eqs (19) and (21), an initial singlet wavefunction is constructed. This wavefunction is superposed by singlet eigenstates involving the ground and excited states of the Hamiltonian $H_e$. Since our desired wavefunction is the ground state that is generated by the relaxation method as shown in Eq. (26) later, the form of Eq. (21) is not so important. But it is significant that the spatial part of the ground-state singlet wavefunction for two-electron system is symmetric with respect to the exchange of two-electron coordinates as shown in Eq. (19).

In this study, we discretize the wave-function in real space. Therefore, the Coulomb potential is approximated by the following softened Coulombic form [62] in order to eliminate the singularity at the origin.

$$\frac{1}{x_{ij}} \approx \frac{1}{\sqrt{a + x_{ij}^2}} \tag{22}$$

where $a$ is a parameter. Javanainen et al. [62] gave $a = 1$ for calculations of above threshold ionization (ATI) spectra for one-dimensional H atom. This potential falls off like the Coulomb potential at large $|x_{ij}|$, but takes an asymptotic form of the Coulomb potential at short $|x_{ij}|$.

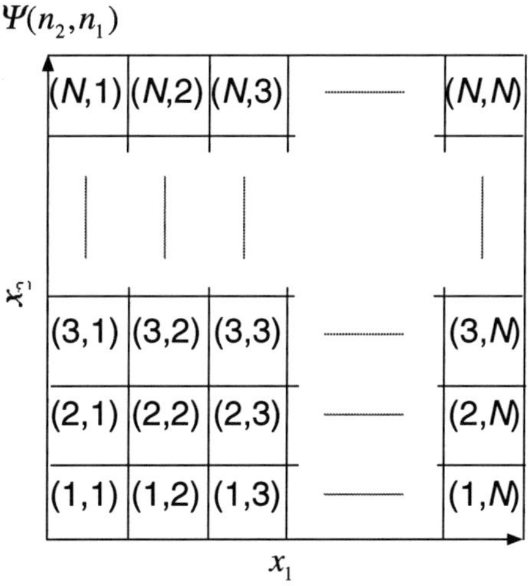


Fig. 4. Descretized coordinate plane for the one-dimensional two-electron system. A finite one-dimensional space is divided into $N$ fragments. The numbers $n_1$ and $n_2$ represent the grid numbers of electrons **1** and **2**, respectively.

After discretizing in space as shown in Fig. 4, the kinetic part on the right-hand side of Eq. (20) is approximated as

$$-\frac{\psi(n_2+1,n_1,t)-2\psi(n_2,n_1,t)+\psi(n_2-1,n_1,t)}{2(\Delta x)^2} \\ -\frac{\psi(n_2,n_1+1,t)-2\psi(n_2,n_1,t)+\psi(n_2,n_1-1,t)}{2(\Delta x)^2} \tag{23}$$

Here, $\psi(n_2, n_1, t)$ represents the wavefunction at the site $(x_2, x_1$, in which $x_1 = (n_i - 1)\Delta x(n_1 = 0, 1, \ldots, N + 1(i = 1, 2))$. Similarly the potential part using the softened Coulombic form Eq. (22) is written by

$$\left[-\frac{1}{\sqrt{|(n_1-n_A)\Delta x|^2+1}}-\frac{1}{\sqrt{|(n_2-n_A)\Delta x|^2+1}}-\frac{1}{\sqrt{|(n_1-n_B)\Delta x|^2+1}}\right. \\ \left.-\frac{1}{\sqrt{|(n_2-n_B)\Delta x|^2+1}}+\frac{1}{\sqrt{|(n_1-n_2)\Delta x|^2+1}}+F(n_1-1)\Delta x+F(n_2-1)\Delta x\right] \\ \psi(n_2,n_1,t), \tag{24}$$

where nuclei **A** and **B** are located at the coordinates $(n_A - 1)\Delta x$ and $(n_B - 1)\Delta x$, respectively. In this work, we adopt the fixed boundary condition as

$$\psi(n_2, n_1 = 0, t) = \psi(n_2 = 0, n_1, t) = \psi(n_2, n_1 - N + 1, t) = \psi(n_2 = N + 1, n_1, t) = 0 \tag{25}$$

The time propagation of $\psi(x_1, x_2, t)$ is carried out by solving Eq. (20) in the sixth-order Runge-Kutta scheme. According to Kosloff et al. [63], the singlet ground-state wavefunction under static electric fields can be obtained by propagating initial wavepackets in imaginary time, i.e., by setting $\tau = it$ in Eq. (20). After sufficient long time propagation, the wave-packets relax to the ground state since all excited states involved in the initial wavepackets decay to zero. The singlet ground-state wavefunction is represented by

$$\psi_g(x_1, x_2) - \lim_{\tau \to \infty} \frac{\psi(x_1, x_2, \tau)}{\int\int |\psi(x_1, x_2, \tau)|^2 dx_1 dx_2}, \tag{26}$$

where $\psi(x_1, x_2, \tau) = \psi(x_1, x_2, t)|_{it=\tau}$. The total energy under static electric field, $E(F)$, is obtained as

$$E(F) = E_e(F) + \frac{1}{\sqrt{|(n_A - n_B)\Delta x|^2 + a}} - F(n_A - 1)\Delta x - F(n_B - 1)\Delta x, \tag{27}$$

in which the first term $E_e(F)$ denotes the electronic energy under the static electric field, the second term represents the nuclear repulsion, and the remaining terms describe the interaction between nuclei and electric field. Using the energies under several external fields and Eq. (15), we calculate the longitudinal static $\gamma$ for one-dimensional $H_2$ model.

## 4. Hyperpolarizability density analysis

In this section, we explain the reduced one-electron and many-electron hyperpolarizability densities using an example of $\gamma$. The reduced one-electron hyperpolarizability density, which is simply referred to as "hyperpolarizability density", can provide an intuitive picture of spatial contributions of electrons to hyperpolarizability, while the many-electron hyperpolarizability density is used for an extraction of quasi-classical picture, which is a "deviation of electrons from their equilibrium positions", for hyperpolarization [33].

### *4.1. Reduced one-electron hyperpolarizability density*

The reduced one-electron density under static electric field $\boldsymbol{F}$ for the one-dimensional system is expanded as [10,64]

$$\rho(\boldsymbol{r}, \boldsymbol{F}) = \rho^{(0)}(\boldsymbol{r}) + \sum_j \rho^{(1)}(\boldsymbol{r}) F^j + \frac{1}{2!}\sum_{jk} \rho^{(2)}(\boldsymbol{r}) F^j F^k + \frac{1}{3!}\sum_{jkl} \rho^{(3)}(\mathbf{r}) F^j F^k F^l + \ldots \tag{28}$$

The induced dipole moment is expressed by

$$\mu^i = \mu_0^i + \sum_j \alpha_{ij} F^j + \sum_{jk} \beta_{ijk} F^j F^k + \sum_{jkl} \gamma_{ijkl} F^j F^k F^l - \ldots \tag{29}$$

where $\mu_0^i$ denotes the ground-state permanent dipole moment. From Eqs (28) and (29), the $\gamma$ is represented by [10,64]

$$\gamma_{ijkl} = -\frac{1}{3!}\int r^i \rho^{(3)}_{jkl}(\boldsymbol{r}) d^3\boldsymbol{r}, \tag{30}$$

where the $\rho^{(3)}_{jkl}(\boldsymbol{r})$, which is defined as the reduced one-electron $\gamma$ densities, respectively, is obtained by

$$\rho^{(3)}_{jkl}(\boldsymbol{r}) = \left.\frac{\partial^3 \rho}{\partial F^j \partial F^k \partial F^l}\right|_{F=0}. \tag{31}$$

This quantity is reduced one-electron second hyperpolarizability density, or simply referred to as $\gamma$ density. This is calculated at each spatial point in the discretized space by using the following numerical differentiation formulae:

$$\rho_{jkl}^{(3)}(\boldsymbol{r}) = \{\rho(\boldsymbol{r}), 2F^i) - \rho(\boldsymbol{r}, -2F^i) - 2(\rho(\boldsymbol{r}, F^i) - \rho(\boldsymbol{r}, -F^i))\}/2(F^i)^3, \quad (32)$$

where $\rho_{jkl}(\boldsymbol{r}, F^i)$ represents the reduced one-electron density at a spatial point $\boldsymbol{r}$ in the presence of the field $F^i$.

In order to explain a method for analysis employing the plots of reduced one-electron (hyper)polarizability density, we consider a pair of localized $\rho_{iii}^{(3)}(\boldsymbol{r})$ shown in Fig. 5 [10]. The positive sign of $\rho_{iii}^{(3)}(\boldsymbol{r})$ implies that the second derivative of the charge density increases with the increase in the field. As can be seen from Eq. (30), the arrow from positive to negative $\rho_{iii}^{(3)}(\boldsymbol{r})$ shows the sign of the contribution to $\gamma$ determined by the relative spatial configuration between the two $\rho_{iii}^{(3)}(\boldsymbol{r})$. Namely, the sign of the contribution to $\gamma$ becomes positive when the direction of the thick arrow coincides with the positive direction of the coordinate system. The contribution to $\gamma$ determined by the $\rho_{iii}^{(3)}(\boldsymbol{r})$ of the two points is more significant, when their distance is larger.

### 4.2. Many-electron (hyper)polarizability density

From the relation between many-electron and reduced one-electron densities, the reduced one-electron $n$th-order (hyper)polarizability density is related to the $M$-electron $n$th-order (hyper)polarizability density as

$$\rho_{jkl}^{(n)}(\boldsymbol{r}) = M \int \rho_{jkl}^{(n)}(\boldsymbol{r}_1, \boldsymbol{r}_2, \ldots, \boldsymbol{r}_m) d^3\boldsymbol{r}_2 \ldots d^3\boldsymbol{r}_M. \quad (33)$$

Therefore, $\gamma$ is also expressed by

$$\gamma_{ijkl} = -\frac{1}{3!} \int \left( \sum_m^M r_m^i \right) \rho_{jkl}^{(3)}(\boldsymbol{r}_1, \boldsymbol{r}_2, \ldots, \boldsymbol{r}_M) d^3\boldsymbol{r}_1 d^3\boldsymbol{r}_2 \ldots d^3\boldsymbol{r}_M, \quad (34)$$

where $\rho_{jkl}^{(3)}(\boldsymbol{r}_1, \boldsymbol{r}_2, \ldots, \boldsymbol{r}_M)$ is defined respectively by

$$\rho_{jkl}^{(3)}(\boldsymbol{r}_1, \boldsymbol{r}_2, \ldots, \boldsymbol{r}_M) = \left. \frac{\partial^3 \rho^{(3)}(\boldsymbol{r}_1, \boldsymbol{r}_2, \ldots, \boldsymbol{r}_M, \boldsymbol{F})}{\partial F^j \partial F^k \partial F^l} \right|_{F=0}. \quad (35)$$

Here, the $M$-electron density under the static electric field $\boldsymbol{F}$ is obtained by

$$\rho(\boldsymbol{r}_1, \boldsymbol{r}_2, \ldots, \boldsymbol{r}_M, \boldsymbol{F}) = \psi^*(\boldsymbol{r}_1, \boldsymbol{r}_2, \ldots, \boldsymbol{r}_M, \boldsymbol{F}) \psi(\boldsymbol{r}_1, \boldsymbol{r}_2, \ldots, \boldsymbol{r}_M, \boldsymbol{F}), \quad (36)$$

where $\psi(\boldsymbol{r}_1, \boldsymbol{r}_2, \ldots, \boldsymbol{r}_M, \boldsymbol{F})$, denotes a spatial wave-function under the static electric field $\boldsymbol{F}$. The $M$-electron hyperpolarizability density is calculated numerically as

$$\begin{aligned} \rho_{jkl}^{(3)}(\boldsymbol{r}_1, \ldots, \boldsymbol{r}_M) = \{&\rho((\boldsymbol{r}_1, \ldots, \boldsymbol{r}_M, 2\boldsymbol{F}) - \rho(\boldsymbol{r}_1, \ldots, \boldsymbol{r}_M, -2\boldsymbol{F}) - 2(\rho(\boldsymbol{r}_1, \ldots, \boldsymbol{r}_M, \boldsymbol{F}) \\ &-\rho(\boldsymbol{r}_1, \ldots, \boldsymbol{r}_M, -\boldsymbol{F}))\}/2(\boldsymbol{F})^3. \end{aligned} \quad (37)$$

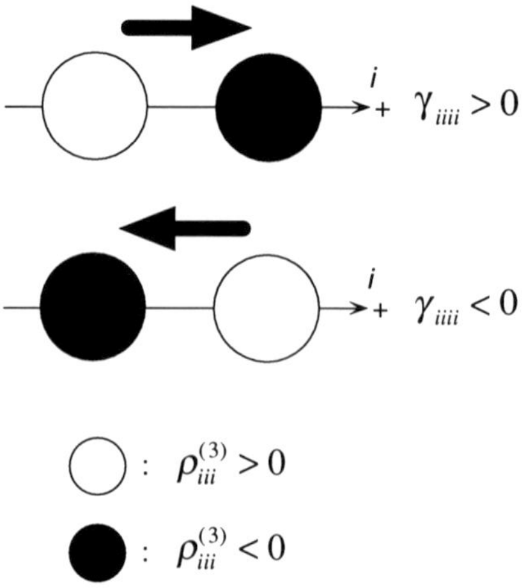


Fig. 5. Schematic diagram of the second hyperpolarizability ($\gamma$) densities ($\rho^{(3)}_{iii}(r)$). The white and black circles denote, respectively, positive and negative $\rho^{(3)}_{iii}(r)$'s. The size of circle represents the magnitude of $\rho^{(3)}_{iii}(r)$ and the arrow shows the sign of $\rho^{(3)}_{iii}(r)$ determined by the relative spatial configuration between these two $\rho^{(3)}_{iii}(r)$'s.

We here explain a method for analysis based on the many-electron hyperpolarizability density. The procedure of evaluating spatial contributions from their densities is basically the same as that in the reduced one-electron hyperpolarizability density analysis mentioned above. The two-electron hyperpolarizability density for the one-dimensional two-electron system (in the direction $x$) is considered for simplicity and for the latter application (See Section 5). Similarly to the case for the reduced one-electron hyperpolarizability density analysis, we consider two pairs of positive and negative localized densities on the two electron coordinate plane. They are located symmetrically with respect to the diagonal line $x_1 = x_2$ since the two electrons cannot be distinguished. Here, let us draw arrows from positive to negative densities similarly to the thick arrow shown in Fig. 5. In this case, we can draw arrows from positive (white circle) to negative (black circle) densities in two manners as shown in Fig. 6(a). In the first manner, two vectors ($\boldsymbol{s}$ and $\boldsymbol{s}'$) are symmetric with respect to a diagonal line $x_1 = x_2$, while for the second one, two vectors ($\boldsymbol{t}$ and $\boldsymbol{t}'$) intersect each other across the diagonal line $x_1 = x_2$. Let's focus on the contribution from vector $\boldsymbol{s}$. We decompose the vector $\boldsymbol{s}$ into vector components $\boldsymbol{s}_1$ and $\boldsymbol{s}_2$, which are along the coordinates $x_1$ and $x_2$, respectively. The contribution from the vector $\boldsymbol{s}$ is obtained by adding the contributions from the vectors $\boldsymbol{s}_1$ and $\boldsymbol{s}_2$. The sign of the contribution becomes positive when the directions of $\boldsymbol{s}_1$ and $\boldsymbol{s}_2$ coincides with the positive direction of the coordinates $x_1$ and $x_2$, respectively. The magnitude of the contribution from the vector $\boldsymbol{s}_i (i = 1, 2)$ increases with the increasing distance $|\boldsymbol{s}_i|$ and with the increasing with magnitude of the two localized densities.

In quantum theory, the position of an electron cannot be determined, and an electron wavefunction is extended in space. In this meaning, a choice of the position of an electron in the two-dimensional plane corresponds to an extraction of a classical picture of the electron. According to this viewpoint, a choice of a vector drawn from the positive to the negative two-electron hyperpolarizability density corresponds to an extraction of the classical picture of polarization relating to the hyperpolarizability specified by the vector. In Fig. 6(b), four classical pictures of polarization corresponding to the vectors ($\boldsymbol{s}, \boldsymbol{s}', \boldsymbol{t}$ and $\boldsymbol{t}'$) are shown as classical displacements of two-electron configurations at the starting and terminal points of these vectors. These classical displacements are divided into two types of the classical ones. The first type of displacements obtained by the vectors $\boldsymbol{s}$ and $\boldsymbol{s}'$ have no interchanges of electrons **1** and **2** in the polarization process, while another type of displacements obtained by the vectors $\boldsymbol{t}$ and $\boldsymbol{t}'$ have an

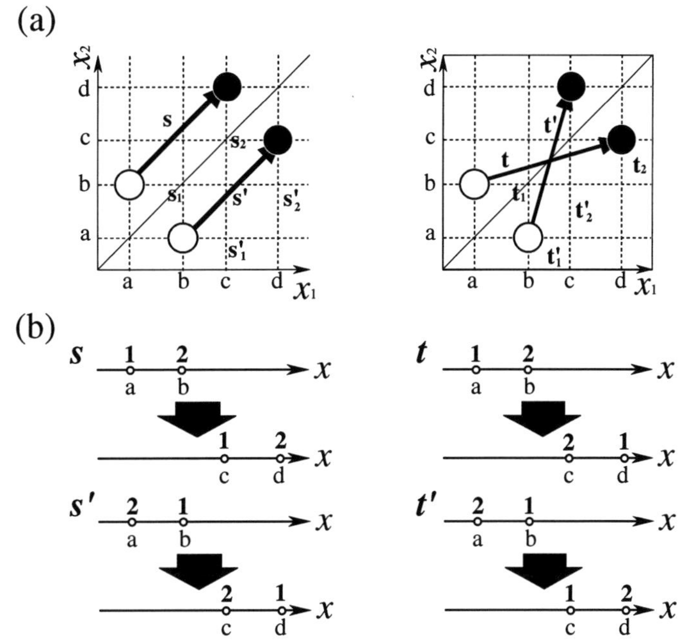

Fig. 6. (a) Two-electron hyperpolarizability density plots corresponding to two methods of drawing vectors $s(s')$ and $t(t')$ from positive to negative two-electron hyperpolarizability densities for the one-dimensional system. The white and black circles indicate positive and negative densities, respectively. The decomposed vectors $s_i(s_i')$ and $t_i(t_i')$ $(i = 1, 2)$ are also shown. (b) Classical pictures of polarization extracted from vectors $s$, $s'$, $t$ and $t'$. Two electrons (**1** and **2**) are represented by white circles, and $a$, $b$, $c$ and $d$ denote the positions of the one-dimensional coordinate.

interchange of them. In general, therefore, plural classical pictures can be extracted from the same pair of positive and negative hyperpolarizability densities. In this study, however, the differences in electrons' number do not need to be considered since the hyperpolarizability relates to only the displacement of the middle point of coordinates of two electrons.

## 5. Variations in $\gamma$ values and their density plots in the dissociation processes of $H_2$ and LiH

In this section, we investigate the variation in longitudinal $\gamma$ for $H_2$ and LiH by the FF method using several *ab initio* MO methods. The $\gamma$ of one-dimensional $H_2$ is also calculated by the FF-MEWP method. Since the basis set dependency of the $\gamma$ for small-size molecules is known to be remarkable, we employ the 6-31G** basis sets augmented by diffuse and polarization functions: 6-31G**+ *sp* ($\zeta_{s,p} = 0.0406$) for H atom and 6-31G**+ *pd* for Li ($\zeta_{p,d} = 0.0164$), which are found to provide semi-quantitative $\gamma$ values. These exponents are determined from the outermost two exponents of 6-31G** by the even-tempered method. The *ab initio* and DF calculations in this paper are performed using Gaussian 98 program package [65].

### 5.1. $H_2$

In $H_2$ system, the single and double excitation CI (SDCI) method is the full CI (FCI) method since $H_2$ has only two electrons. As seen from the variation feature of longitudinal $\gamma$ for $H_2$ shown in Fig. 7, the $\gamma$

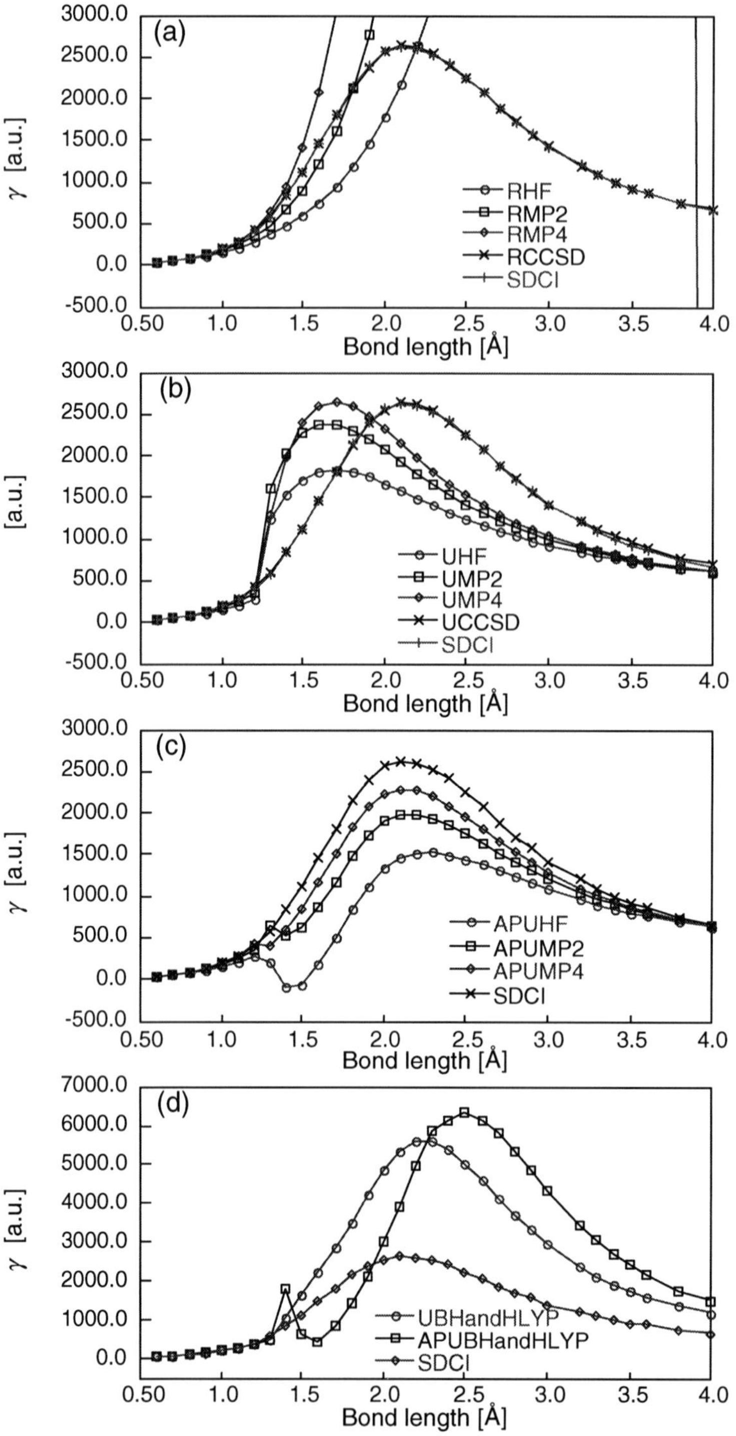


Fig. 7. Variation in $\gamma$ value of $H_2$ by (a) restricted (RHF, RMP2, RMP4, RCCSD), (b) unrestricted (UHF, UMP2, UMP4, UCCSD), (c) approximate projection (APUHF, APUMP2, APUMP4) and (d) density functional (UBHandHLYP, APUBHandHLYP) methods. The SDCI (full CI in this case) result is also shown for comparison.

significantly increases from the equilibrium bond distance ($R = 0.74$ Å) to intermediate bond distance, takes a maximum value ($\cong 2600$ a.u.), which is 34 times as large as that at the equilibrium bond distance, at $R = 2.1$ Å and then decreases to the sum of $\gamma$ for H atom in the infinite bond distance.

Figure 7(a) shows the results for $H_2$ calculated by the spin-restricted HF based methods. The RHF $\gamma$ value is shown to only increase remarkably as the bond distance increases. It is found that the electron correlation effects by the MP$n$ ($n$ = 2, 4) methods do not correct this remarkable increase behavior but more enhance the increase behavior. Such incorrect behavior is predicted to originate in the triplet instability of the RHF solution in $R \geqslant 1.2$ Å [35]. The RCCSD method is shown to rectify this incorrect behavior and to give a behavior which coincides with that by the SDCI method except for the case beyond $R > 2.8$ Å. We here briefly mention the relation between UHF and RHF solutions as well as spin projected UHF solution [35]. The RHF solution of $H_2$ exhibits triplet instability and is reorganized into singlet UHF solution, where the split HOMOs for the up and down spins are given by

$$\chi_{HO} = \cos(\omega/2)\sigma + \sin(\omega/2)\sigma^*, \tag{38}$$

$$\eta_{HO} = \cos(\omega/2)\sigma - \sin(\omega/2)\sigma^*, \tag{39}$$

where $\sigma$ and $\sigma^*$ are the bonding and anti-bonding RHF $\sigma$ orbitals of $H_2$. The orbital overlap between these split MOs in the UHF solution $|\chi_{HO}\eta_{HO}|$ is given by $\cos\omega$. The spin density on the hydrogen atom and total spin angular momentum $\langle S^2 \rangle$ are not zero in the UHF solution. The UHF solution exhibits a smooth energy dissociation curve although it suffers from spin contamination. The spin-projected UHF (PUHF) solution eliminating the triplet component from UHF is expressed by the 2x2 UHF natural orbital (UNO)-CI form as [35]

$$\Psi(\mathrm{PUHF}) = |\{\chi_{HO}\overline{\eta}_{HO} + \eta_{HO}\overline{\chi}_{HO}\}| = \cos^2\omega\,|\sigma\overline{\sigma}| - sin^2\omega\,|\sigma^*\overline{\sigma}^*|\,. \tag{40}$$

In the two-electron system, APUHF and PUHF are equivalent to each other. The reoptimization of $\omega$ provides the UNO CASSCF solution. The $\gamma$ values by the UHF based methods are shown in Fig. 7(b). The UHF $\gamma$ value exhibits a cusp at $R = 1.2$ Å (triplet instability threshold) and provides a maximum $\gamma$ value ($\cong 1800$ a.u.) at $R = 1.7$ Å, which is shorter than that ($R = 2.1$ Å) by the SDCI method. It is shown that the UMP$n$ ($n$ = 2, 4) correlations more enhance the $\gamma$ values at the HF level, while cannot move the maximum point. The cusp at $R = 1.2$ Å are not shown to be removed by including higher-order UMP-like electron correlations though the cusp becomes small at the UMP4 level. The UCCSD method is shown to correct such deficient behavior and to reproduce the behavior by the SDCI method.

In Fig. 7(c), we investigate the effects of spin-contamination correction on the variation in $\gamma$. Although the APUHF exhibits a cusp at $R = 1.2$ Å similarly to the UHF method, the $\gamma$ value takes a maximum value at a similar bond distance to that by the SDCI method. In addition, the APUMP$n$ ($n$ = 2, 4) is found to reduce the deviation from the SDCI result. This feature suggests that the spin-contamination correction can significantly improve the variation of $\gamma$ in the bond dissociation, i.e., from weak to strong correlation via the intermediate correlation regime. However, the small cusp is found to still exist at $R = 1.2$ Å in the case of $\gamma$ calculated by the APUMP4 method. Since the APUHF ( = PUHF for a two-electron system) result can be equivalent with that by the 2x2 UNO CASSCF method by reoptimizing the $\omega$ in Eq. (38), and the UNO CASSCF method is found to provide a smooth variation which is equivalent to that by the SDCI method at $R = 1.2$ Å, the FF-APUHF and FF-APUMP4 methods are not found to completely reproduce the correct non-dynamical correlation correction around the triplet instability threshold ($R = 1.2$ Å), which is obtained by the reoptimization of $\omega$ Eq. (40). In contrast, judging from results in the intermediate and strong correlation regime ($R > 1.2$ Å), the FF-AP scheme is predicted to effectively include non-dynamical correlation not only on the ground state but also the excited states in that region, and further the FF-APUMP$n$ method is expected to describe the dynamical correlation both on the ground and excited states. Recently, the density functional (DF) method is known to well reproduce the energies, properties and structures for molecules calculated by the *ab initio* MO electron-correlation methods. To check the applicability of the DF method to the $\gamma$ for molecules in the intermediate and strong correlation regime, we examine the $\gamma$ of $H_2$ by the BHandHLYP method Fig. 7(d). Although a cusp at $R = 1.2$ Å also appears and the $\gamma$ value by the BHandHLYP method is shown to be more significantly overestimated than that by the SDCI method, the bond distance, at which $\gamma$ takes a maximum value, obtained by the SDCI method is reproduced by the BHandHLYP method. This suggests that the BHandHLYP result somewhat includes the non-dynamical effects on $\gamma$ though the DF methods with the present exchange correlation functionals are known to have some deficiencies in reproducing reliable magnitudes of (hyper)polarizabilities [66].

Figure 8 shows $\gamma$ density distributions of $H_2$ at (a) $R = 1.2$ Å, (b) $R = 1.6$ Å, (c) $R = 2.1$ Å and (d) $R = 3.2$ Å calculated by the QCISD method, which is known to well reproduce the CCSD results. The $\gamma$ densities are plotted on a plane located 0.5 a.u. above the molecular $(z - x)$ plane. At (a) $R = 1.2$ Å, there are found to be two types of positive contributions, in which internal contributions are much larger than the outer ones. At (b) $R = 1.6$ Å, only a large positive contribution to $\gamma$ is observed. At (c) $R = 2.1$ Å, where the $\gamma$ value takes maximum, the positive contribution is further enhanced, while at (d) $R = 3.2$ Å, which corresponds to strong correlation (magnetic) regime, the whole $\gamma$ density distributions are reduced and the contributions tend to be partitioned to that for each H atom.

### 5.2. *One-dimensional $H_2$ model calculated by the FF-MEWP method*

In order to better elucidate the relation between spatial contribution of two electrons and $\gamma$ for $H_2$, the FF-MEWP method is applied to a one-dimensional $H_2$ model (Fig. 3). Figure 9 shows the variations in the longitudinal $\gamma$ in the dissociation process of the one-dimensional $H_2$ (Fig. 3). Similarly to the *ab initio* results on real $H_2$ shown in Fig. 7, the $\gamma$ increases slowly at the bond distance less than 1.6 Å and takes a maximum at 2.1 Å and then decreases. For the validity of the MEWP results, we here consider the effects of the softened Coulombic potential Eq. (22) on the hyperpolarizabilities. The use of this potential has two types of influences on electronic structures. One is caused by the effects on the electron-nucleus interaction, and the other is done by the effects on the electron-electron interaction. Since electrons distributed far from nuclei are known to primarily contribute to the hyperpolarizability, the use of the softened Coulombic potential is predicted to have a primary influence on the hyperpolarizability by reducing the Coulomb repulsion for an electron pair with short interelectronic distance nearly equal to or less than spatial grid interval $\Delta x$. As mentioned above, however, the hyperpolarizability is found to be essentially characterized by electron distributions from extended and diffuse spatial regions. Therefore, qualitative features of hyperpolarizability could be meaningfully discussed if we used a sufficiently small grid $\Delta x$ and a sufficiently large coordinate plane. Actually, from the comparison of Figs 7 and 9, the characteristics of $\gamma$ in the one-dimensional model $H_2$ obtained here can reproduce those of the longitudinal $\gamma$ for real $H_2$.

The maximum $\gamma$ at $R \cong 2.1$ Å implies that both theoretical and experimental searches of species with chemical bonds in the intermediate correlation regime are important in relation to molecular design of nonlinear optical materials, e.g., $\pi$-conjugated compounds with labile chemical bonds [5].

The two-electron and reduced one-electron $\gamma$ densities at 3.0 a.u. ($\cong$ 1.6 Å), 4.0 a.u. ($\cong$ 2.1 Å) and 6.0 a.u. ($\cong$ 3.2 Å) are shown in Fig. 10. Since the feature of one-electron $\gamma$ density at each bond distance by

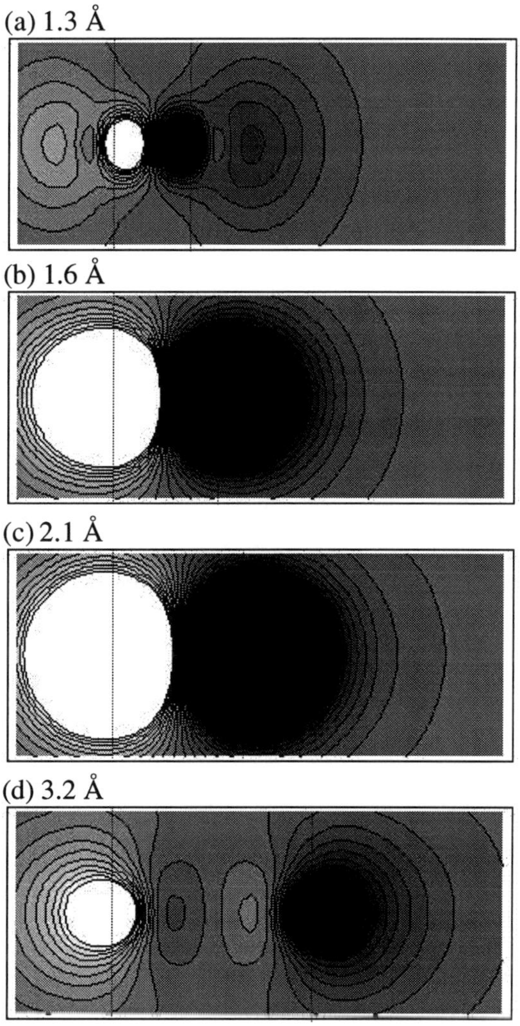


Fig. 8. The $\gamma$ density plots of $H_2$ at (a) $R = 1.2$ Å, (b) $R = 1.6$ Å, (c) $R = 2.1$ Å and (d) $R = 3.2$ Å calculated by the UQCISD method. The $\gamma$ densities are plotted on a plane located 0.5 a.u. above the molecular $(z - x)$ plane. The contours are drawn from $-10$ a.u. to 10 a.u. (contour step: 1 a.u.). The positions of H atoms are shown by the vertical dotted lines.

the MEWP method is shown to be in agreement with that by the UQCISD method (see Fig. 8) though the $\gamma$ density distribution with two types of positive contributions by the UQCISD method (Fig. 8(a) $R = 1.2$ Å) corresponds to that at $R = 3.0$ a.u. ($\cong 1.6$ Å) by the MEWP method as shown in Fig. 10. This difference in the bond distance seems to originate in the difference in the spatial dimension and the softened Coulomb potential (Eq. (22)) used in the MEWP method. The two-electron $\gamma$ density of one-dimensional $H_2$ model is useful for predicting the classical picture of hyperpolarization of real $H_2$. We can see four pairs of positive and negative densities constructed from internal and external two pairs which give negative and positive $\gamma$ values, respectively. This feature can be also understood by the negative contribution in the bond region and positive one in the outer region in the reduced one-electron $\gamma$ density plots. As seen from the explanation in Section 2, the difference between internal and external contributions in the two-electron $\gamma$ density is predicted to be ascribed to the two types of virtual excitation processes (types II and III) in the one-dimensional $H_2$. In the present case, the $\gamma$ is composed of types II

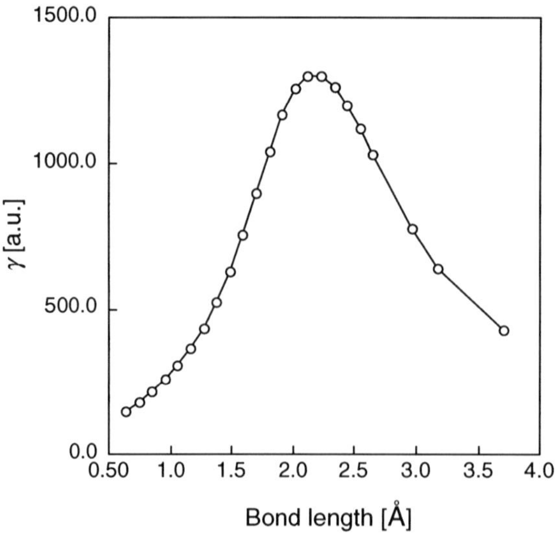


Fig. 9. Variations in $\gamma$ with bond distance $R$ for the one-dimensional $H_2$ calculated by the FF-MEWP method.

and III contributions. It is presumed for $H_2$ that the type II paths mainly include the ground and low-lying ionic excited states, while the type III paths include higher-lying ionic excited states. Therefore, the internal and external contributions seem to correspond to type II and III processes, respectively. With the increase of the bond distance, the internal two-electron $\gamma$ densities are stretched and divided along the diagonal line $x_2 = -x_1 + 20$, while the external extremum region is not much divided even at the bond distance 4.0 a.u.

This change in the two-electron coordinate system corresponds to the fact that a classical picture of polarization in the internal region rapidly turn from the ionic to the radical structure at small bond distance less than 3.0 a.u., while that in the external region continues to hold a polarization in the ionic structure even at larger bond distance ($\cong$ 4.0 a.u.). This tendency of external region seems to be realized by the virtual excitation paths (type III) including higher-lying excited sates which are more spread in space than lower excited states primarily contributing to the internal region.

The slow development of $\gamma$ at the bond distance less than 3.0 a.u. seems to be caused by a reduction of the type III (positive contribution) by the type II (negative contribution), which are predicted to has the maximum at small bond distance less than 3.0 a.u. since the excited states involved mainly in the virtual excitation paths for seem to be the same as those of type II for $\tilde{\gamma}$.

### 5.3. LiH

As an example of heteropolar molecule, we consider the variation in $\gamma$ of LiH in the bond dissociation. The variations in $\gamma$ by the RCCSD(T) Fig. 11(a) and UCCSD(T) Fig. 11(b) methods, which are found to be in good agreement with that by the FCI (four electron excitation CI in this case), show more complicated increase and decrease behavior, which involves even the change in sign of $\gamma$ in contrast to $H_2$. It is found that the positive $\gamma$ value of LiH in the equilibrium bond distance increases gradually, takes a local maximum at $R = 2.8$ Å, and then decreases until $R = 3.5$ Å, where the $\gamma$ takes the negative minimum value. Beyond $R = 3.5$ Å, the $\gamma$ value is shown to increase again, to take a positive maximum at $R = 4.4$ Å and to decrease toward to the sum of $\gamma$ for independent H and Li atoms.

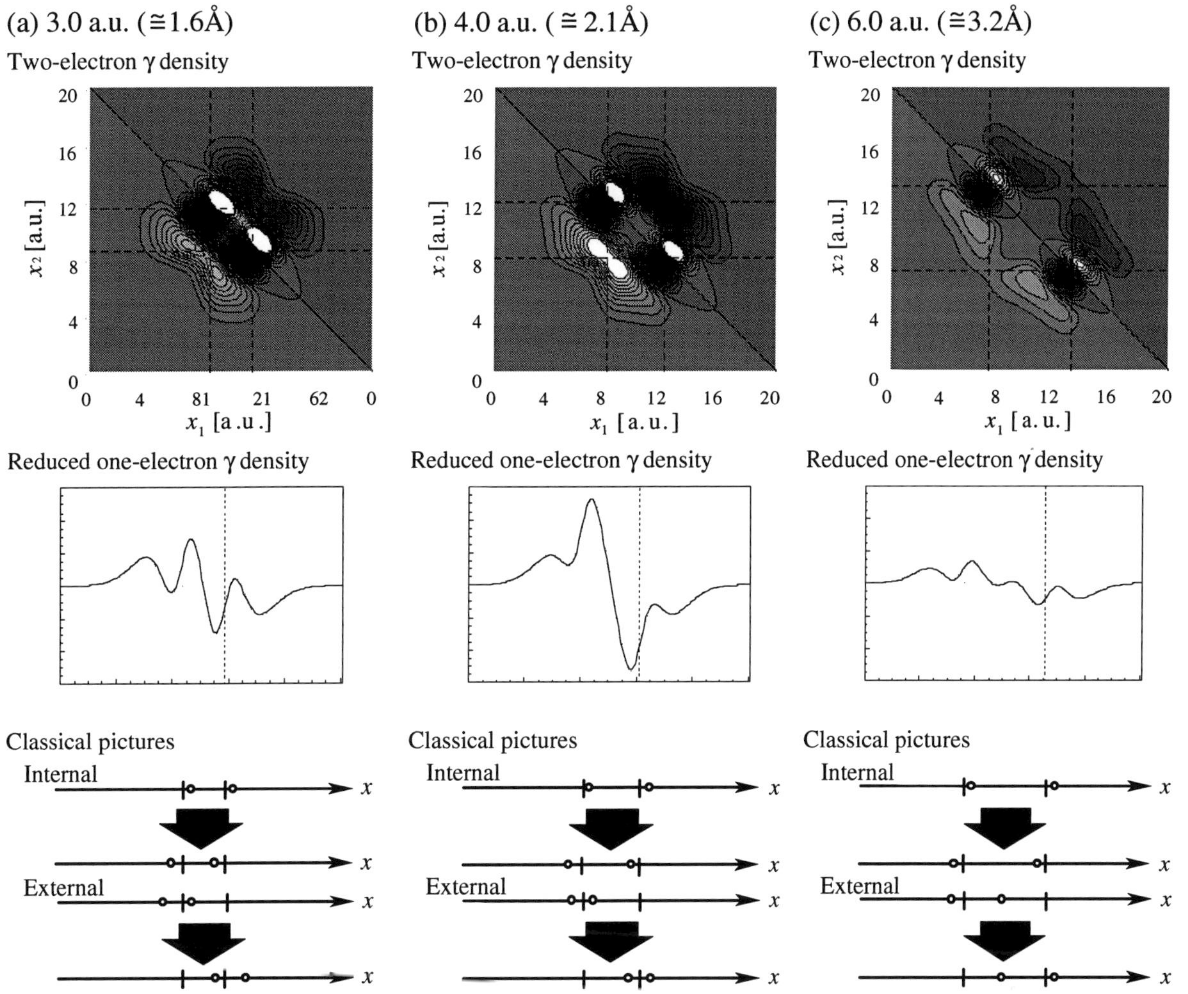


Fig. 10. Two-electron $\gamma$ densities, reduced one-electron $\gamma$ densities and classical pictures of polarization primarily contributing to $\gamma$ for the one-dimensional $H_2$ model at each bond distance: (a) 1.6 a.u., (b) 3.0 a.u., (c) 4.0 a.u. and (d) 6.0 a.u. Contour lines are drawn from −50 to 50 a.u. (contour step : 5 a.u.). Classical pictures corresponding to internal and external two-electron $\gamma$ densities are shown. In classical pictures, ticks and white circles indicate the positions of nuclei and electrons, respectively. Thick arrows represent the displacements of two-electron configurations.

The results by the RHF based methods are shown in Fig. 11(a). All these results show smooth variations in the whole region of bond distance, while the variation feature is in disagreement with that by the UCCSD(T) method Fig. 11(b). The RHF result shows a gradual increase of $\gamma$ (with a local maximum at $R = 3.6$ Å) and the subsequent decrease. It is found that the RMP2 and RMP4 correlation corrections cannot rectify such incorrect variation but enhance the magnitude of variation in $\gamma$. Such incorrect behavior is predicted to originate in the triplet instability of the RHF solution in the intermediate and strong correlation regime. Although the RCCSD(T) results is shown to be in good agreement with the UCCSD(T) one, a slight difference is detected beyond about $R = 4.0$ Å. This indicates that even the correlation correction at the RCCSD(T) level is hard to completely remove the effect of triplet instability of the RHF solution of LiH in the large bond distance region.

Figure 11(b) shows the results by the UHF based methods. All the UHF based results show a cusp

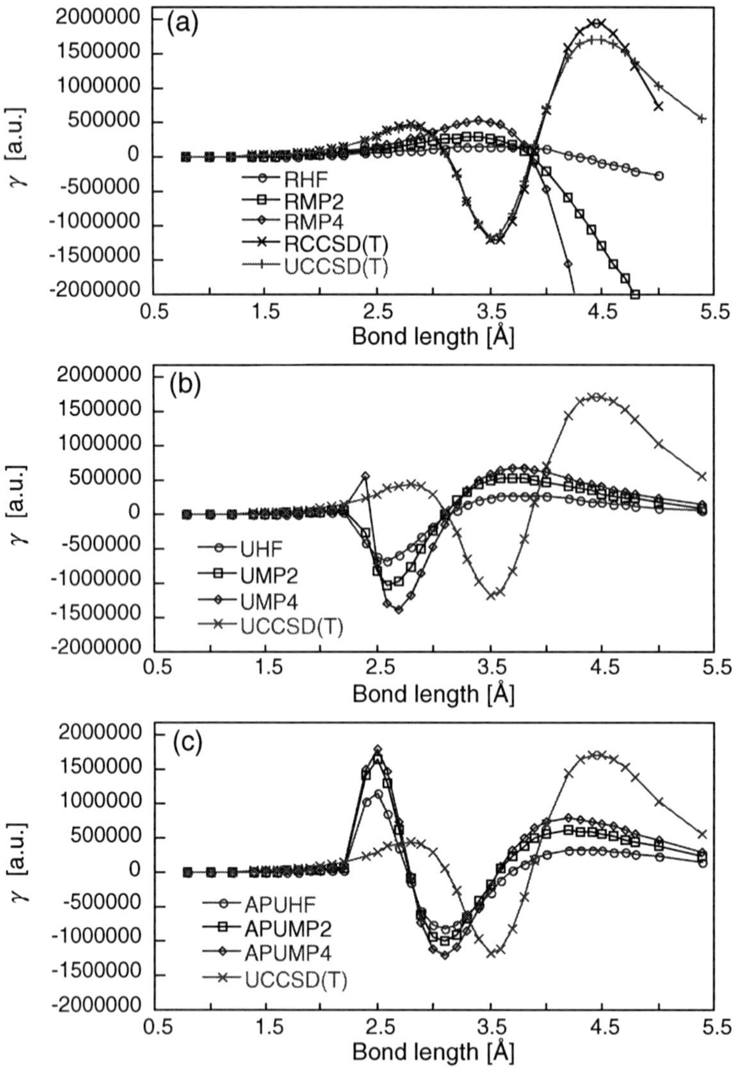


Fig. 11. Variation in $\gamma$ value of LiH by (a) restricted (RHF, RMP2, RMP4, RCCSD(T)), (b) unrestricted (UHF, UMP2, UMP4, UCCSD(T)) and (c) approximate projection (APUHF, APUMP2, APUMP4) methods. The UCCSD(T), which well reproduces the full CI results in this case, is also shown for comparison.

at $R = 2.1$ Å. This feature represents a deficiency in the response feature of the UHF wavefunction of LiH at the triplet instability threshold similarly to the $H_2$ case Fig. 7(b). Beyond this cusp point, the $\gamma$ value by the UHF method is shown to decrease, to take a negative local minimum, and then to increase. It is shown that the change in sign of $\gamma$ from negative to positive occurs at $R = 3.2$ Å and then the $\gamma$ value takes a positive maximum at about $R = 3.8$ Å. It is noted that the number (2) of local extrema of $\gamma$ by the UHF method is less than that Eq. (3) by the UCCSD(T) method. The UMP2 correlation correction shows the enhancement of the first (negative) and the second (positive) peaks of $\gamma$, which are significantly smaller than those by the UCCSD(T) method, respectively. Namely, we can say that these UHF and UMP2 results cannot reproduce the first positive peak and the positions of the remaining peaks (negative and positive peaks) by the UCCSD(T) method. Although the UMP4 correlation correction is shown to give a positive value just after a cusp at $R = 2.4$ Å, the variation of $\gamma$ seems not to be smooth around that region.

The effects of approximate spin-projection (AP) on the UHF and UMP$n$ ($n$ = 2, 4) are shown in Fig. 11(c). The APUHF and APUMP$n$ method is shown to generate the first positive peaks though they are shown to overshoot that by the UCCSD(T) method. It is also found that these AP methods change the second (negative) and the third (positive) peak positions by the UHF and UMP$n$ methods to more approach those by the UCCSD(T) method, respectively, though the relative relation among the magnitudes of these peaks by the AP methods is not found to well reproduce that by the UCCSD(T) method. In order to improve the qualitative features of the APUMP4 results, higher-order dynamical correlation corrections would be required.

We next investigate the variation in $\gamma$ densities located at 0.5 a.u. above the molecular plane, by the UHF, UMP2 and UQCISD methods (Fig. 12). Since the UQCISD method is found to provide a similar behavior to that UCCSD(T) method in the present case, we use the $\gamma$ density by the UQCISD method to analyze the spatial contributions of electrons to $\gamma$ by the UCCSD(T) method. The variations in $\gamma$ densities by the UHF and UMP2 methods are found to be similar to each other. At $R = 2.6$ Å Fig. 12(a), where the $\gamma$ value by the UHF method takes a negative local minimum, the dominant spatial contributions by the UHF and UMP2 methods are found to be caused by the virtual charge transfer (CT) between the H-Li bond region (with negative $\gamma$ density) and the external region of Li with positive $\gamma$ density. As the bond distance increases up to $R = 3.5$ Å (see Fig. 12(a)–(e)), the positive $\gamma$ density around H atom is shown to be increased in the UHF and UMP2 results. For the UHF and UMP2 results at $R = 3.1$ Å Fig. 12(c), the positive contributions between the H atom region (with positive $\gamma$ density) and bond region (with negative $\gamma$ density) are found to cancel out the negative contributions between the bond region and the external region (with positive $\gamma$ density) of Li atom. For the UHF and UMP2 results at $R = 3.5$ Å Fig. 12(e), the magnitude of $\gamma$ density in the external region of Li atom is shown to decrease, so that the contribution to total $\gamma$ by the UHF and UMP2 methods is primarily caused by the positive contribution originating from the virtual CT between the H atom (with positive $\gamma$ density) and bond region (with negative $\gamma$ density). In the large bond distance region ($R = 4.5$ Å shown in Fig. 12(g)) for the UHF and UMP2 results, the magnitude of $\gamma$ densities are found to be decreased in the whole region, the feature of which is the origin of the decrease in total $\gamma$ in the bond dissociation limit by the UHF and UMP2 methods. The variation in $\gamma$ density from $R = 2.6$ Å to 3.1 Å by the UQCISD method (Figs. 12(a)–(c)) are not observed in those by the UHF and UMP2 methods. The variation of $\gamma$ in this region corresponds to the first (positive) peak of $\gamma$ by the UCCSD(T) method shown in Fig. 11(b), and the primary positive contribution is found to come from the virtual CT between the bond region (with positive $\gamma$ density) and the external region of Li atom (with negative $\gamma$ density). For the UQCISD results from $R = 2.8$ Å to 3.1 Å, the increase in the magnitude of negative $\gamma$ density in the H atom region, the increase in the positive $\gamma$ density in the bond region and the decrease in the magnitude of negative $\gamma$ density in the external region of Li atom are observed. This feature causes the cancellation between positive and negative contributions, providing the decrease in the magnitude of $\gamma$ in that region. For the UQCISD result at $R = 3.5$ Å Fig. 12(e), there is shown to be a large negative contribution between the bond region and the external region of Li atom. It is noted that this distribution is similar to that at $R = 2.6$ Å by the UHF and UMP2 methods (see Fig. 12(a)). The feature of $\gamma$ density distribution at $R = 4.5$ Å Fig. 12(g) by the UQCISD method is also shown to be similar to that at $R = 3.5$ Å by the UHF and UMP2 methods Fig. 12(e). These similarities indicate that the first (negative) and second (positive) peaks of $\gamma$ values by the UHF and UMP$n$ methods correspond to the second and the third peaks of $\gamma$ values by the UCCSD(T) method. However, the decrease in the magnitude of $\gamma$ density in the Li atom region at $R = 4.5$ Å Fig. 12(g) by the UQCISD method is shown to be more remarkable than that at $R = 3.5$ Å by the UHF and UMP2 methods. This corresponds to the fact that the last positive peaks at $R = 3.5$ Å by the UHF and UMP2 methods are significantly smaller than that at $R = 4.5$ Å by the UQCISD method. As a result, the UHF and UMP2 methods are suggested to reproduce qualitative $\gamma$ density distributions (spatial contributions of electrons to $\gamma$) in the second and the third peaks of $\gamma$ by the UQCISD method in the relatively strong correlation regime though the peak position and relative magnitudes of peaks by the UHF and UMP$n$ methods are different from those by the UQCISD method. The AP method somewhat improves these two peak position and their peak values, though the improved values are not still far from the UQCISD values, and also generates the first peak, which cannot be generated by the UHF and UMP2 methods. This suggests that the AP method tend to improve the non-dynamical correlation effects involved in the UHF method on $\gamma$ LiH.

## 6. Summary

We investigated in this study the variation in longitudinal $\gamma$ in the bond-dissociation process of small-size molecule, i.e., $H_2$ and LiH, in order to elucidate the feature of $\gamma$ for systems in intermediate and strong correlation regime. From the results by high-order electron correlated *ab initio* MO methods, e.g., UCCSD(T), the magnitude of $\gamma$ value in equilibrium bond distance, i.e., weak correlation, regime tends to be remarkably enhanced in the intermediate and strong correlation regime with labile chemical bonds. For $H_2$ molecule, the $\gamma$ value exhibits only an increase-decrease behavior with a peak, while for LiH molecule three peaks with significant magnitude of $\gamma$ appear and even the change in the sign of $\gamma$ is observed in the intermediate and strong correlation regime. These features suggest that the molecular systems with chemical bonds in the intermediate and strong correlation regime are candidates for novel third-order NLO systems, in which the magnitudes of $\gamma$ values can be significantly enhanced and/or their signs are changed by slight chemical modifications, e.g., change in the metal spieses and ligands in transition-metal compounds.

We also examined the applicability of various *ab initio* MO methods to $\gamma$ for molecules in the intermediate and strong correlation regime. The RHF and RMP$n$ methods are found to fail in giving qualitative variation in $\gamma$ in the bond-dissociation region due to the triplet instability of the RHF solution in that region. Although the RCCSD(T) method can remarkably improve the variation in $\gamma$, the variation shape of $\gamma$ is shown to somewhat deviate from the FCI result of LiH in the strong correlation regime. In contrast, the unrestricted CC methods, e.g., UCCSD(T), are found to be used for at least semi-quantitative study on the $\gamma$ for small-size systems. In contrast, the UHF and UMP$n$ ($n = 2, 4$) methods are found to fail in reproducing qualitative shape of variation in $\gamma$ in the whole bond-dissociation region though the $\gamma$ density distribution in the strong correlation regime is similar to that by the UQCISD method. The approximate spin-projected MP$n$ (APUMP$n$) methods tend to somewhat improve the variation in $\gamma$ but the correction at the APUMP4 level is predicted to be still insufficient for getting the qualitative variation in $\gamma$ judging from LiH results. Also from our previous studies [54], the CASSCF and CASSCF PT2 methods without extended active space seem to fail in providing convergent $\gamma$ in the present case. Probably, the multi-reference CC (MRCC) with relatively small active space or CCSDT and CCSDTQ methods based on the UHF solution will be necessary for obtaining reliable results of $\gamma$ for systems in the whole region from weak to strong correlation. For larger systems, however, the MRCC method is hard to apply to attractive NLO systems with extended $\pi$ conjugation. Recently, the DF method based on the UHF natural orbital (UNO) CASSCF, which is referred to as the CASDFT, has been developed [67]. This method has the advantage of effectively involving non-dynamical and dynamical correlation effects

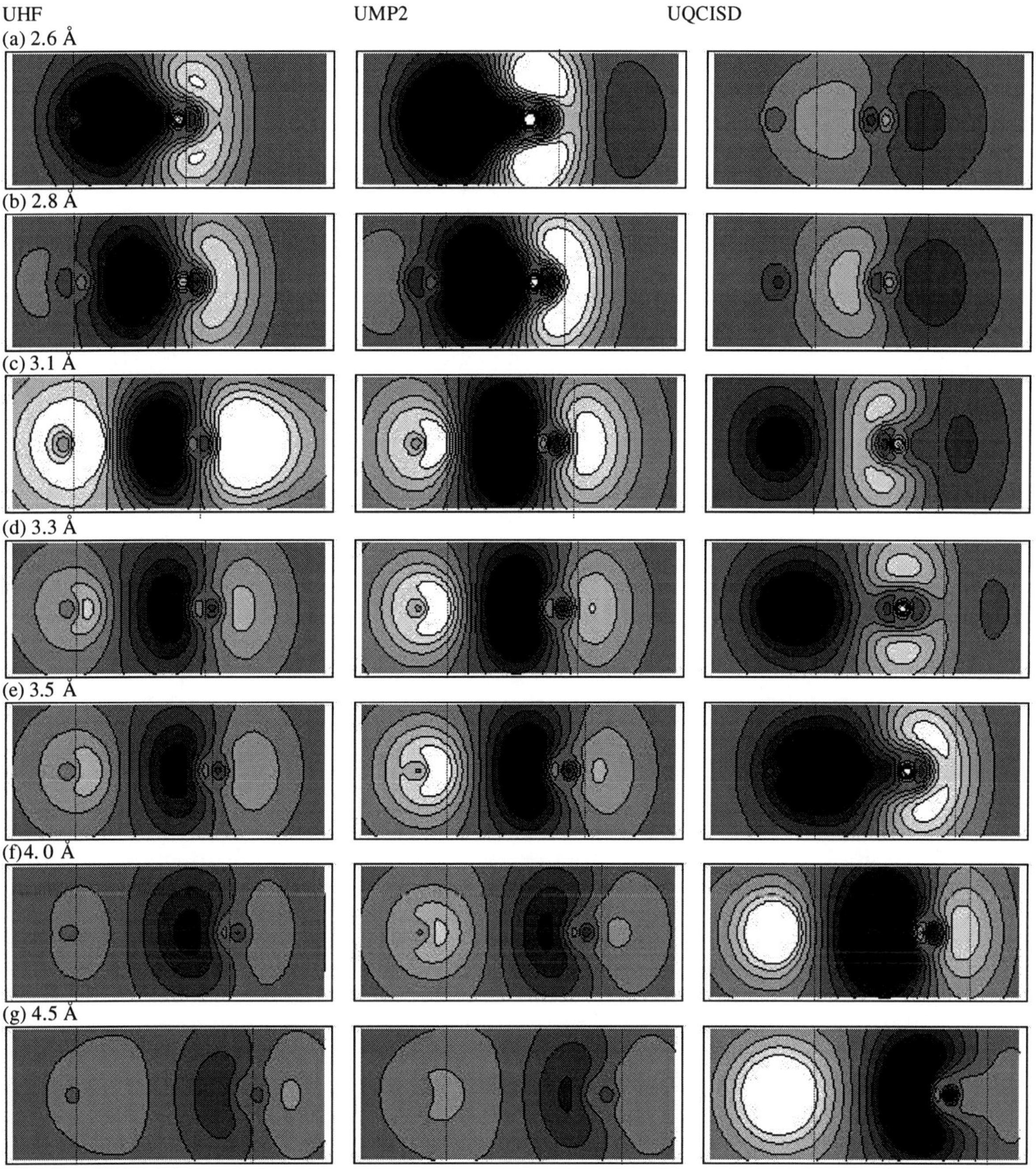


Fig. 12. The $\gamma$ density plots of LiH (see Fig. 2 for the coordinate) at (a) $R = 2.6$ Å, (b) $R = 2.8$ Å, (c) $R = 3.1$ Å, (d) $R = 3.3$ Å, (c) $R = 3.5$ Å, (f) $R = 4.0$ Å and (g) $R = 4.5$ Å calculated by the UHF, UMP2 and UQCISD method. The $\gamma$ densities are plotted on a plane located 0.5 a.u. above the molecular $(z - x)$ plane. The contours are drawn from $-10000$ a.u. to 10000 a.u. (contour step: 2000 a.u.). The positions of H (left-side) and Li (right-side) atoms are shown by the dotted lines.

using much smaller computational resources than the MRCC. The FF method using such CASDFT is expected to provide reliable $\gamma$ values for larger systems in the intermediate and strong correlation regime.

## Acknowledgements

This work was supported by Grant-in-Aid for Scientific Research (No. 14340184) from Japan Society for the Promotion of Science (JSPS).

## References

[1] D. J. Williams, ed., *Nonlinear Optical Properties of Organic and Polymeric Materials*; ACS Symposium Series 233; American Chemical Society: Washington, D.C., 1984.
[2] P.N. Butcher and D. Cotter, *The Elements of Nonlinear Optics*, Cambridge University Press, Cambridge, 1990.
[3] N.P. Prasad and D.J. Williams, *Introduction to Nonlinear Optical Effects in Molecules and Polymers*, Wiley: New York, 1991.
[4] (a) D.R. Kanis, M.A. Ratner and T.J. Marks, *Chemical Reviews* **94** (1994), 195; (b) J.L. Brédas, C. Adant, P. Tackx and A. Persoons, *Chemical Reviews* **94** (1994), 243.
[5] (a) M. Nakano and K. Yamaguchi, *Analysis of nonlinear optical processes for molecular systems*, in Trends in Chemical Physics Vol. 5, pp. 87–237, Research trends, Trivandrum, India, 1997. (b) A. Nakamura, N. Ueyama and K. Yamaguchi, eds, *Organometallic Conjugation, Structures, Reactions and Functions of d-d and d-$\pi$ Conjugated Systems*, KODANSHA, Springer, Tokyo, 2002. (c) M. Nakano and K. Yamaguchi, *Polarizabilities and hyperpolarizabilities of dendritic systems*, in: *Advances in Multi-Photon Processes and Spectroscopy*, (Vol. 15), World Scientific, Singapore, 2003, pp. 1–146.
[6] H.S. Nalwa and S. Miyata, eds., *Nonlinear Optics of Organic Molecules and Polymers*, CRC Press, Boca Raton, 1997.
[7] B. Champagne and D.M. Bishop, *Calculations of nonlinear optical properties of solid state, in Advances in Chemical Physics*, **126** (2003), 41.
[8] (a) M. Nakano, M. Okumura, K. Yamaguchi and T. Fueno, *Mol. Cryst. Liq. Cryst.* **182A** (1990), 1; (b) M. Nakano and K. Yamaguchi, *Chem. Phys. Lett.* **206** (1993), 285.
[9] M. Nakano, K. Yamaguchi and T. Fueno, *Nonlinear Optics.* **6** (1993), 289.
[10] M. Nakano, I. Shigemoto, S. Yamada and K. Yamaguchi, *J. Chem. Phys.* **103** (1995), 4175.
[11] I. Shigemoto, M. Nakano, S. Yamada, S. Kiribayashi and K. Yamaguchi, *Mol. Cryst. Liq. Cryst.* **286** (1996), 159.
[12] S. Yamada, M. Nakano, I. Shigemoto and K. Yamaguchi, *Chem. Phys. Lett.* **254** (1996), 158.
[13] M. Nakano, S. Kiribayashi, S. Yamada, I. Shigemoto and K. Yamaguchi, *Chem. Phys. Lett.* **262** (1996), 66.
[14] S. Yamada, M. Nakano, I. Shigemoto, S. Kiribayashi and K. Yamaguchi, *Chem. Phys. Lett.* **267** (1997), 438.
[15] S. Yamada, M. Nakano and K. Yamaguchi, *Chem. Phys. Lett* **276** (1997), 375.
[16] M. Nakano, S. Yamada and K. Yamaguchi, *Int. J. Quant. Chem.* **70** (1998), 269.
[17] M. Nakano, S. Yamada and K. Yamaguchi, *Bull. Chem. Soc. Jpn.* **71** (1998), 845.
[18] M. Nakano, S. Yamada and K. Yamaguchi, *J. Phys. Chem. A* **103** (1999), 3103.
[19] M. Nakano, S. Yamada and K. Yamaguchi, *Chem. Phys. Lett.* **306** (1999), 187.
[20] I. Shigemoto, M. Nakano, S. Yamada and K. Yamaguchi, *Synthetic Metals* **102** (1999), 1562.
[21] S. Yamada, M. Nakano and K. Yamaguchi, *J. Phys. Chem. A* **103** (1999), 7105.
[22] M. Nakano, S. Yamada and K. Yamaguchi, *Chem. Phys. Lett.* **321** (2000), 491.
[23] H. Fujita, M. Nakano, M. Takahata and K. Yamaguchi, *Chem. Phys. Lett.* **358** (2002), 435.
[24] H. Kisida, H. Matsuzaki, H. Okamoto, T. Manabe, M. Yamashita, Y. Taguchi and Y. Tokura, *Nature* **405** (2000), 929.
[25] (a) D.M. Bishop and L.M. Cheung, *Phys. Rev. A* **20** (1979), 1310; (b) D.M. Bishop, *J. Pipin, Phys. Rev. A* **36**(1987), 2171; (c) D.M. Bishop and B. Lam, *J. Chem. Phys.* **89** (1988), 1571; (d) D.M. Bishop, J. Pipin and M. Rérat, *J. Chem. Phys.* **92** (1990), 1902; (e) D.M. Bishop, J. Pipin and S.M. Cybulski, *Phys. Rev. A* **43** (1991), 4845.
[26] (a) G. Maroulis and A.J. Thakkar, *J. Chem. Phys.* **90** (1989), 366; (b) G. Maroulis, *J. Chem. Phys.* **101** (1994), 94, 4949; (c) G. Maroulis, *J. Chem. Phys.* **111** (1999), 6846; (d) G. Maroulis, *J. Chem. Phys.* **118** (2003), 2673.
[27] M. Rérat, M. Mérawa and C. Pouchan, *Phys. Rev. A* **46** (1992), 5471.
[28] M.K. Harbola and A. Banerjee, *Phys. Rev. A* **54** (1996), 283.
[29] T. Voegel, J. Hinze and F. Tobin, *J. Chem. Phys.* **70** (1979), 1107.
[30] C.A. Nicolaides, Th. Mercouris and G. Aspromallis, *J. Opt. Soc. Am. B* **7** (1990), 494.
[31] S.I. Themelis and C.A. Nicolaides, *Phys. Rev. A* **46** (1992), R21.
[32] N. E.-B. Kassimi and A.J. Thakkar, *Phys. Rev. A* **50** (1994), 2948.
[33] M. Nakano, H. Nagao and K. Yamaguchi, *Phys. Rev. A* **55** (1997), 1503.
[34] H.D. Cohen and C.C.J. Roothaan, *J. Chem. Phys.* **S34** (1965), 43.
[35] S. Yamanaka, M. Okumura, M. Nakano and K. Yamaguchi, *J. Mol. Structure* **310** (1994), 205.

[36] (a) M. Nakano, K. Yamaguchi and T. Fueno, *Chem. Phys. Lett.* **185** (1991) , 550; (b) M. Nakano, S. Yamada, I. Shigemoto and K. Yamaguchi, *Chem. Phys. Lett.* **250** (1996), 247; (c) M. Nakano, H. Fujita, M. Takahata and K. Yamaguchi, *Chem. Phys. Lett.* **356** (2002), 462.
[37] H. Nagao, M. Nakano, S. Yamanaka, Y. Shigeta, S. Yamada, D. Yamaki, I. Shigemoto, S. Kiribayashi and K. Yamaguchi, *Int. J. Quantum Chem.* **60** (1996), 79.
[38] Y.R. Shen, *The Principles of Nonlinear Optics*, Wiley, New York, 1984.
[39] C.P. de Melo and R. Silbey, *J. Chem. Phys.* **88**, 2567.
[40] K. Yamaguchi, Y. Takahara, T. Fueno and K.N. Houk, *Theor. Chim. Acta* **73** (1988), 337.
[41] K. Yamaguchi, R. Carbo and M. Klobukowski, eds., *Self-Consistent Field. Theory and Applications*, Elsevier, Amsterdam, 1990.
[42] K. Takatsuka, K. Yamaguchi and T. Fueno, *Theor. Chim. Acta* **48** (1978), 175.
[43] J. Cizek and J. Paldus, *J. Chem. Phys.* **47** (1967), 3976.
[44] D.J. Thouless, *The Quantum Mechanics of Many-Body Systems*, Academic Press, New York, 1961.
[45] M.-A. Ozaki, *J. Math. Phys.* **26** (1985), 1521.
[46] J.M. Bofill and P. Pulay, *J. Chem. Phys.* **90** (1989), 3657.
[47] K. Yamaguchi, M. Okumura, W. Mori, J. Maki, K. Takada, T. Noro and K. Tanaka, *Chem. Phys. Lett.* **210** (1993), 201.
[48] K. Yamaguchi, T. Tsunekawa, Y. Toyoda and T. Fueno, *Che. Phys. Lett.* **143** (1988), 371.
[49] K. Yamaguchi, M. Okumura, J. Maki, T. Noro, H. Namimoto, M. Nakano, T. Fueno and K. Nakasuji, *Chem. Phys. Lett.* **190** (1992), 353.
[50] M. Okumura, S. Yamanaka, W. Mori and K. Yamaguchi, *J. Mol. Structure* **310** (1994), 177.
[51] (a) K. Yamaguchi, *Int. J. Quantum Chem. Symp.* **14** (1980), 269; (b) K. Yamaguchi, *Chem. Phys. Lett.* **33** (1975), 33; (c) *ibid.* **35** (1975), 230.
[52] B.O. Roos, P. Linse, P.E.M. Siegbahn and M.R.A. Blomberg, *Chem. Phys.* **66** (1982), 197.
[53] K. Andersson, P.-A. Malmqvist and B.O. Roos, *J. Chem. Phys.* **96** (1992), 1218.
[54] S. Yamada, M. Nakano and K. Yamaguchi, *Int. J. Quant. Chem.* **71** (1999), 329.
[55] D. Jacquemin, D. Beljonne, B. Champagne, V. Geskin, J.L. Bredas and J.M. Andre, *J. Chem. Phys.* **115** (2001), 6766.
[56] (a) B. Kirtman and M. Hasan, *Chem. Phys. Lett.* **157** (1989) , 13; (b) D.R. Kanis, M.A. Ratner and T.J. Marks, *J. Am. Chem. Soc.* **112** (1990), 8203; (c) D.R. Kanis, M.A. Ratner, T.J. Marks and M.C. Zerner, *Chem. Mater.* **3** (1991), 19; (d) D.R. Kanis, M.A. Ratner and T.J. Marks, *J. Am. Chem. Soc.* **114** (1992), 10338; (e) D.R. Kanis, M.A. Ratner and T.J. Marks, *Int. J. Quant. Chem.* **43** (1992), 61.
[57] G.J.B. Hurst, M. Dupuis and E. Clementi, *J. Chem. Phys.* **89** (1988), 385.
[58] (a) G. Maroulis and A.J. Thakkar, *J. Chem. Phys.* **88** (1988), 7623; *ibid.* **93** (1990), 652; *ibid.* **93** (1990), 4164; (b) D.M. Bishop and G. Maroulis, *J. Chem. Phys.* **82** (1985), 2380; (c) G. Maroulis, *J. Chem. Phys.* **94** (1991), 1182; (d) H. Sekino and R. Bartlett, *J. Chem. Phys.* **94** (1991), 3665; (e) T. Kobayashi, K. Sasagane, F. Aiga and K. Yamaguchi, *J. Chem. Phys* **110** (1999), 11720; (f) H. Larsen, J. Olsen, C. Hättig, P. Jørgensen, O. Christiansen and J. Gauss, *J. Chem. Phys.* **111** (1999), 1917.
[59] N. Matsuzawa and D.A. Dixon, *J. Phys. Chem.* **98** (1994), 2545.
[60] P. Calaminici, K. Jug and A.M. Köester, *J. Chem. Phys.* **109** (1998), 7756.
[61] (a) R. Feynman, *Phys. Rev.* **56** (1939), 340; (b) S.T. Epstein, *Am. J. Phys.* **22** (1954), 613.
[62] (a) J. Javanainen, J.H. Eberly and Q. Su, *Phys. Rev. A* **38** (1988), 3430; (b) Q. Su and J.H. Eberly, *ibid.* **44** (1991), 5997.
[63] R. Kosloff and H. Tal-Ezer, *Chem. Phys. Lett.* **127** (1986), 223.
[64] P. Chopra, L. Carlacci, H.F. King and P.N. Prasad, *J. Phys. Chem.* **93** (1989), 7120.
[65] M.J. Frisch, G.W. Trucks, H.B. Schlegel, G.E. Scuseria, M.A. Robb, J.R. Cheeseman, V.G. Zakrzewski, J.A. Montgomery, Jr., R.E. Stratmann, J.C. Burant, S. Dapprich, J.M. Millam, A.D. Daniels, K.N. Kudin, M.C. Strain, O. Farkas, J. Tomasi, V. Barone, M. Cossi, R. Cammi, B. Mennucci, C. Pomelli, C. Adamo, S. Clifford, J. Ochterski, G.A. Petersson, P.Y. Ayala, Q. Cui, K. Morokuma, P. Salvador, J.J. Dannenberg, D.K. Malick, A.D. Rabuck, K. Raghavachari, J.B. Foresman, J. Cioslowski, J.V. Ortiz, A.G. Baboul, B.B. Stefanov, G. Liu, A. Liashenko, P. Piskorz, I. Komaromi, R. Gomperts, R.L. Martin, D.J. Fox, T. Keith, M.A. Al-Laham, C.Y. Peng, A. Nanayakkara, M. Challacombe, P.M.W. Gill, B. Johnson, W. Chen, M.W. Wong, J.L. Andres, C. Gonzalez, M. Head-Gordon, E.S. Replogle and J.A. Pople, *Gaussian 98*, Gaussian, Inc., Pittsburgh PA, 2001.
[66] (a) S.J.A. van Gisbergen, P.R.T. Schipper, O.V. Gritsenko, E.J. Baerends, J.G. Snijders, B. Champagne and B. Kirtman, *Phys. Rev. Lett.* **84** (1999), 694; (b) B. Champagne, E.A. Perpéte, S.J.A. van Gisbergen, E. Baerends, J.G. Snijders, C. Soubra-Ghaoui, K.A. Robins and B. Kirtman, *J. Chem. Phys.* **108** (1999), 10489.
[67] R. Takeda, S. Yamanaka and K. Yamaguchi, *Chem. Phys. Lett.* **366** (2002), 321.

Journal of Computational Methods in Sciences and Engineering 4 (2004) 703–720
IOS Press


# Static polarizability of molecular materials: Environmental and vibrational contributions

Francesca Terenziani[a], Anna Painelli[a,*] and Zoltan G. Soos[b]
[a]*Dipartimento di Chimica GIAF Università di Parma, 43100 Parma, INSTM UdR Parma, Italy*
[b]*Department of Chemistry, Princeton University, Princeton, NJ 08544, USA*

**Abstract.** Modeling the dielectric behavior of molecular materials made up of large $\pi$-conjugated molecules is an interesting and complex task. Here we address linear polarizabilities, and the related dielectric constant, of molecular crystals and aggregates made up of closed-shell $\pi$-conjugated molecules with either a non-polar or largely polar ground-state, and also examine the behavior of mixed-valence (or charge-transfer) organic salts. We recognize important collective phenomena due to supramolecular interactions in materials with large molecular polarizabilities, and underline large vibrational contributions to the polarizability in materials with largely delocalized electrons.

Keywords: Static polarizability and dielectric constant, electron-vibration coupling, molecular materials, charge-transfer organic salts

## 1. Introduction

A static electric field $F$ applied to an insulating material induces a dipole, $\mu_{\text{ind}} = \alpha F + \beta F^2/2 + \gamma F^3/6 + \ldots$, whose magnitude depends on the polarizability, $\alpha$, and hyperpolarizabilities, $\beta$, $\gamma\ldots$, of the system. Large (hyper-)polarizabilities imply easy redistribution of electronic or nuclear charge by the field. For isolated molecules in the gas phase, the formal theory of static polarizabilities and hyperpolarizabilities is well known, and accurate results can be obtained from quantum chemical calculations on simple molecules [1]. For complex molecular structures the calculation becomes challenging, particularly in large $\pi-$conjugated molecules, where important contributions to (hyper-)polarizabilities are expected from molecular vibrations [2]. For molecules in solution the polarization of the solvent screens the applied fields and the calculation of (hyper-)polarizabilities becomes fairly involved [3].

In the solid state, and particularly in crystals, the definition of the dipole moment itself is challenging, and a formal theory for the polarization, $P$, the dipole moment for unit cell, was developed only in the last decade [4]. The problem is that the polarization of the material apparently depends on the choice of the unit cell [5]. This problem finds a natural solution in the so-called Berry-phase formulation of $P$, that relates the macroscopic polarization of the crystal to the ground state (gs) wavefunction rather than to the gs charge-density [6–8,4]. A key concept in the modern theory of polarization is that $P$ is by itself ill-defined and not accessible experimentally: only *variations* of $P$ can in fact be defined as independent of boundary-conditions, and are experimentally accessible [8]. The Berry-phase formulation of polarizability in insulating materials constitutes an important advance in understanding the electrical

*Corresponding author. E-mail: anna.painelli@unipr.it.

1472-7978/04/$17.00 

properties of crystalline materials, but does not solve the problem of the definition of a dipole moment operator in systems with periodic boundary conditions [9,10]. A complete and coherent picture for the polarizability (and hyperpolarizability) of crystalline materials is still lacking.

The Berry-phase definition of $P$ is particularly important in covalent insulators, i.e. in crystals such as oxides or silicon, where the charge density is sizeable even between atoms. In organic molecular crystals, by contrast, the overlap between electronic wavefunctions on different molecules is small and there is an unambiguous natural way of partitioning the crystal into an assembly of neutral, localized unit cells (the so-called Clausius-Mossotti limit) [8,5]. In the zero overlap approximation for charge-densities on different molecules, the dipole moment operator for the crystal is simply the sum of the dipole moments of the basic unit cells, and standard approaches to linear and non-linear polarizabilities apply: the formal problem of calculating the crystal polarization and the relevant (hyper)polarizability is trivially solved in this limit. The actual calculation of the electrical responses of molecular crystals is difficult, however, particularly for crystals of highly polarizable molecules. In these materials in fact important collective effects are expected [11] and, even if the dipole moment can be calculated as the sum of local dipole moments, the polarizability and the hyperpolarizabilities are not the sum of local contributions.

In this paper we address the calculation of the linear polarizability $\alpha$ in crystalline materials, as related to the experimentally accessible static dielectric constant. Sections 2 and 3 are devoted to the analysis of organic molecular crystals with little or no intermolecular overlap. In these insulators the zero-overlap approximation for charge densities allows to adopt standard expressions for the polarizability tensor, whose Cartesian components $\alpha_{i,j}$ measure the curvature of the gs energy $E_0(F)$ with respect to the field:

$$\alpha_{ij} = -\left(\frac{\partial^2 E_0(F)}{\partial F_i \partial F_j}\right)_0 = 2{\sum_R}' \frac{\langle G|\mu_i|R\rangle\langle R|\mu_j|G\rangle}{E_R} \tag{1}$$

The sum-over-state (SOS) expression is the second order correction. It involves the excited states $R$ at energy $E_R$ above the gs at $F = 0$ and the dipole operator $\mu$. The first expression for $\alpha$ above is the finite-field (FF) result, that requires only the gs, albeit at finite fields. We emphasize the long-range nature of electrostatic interactions that lead to significant perturbations to $\alpha$ as due to the environment even in the absence of direct intermolecular overlap. Section 2 reviews a recent approach to the detailed calculation of $\alpha$ in non-polar chromophores with an extended and strongly polarizable $\pi$-system. This approach, based on the FF calculation of $\alpha$ for molecules experiencing the electric field generated by the surrounding molecules, relies on quantum chemical models for the isolated molecular fragments and leads to quantitative estimates of the dielectric constant for crystals and films that nicely compare with experimental data.

Section 3 presents an instructive toy model for clusters of polar and polarizable chromophores, based on a two-state picture for each chromophore. Whereas this simple model is hardly quantitative, it correctly grasps the basic physics of supramolecular interactions in these materials. Very schematically, Fig. 1 depicts a polar chromophore in an environment of polar molecules that can be either an ordered crystal or a solution. Electrostatic interactions among dipoles and induced dipoles require a self-consistent treatment that accounts for large local fields due to the dipoles and induced dipoles themselves. Huge collective effects are possible in systems of polar-polarizable molecules, and are indeed expected for certain ranges of parameters. We note that the dipoles in Fig. 1 may represent both electronic and nuclear degrees of freedom or more complicated charge distributions. Quite generally, the zero-overlap approximation for molecular aggregates or crystals leads to such electrostatic problems.

Models with a limited number of states are then particularly instructive for assessing the inevitable approximations in molecular or extended systems. The Born-Oppenheimer (BO) separation of electronic

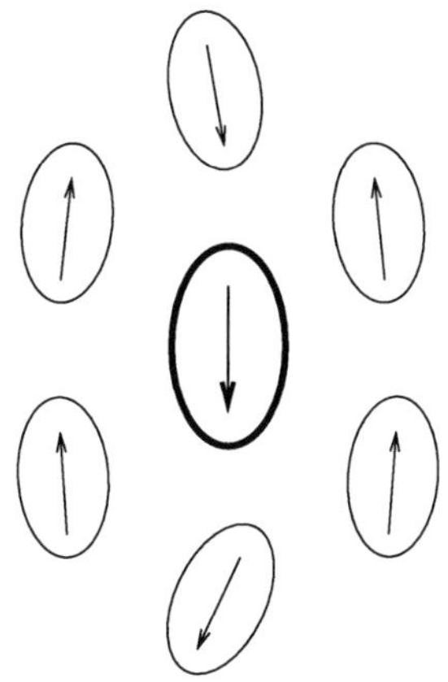

Fig. 1. Schematic view of dipolar molecule interacting with the surroundings, not drawn to minimize energy. The arrows represent both dipoles and induced dipoles.

and nuclear degrees of freedom is ubiquitous for molecules and generates potential energy surfaces (PES). In Section 4 we will assess the BO approximation for $\alpha$ in models whose exact eigenstates are accessible, and will present a very simple expression for the vibrational contributions to $\alpha$ that applies in the limit of low vibrational frequencies.

Based on these results, in Section 5 we briefly address the role of vibrational degrees of freedom in charge-transfer (CT) salts with a mixed (donor acceptor, DA) stack motif and variable ionicity. A first description of these materials was given in terms of non-overlapping DA pairs [12], so that collective effects in $\alpha$ are easily predicted. However, any realistic model for CT salts must account for charge delocalization along the stack. Since the Clausius-Mossotti approximation no longer applies, we resort to Berry-phase polarization and consider $P$ at instabilities. We relate the peak in the dielectric constant observed in some CT salts at a Peierls instability to large charge-fluxes induced along the stack by lattice vibrations.

## 2. Electronic polarization in organic molecular crystals and thin films

The polarization energy of a charge $q$ in a cavity of radius $a$ in medium with dielectric constant $\kappa$ is $q^2(1 - 1/\kappa)/2a$, or $\sim$1 eV for typical $a \sim 5$ Å and $\kappa \sim 3$ in organic molecular crystals [13]. The major role of polarization has long been appreciated in organic semiconductors [14,15]. The transport gap $E_t$ for creating an electron-hole pair at infinite separation is reduced from the gas phase value by $P = P_+ + P_-$, the separate stabilization of the cation and anion. The major ($\sim$90%) part of $P$ is thought to be electronic [15], with lattice relaxation or polarons accounting for the rest, and our discussion of $P$ in this Section is restricted to electronic polarization.

We ask how the charge density $\rho(r)$ in the solid differs from the gas-phase density $\rho_G(r)$ of molecules at the same nuclear positions. The crystal induces the difference, $\Delta\rho(r) = \rho(r) - \rho_G(r)$. The lowest-order correction can again be viewed as polarization and rationalized by small overlaps in organic crystals. Organic molecular solids typically have van der Waals contacts and slightly shifted electronic or vibrational excitations, as suggested by the oriented gas model that is the starting point of molecular exciton theory. First-order corrections to energies clearly depend only on $\rho_G(r)$, which can be inferred from gas-phase experiment or theory. Electronic polarization is the first-order correction to $\rho_G(r)$, or to the wavefunction, and is accordingly more difficult to compute.

Small overlap at van der Waals separations suggests starting with zero intermolecular overlap and introducing finite band widths later. In the zero-overlap approximation, charge redistribution $\Delta\rho(r)$ is clearly confined to molecules and intermolecular interactions are purely electrostatic [16]. In this sense, the specific model of polar-polarizable molecules introduced in the next Section represents any aggregate. Conceptually, we have a crystal of quantum mechanical molecules with classical interactions that may be viewed as a non-uniform electric field $F(r)$. Since induced dipoles contribute to $F(r)$, a self-consistent solution is required and made practical by translational symmetry. For example, consider a crystal in a uniform $F_{\text{ext}}$. Since all unit cells have the same total field, we apply constant $F$ and resolve $F$ into local and external fields at the end [17]. The calculation of $P_+$ or $P_-$ , the polarization energy of an isolated charge, is done in two steps [16]. First, the *extensive* quantity $\Delta\rho_N(r)$ is found for the neutral lattice using translational invariance. Then the *intensive* quantity $\Delta\rho_\pm(r, R)$ is found relative to $\Delta\rho_N(r)$ in spheres of radius $R$ about the charge without redistributing charge on molecules whose center is outside the sphere. Convergence to $P_\pm$ rigorously goes as $1/R$ at large $R$ with a slope that depends on the dielectric tensor $\kappa$. Representative organic crystals show proper convergence by $R \sim 100$ Å. This corresponds to clusters of thousands of molecules and underscores the long-range nature of polarization. The polarization energy of CT states also requires two steps, again starting with $\Delta\rho_N(r)$. The sphere $R$ now encloses a cation and anion at fixed separation in the crystal lattice. The polarization energy converges faster, as $1/R^3$, since a dipole is enclosed.

The zero-overlap approximation reduces electronic polarization to finding $\alpha$ for molecules in a non-uniform field $F(r)$. This major simplification yields a continuum problem for the functional derivative, $\partial\rho(r)/\partial V(r')$, of the gs charge density at $r$ with respect to the potential at $r'$. To be practical for crystals or thin films of large molecules, we introduce a discrete approximation for $\partial\rho/\partial V$ by restricting $r$ and $r'$ at molecule $a$ to the atomic positions $r_i^a$ and evaluating $V(r)$ and $F = -\nabla V$ at $r_i^a$. As done routinely in electrostatic (Madelung) calculations [18], we represent the gas phase $\rho_G(r)$ by atomic charges $\rho_i^{a(0)}$ at $r_i^a$. Electronic polarization then yields new atomic charges and induced atomic dipoles that are given by the following linear equations [16],

$$\rho_i^a = \rho_i^{a(0)} - \sum_j \Pi_{ij}^a \phi_j^a \tag{2}$$

$$\mu_i^a = \mu_i^{a(0)} - \tilde{\alpha}_i^a F_i^a \tag{3}$$

Here the sum is over atoms $j$ of molecule $a$, the electrostatic potential is $\phi_i^a = V(r_i^a)$, and $\Pi$ is the atom-atom polarizability tensor,

$$\Pi_{ij} = -\left(\frac{\partial\rho_i}{\partial\phi_j}\right)_0 = -\left(\frac{\partial^2 E}{\partial\phi_i\partial\phi_j}\right)_0 \tag{4}$$

that governs how charge redistributes within a molecule. The dipoles $\mu_i^a$ contribute to the potential and change according to $F(r_i^a)$. The approximations of zero-overlap and discrete atomic moments lead to eight linear equations per atom, namely the scalars $\rho_i^a$, $\phi_i^a$ and the vectors $\mu_i^a$, $F_i^a$. They are solved iteratively starting with the oriented-gas potential $\phi_i^{a(0)}$ produced by $\rho_i^{a(0)}$. Clusters of thousands of molecules with about 50 atoms each are accessible on workstations [16,19].

We comment on general aspects of charge redistribution without going into computational details that are found in refs. 16 and 19. First, semiempirical theory such as INDO/S [20] is particularly convenient for $\Pi_{ij}$ since $\phi_i^a$ is then simply a site energy; this approximation, typical in solid-state models, is neither

Table 1
Principal components of the dielectric tensor and indices of refraction, $\kappa = n^2$, of crystalline anthracene. The calculated $\kappa$ are based on the indicated molecular $\alpha$. The unique axis is $\kappa_{bb}$; $\kappa_{11}$, $\kappa_{33}$ are in the $ac$ plane and $\theta$ is the angle between $\kappa_{11}$ and $a$

| Molecular Inputs | $\kappa_{11}$ | $\kappa_{bb}$ | $\kappa_{33}$ | $\theta$ |
|---|---|---|---|---|
| Expt. (Ref. [23]) | 2.49(10) | 3.07(10) | 4.04(20) | 28(2)° |
| TZVP-FIP (Ref. [24]) | 2.35 | 3.09 | 4.31 | 31.8° |
| B3LYP/6-311++G(d,p) | 2.23 | 2.91 | 4.03 | 31.6° |
| Expt. gas-phase $\alpha$ | 2.69 | 3.14 | 3.39 | 31.6° |
| Oriented gas | 1.36 | 2.39 | 3.90 | |

the finite-field nor the SOS method of Eq. (1), but can be compared to them in exact models. Second, the best gas-phase polarizability $\alpha$ from experiment or theory is retained and partitioned between charge redistribution and a remainder, the $\tilde{\alpha}_i^a$ above. An intuitive partitioning of $\alpha$ in acenes, as done in the submolecular method [21,15], leads to self-consistent equations for induced dipoles in external fields. Similarly, the gs dipole $\mu$ of polar molecules can be partitioned between charge redistribution and the $\mu_i^{a(0)}$ above. In the spirit of molecular exciton theory, we focus on how $\Delta\rho(r)$ changes in the actual crystal lattice for molecules whose electronic structure is given by hypothesis. Third, polarization energies are bilinear expressions [16] in self-consistent atomic charges and induced dipoles coupled to gas-phase potentials or, vice versa, self-consistent potentials and fields coupled to gas-phase charges or dipoles. Since potentials and fields produced by the best gas-phase charge distribution $\rho_G(r)$ can be evaluated at atomic positions of the crystal lattice, first-order corrections to polarization energies based on discrete $\rho_i^{a(0)}$ can readily be found [22]. Such corrections are important in acenes or other systems with electron-hole symmetry and hence $\rho_i^{a(0)} \sim 0$ in the neutral molecule.

The optical dielectric tensor, or indices of refraction, of anthracene crystals has been measured [23] separately and taken together yield the principal values $\kappa_{ii}$ and uncertainties in Table 1. The principal axes are the crystallographic $b$ axis, by symmetry, and $\theta$ is the angle between $\kappa_{11}$ and $a$ in the $ac$ plane. All calculations are based on the same, INDO/S-based $\Pi_{ij}$, but have different molecular $\alpha$ inputs as indicated in the Table, including the experimental polarizability. The triple-zeta basis with field-induced polarization functions from ref. 24 and density functional theory (B3LYP) with a large basis are nearly quantitative, while the oriented-gas value based on $\rho_G(r)$ is clearly not. The dielectric tensor of the crystal includes charge redistribution and induced dipoles whose interactions are treated self-consistently via Eqs (2) and (3).

Polarization energies $P_+$ or $P_-$ are surface rather than bulk measurements [25]. As sketched in Fig. 2, photoelectron spectroscopy (UPS) involves $P_+$ for a cation that is mainly at the surface while inverse photoelectron spectroscopy (IPES) involves $P_-$. To minimize charging, thin organic films on metallic substrates are used. Recent interest in organic electronic devices is made possible by advances in forming and characterizing crystalline thin films that compensate for the limited mobility of charges. Charge injection then involves polarization at the metal-organic interface. A constant potential surface with image charges is the simplest model of the metal-organic interface. The zero-overlap procedure for electronic polarization, Eqs (2) and (3), is readily extended to surfaces or interfaces [26]. We take $N$ molecular layers based on the crystal, place them in van der Waals contact with the metal, and introduce image charges and dipoles for neutral molecules as well as for ions at specified locations. Instead of spheres of radius $R$ that enclose ions, we use pill-boxes of thickness $2N$ and variable radius. Convergence again requires thousands of molecules and can be monitored with respect to $N$ as well as pill-box radius.

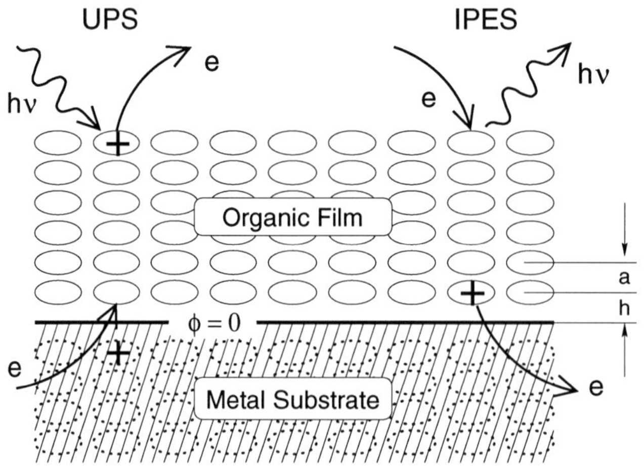


Fig. 2. Idealized model for electronic polarization in crystalline thin films on a metallic substrate at separation $h$. The $N$ layers of the film appear as image charges. UPS and IPES generate a cation and anion at the surface, while charge injection generates ions in the interface layer.

Perylenetetracarboxylic dianhydride (PTCDA) is a good hole conductor and excellent film former, with molecules lying almost flat on the surface [27]. It is a prototypical molecule for organic devices. The crystal value for $P = P_+ + P_-$ is 1.82 eV, which is in the expected range [16]. The calculated monolayer value is $P_{\text{mono}} = 1.93$ eV [26]. The large polarizability of image charges on one side offsets the vacuum on the other side. Accordingly, thick films have reduced $P_{\text{surf}} = 1.41$ eV. The 500 meV change between monolayers and thick films agrees quantitatively with UPS, IPES, and tunneling spectra of PTCDA monolayers and films on gold and silver [26]. The calculated $P$ at the metal-organic interface, with image charges on one side and crystal on the other, increases to $P_{\text{inter}} = 2.21$ eV. The separate values of $P_+$ or $P_-$ are relevant for matching energy levels to facilitate the injection of holes or electrons. Efficient injection is a major challenge whose pursuit is largely empirical at present, usually without any consideration of polarization.

Pentacene is another widely used molecule, mainly as thin film transistors [28]. Its herringbone structure is more common than PTCDA stacks. Pentacene films have the long axis almost normal to the surface, with high conductivity parallel to the surface. The two inequivalent molecules per unit cell are calculated [19] to have slightly (70 meV) different $P = 2.01$ eV in the crystal. $P_{\text{mono}}$ differs by only 6 meV from the crystal, while $P_{\text{surf}}$ is 0.23 eV less and $P_{\text{inter}}$ is 0.13 eV greater [19]. The contrasting structures of pentacene and PTCDA films lead to different electronic polarizabilities that can now be estimated in the well-defined limit of zero overlap rather than just expected on general grounds in anisotropic solids. Similarly, accurate polarization energies are needed for quantitative analyses of transport gaps or of CT states seen in electroabsorption.

In the context of polarizabilities, we close this Section by noting that since the pentacene long axis is almost normal to the metal, the images charges for a cation or anion produce fields along the direction of largest $\alpha$ and redistribute charge to the ends of the molecule. The induced dipoles of adjacent molecules are parallel and close to each other. These repulsive interactions are relieved by redistributing charge toward the molecule's center, as actually found in the self-consistent solution [19]. The field $F(r_i^a)$ at a pentacene ion on a metallic surface vanishes for atoms near the center and ranges from $\pm 10^7$ V/cm for atoms close to and far from the metal. Electronic polarization in organic molecular crystals produces highly non-uniform fields, as found explicitly in pentacene films and studied below in clusters of polar-polarizable molecules.

## 3. Electronic polarization in clusters of polar-polarizable molecules

Important intermolecular effects are expected in materials where several polar and highly polarizable molecules interact. In these materials in fact the local fields generated by the polar molecules strongly affect the charge distribution on the molecules themselves, leading to a non-trivial self-consistent problem. In so-called push-pull chromophores an electron-donor (D) and an electron-acceptor (A) group are linked by a $\pi$-conjugated bridge. These intrinsically polar molecules are also highly polarizable due to the presence of delocalized $\pi$-electrons. They are also often largely hyper-polarizable, and represent the molecules of choice for second-order NLO applications [29]. As originally recognized by Oudar and Chemla [30], the low-energy physics of push-pull chromophores is well described by a two-state picture. The gs resonates between two limiting structures: $|DA\rangle$ and $|D^+A^-\rangle$, that can be taken as the two basis states of a Mulliken model [31]. These two states are separated by an energy $2z_0$ and are mixed by a matrix element $-\sqrt{2}t$, that, with no loss of generality, will be fixed as the energy unit. The relevant gs is written as $|G\rangle = \sqrt{1-\rho}|DA\rangle + \sqrt{\rho}|D^+A^-\rangle$, where the ionicity $\rho$ only depends on $z_0$. $\rho$ is proportional to the gs dipole moment: $\mu_G = \mu_0\rho$ and $\mu_0 = \langle D^+A^-|\hat{\mu}|D^+A^-\rangle$. The excited state, $|E\rangle$, is orthogonal to $|G\rangle$ and its dipole moment is $\mu_0(1-\rho)$. The excitation energy is $\omega_{CT} = 1/\sqrt{\rho(1-\rho)}$ and the corresponding transition dipole moment is $\mu_0\sqrt{\rho(1-\rho)}$.

This simple model sets the basis for current understanding of NLO responses of push-pull chromophores and, if extended to account for the coupling to molecular vibrations and solvation effects, it offers a good description of the spectroscopic properties of push-pull chromophores in solution [32, 33]. Here we adopt this model to investigate the role of intermolecular interactions in clusters of polar-polarizable chromophores. We consider a cluster of Mulliken molecules with purely electrostatic interactions. The relevant Hamiltonian is [34]:

$$\mathcal{H}_{\text{int}} = \sum_i (2z_0\hat{\rho}_i - \sqrt{2}t\hat{\sigma}_{x,i}) + \sum_{i,j>i} V_{ij}\hat{\rho}_i\hat{\rho}_j \tag{5}$$

The first term in (5) describes the on-site problem, with $\hat{\rho}_i$ measuring the polarity (i.e. the weight of the zwitterionic state) of the $i$-th chromophore, and $\sigma_{x/z,i}$ is the $x/z$-Pauli matrix for the $i$-site. The second term accounts for electrostatic intermolecular interactions with $V_{ij}$ measuring the interaction between zwitterionic species located on sites $i$ and $j$.

The Hamiltonian (5) is fairly general. Here we consider 1-dimensional clusters of $N$ equivalent molecules with the three geometries sketched in Fig. 3. We model each zwitterionic molecule as a segment of length $l$ carrying $\pm e$ charges at the D/A ends, so that, for unscreened interactions, $V_{ij}$ is fixed by $v = e^2/l$, the interaction between two charges at unit distance, and $r$, the interchromophore distance. In any case the specific expression for $V_{ij}$ does not alter the basic physics of the model. The above Hamiltonian is easily written and diagonalized on the $2^N$ basis obtained from the direct product of the two basis functions, $|DA\rangle$ and $|D^+A^-\rangle$, on each site. By exploiting the translational symmetry we are able to find exactly at least the lowest 30 eigenstates for systems with up to 16 sites.

Figure 4 shows the evolution of the chromophore polarity with the inverse interchromophore distance, $w = l/r$ for the three lattices sketched in Fig. 3. All results are obtained for $v = 1$; $z_0$ is fixed to 1 in upper panels, to show the behavior of a chromophore with a neutral (N) ground state in the gas phase ($\rho = 0.15$ at $w = 0$). The bottom panels ($z_0 = -1$) instead describe the behavior of zwitterionic (I) chromophores ($\rho = 0.85$ at $w = 0$). Interchromophore interactions disfavor charge separation in A geometry, and $\rho$ decreases with $w$ in the leftmost panels in Fig. 4, whereas just the opposite occurs for geometry B and C (Fig. 4, middle and right panels). The behavior of an A cluster of I chromophores

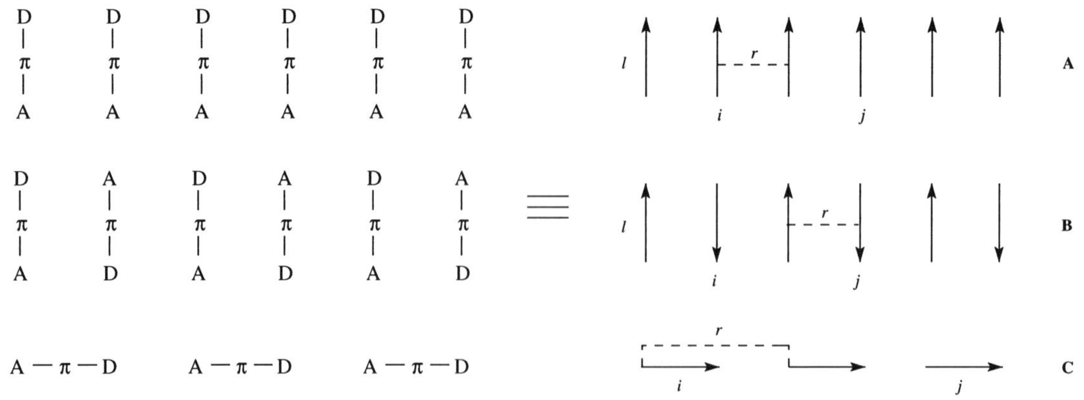


Fig. 3. Schematic view of the three one-dimensional clusters of polar and polarizable molecules considered in this work, the left panel shows the relative orientation of molecular dipole.

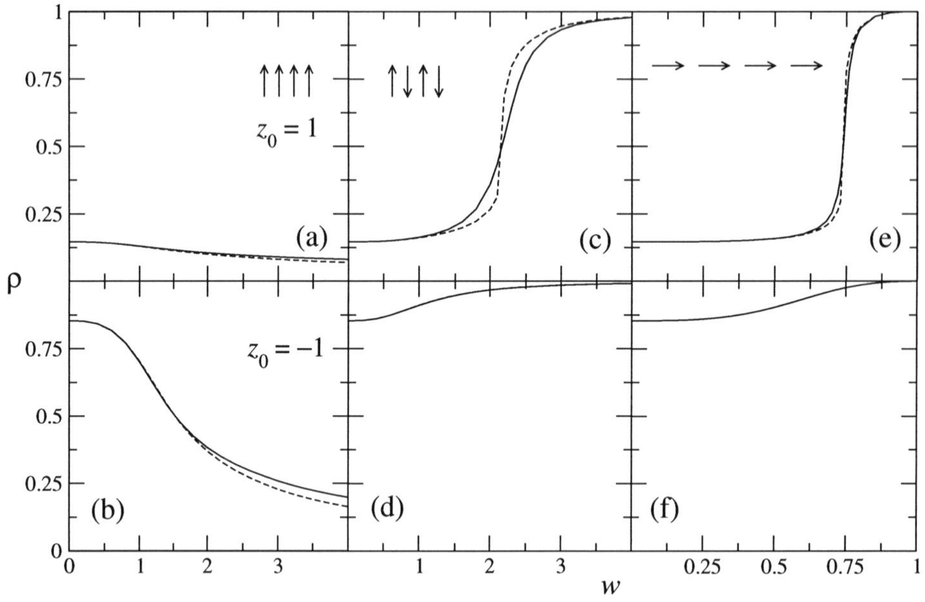


Fig. 4. Molecular ground state polarity, $\rho$, as a function of the strength of the interchromophore interaction, $w$, calculated for clusters of 16 molecules, with $v = 1$. Left, middle and right columns refer to geometries A, B and C, respectively; top and bottom rows correspond to $z_0 = 1$ and $-1$, respectively. Continuous and dashed lines refer to exact and mean-field results, respectively. Finite size effects are negligible in all cases.

(Fig. 4(b)) and of B and C clusters of N molecules (Figs 4(c) and (e), respectively) are particularly interesting. In the first case the isolated chromophore is zwitterionic, but, with increasing $w$ (i.e. by decreasing the interchromophore distance) the molecular polarity decreases down to the cyanine limit ($\rho = 0.5$) reaching the N regime for an interchromophore distance of about 0.7 times the dipole length. Similarly, a neutral isolated chromophore can be driven to the I regime for large enough interactions in either B and C geometries when the interchromophore distance is about one half of the dipole length (B cluster) or about 1.4 times the dipole length (C cluster).

To make contact with experiment, we note that as a first estimate, push-pull chromophores have

$\sqrt{2}t \sim 1$ eV. Then $v = e^2/l = 1$ corresponds to typical molecular lengths ($l \sim 15$ Å). For these parameters, an I chromophore in A geometry crosses the I-N interface at $w \sim 1.5$ (panel (b)), i.e. for intermolecular distance of $\sim 10$ Å, a reasonable situation. Similarly, curves in panels (c) and (e) indicate that the N-I interface is crossed for interchromophore distance of $\sim 7$ Å and $\sim 20$ Å for B and C cluster, respectively. Again the inversion of the polarity occurs for reasonable intermolecular distances. As a matter of fact, it has already been predicted [35] that the polarity of a polar and polarizable molecule varies and eventually inverts due to environmental interactions. However, most of the approaches presented in the literature on the properties of interacting push-pull chromophores disregard the molecular polarizability and do not allow the molecular polarity to readjust in response to supramolecular interactions. On general ground, push-pull chromophores have large transition dipole moments (~5–10 D) and their permanent dipole moment undergoes a large variation (~20–30 D) upon excitation [33]: in samples with a medium-large concentration of chromophores (intermolecular distances ~5–10 Å) interchromophore interactions are a sizable fraction of, or even larger than typical excitation energies (~1–3 eV) [33].

The mf treatment in the previous Section becomes trivial when applied to the above Hamiltonian. In the two-state model in fact a single parameter, $\rho_i$, fully defines both the molecular gs and local electric fields at the molecular positions. For lattices of equivalent molecules, as in Fig. 3, a single-parameter self-consistent problem results from the mf approach. In particular, within mf the lattice reduces to a collection of non-interacting molecules, each one described by the same two-state Hamiltonian as the isolated molecule, but with a renormalized energy gap between $|DA\rangle$ and $|D^+A^-\rangle$, $z_0 \rightarrow z_0 + m\rho$, where $m = \sum_j V_{ij}/2$. Dashed lines in Fig. 4 show $\rho$ as obtained from the self-consistent solution of the mf problem, and demonstrate that mf offers a quite satisfactory description of the behavior of interacting molecules, at least for not too large interactions.

Within mf it is easy to recognize a qualitative difference between the I to N crossover in repulsive lattices (A) and the N to I crossover in attractive lattices (B and C). With increasing supramolecular interactions in repulsive lattice (A, $m > 0$) the slope of the $\rho(z_0)$ curve becomes less negative, whereas it becomes more negative in attractive lattices (B and C, $m < 0$). For large negative $m$ ($m < -2$) a divergent $\frac{\partial \rho}{\partial z_0}$ is expected, marking the occurrence of a discontinuous crossover from the N to the I regime [37]. The N-I crossover is located at $z_0 \sim -m/2$, and for large $z_0$ ($> 1$) S-shaped $\rho(w)$ curves are calculated within mf. The appearance of a discontinuous crossover in mf treatments of C-lattices was discussed many years ago, and offered a first description of the neutral-ionic phase transition observed in CT crystals with a mixed stack motif [12]. The behavior of the system in the proximity of a discontinuous interface is very interesting, but is beyond the scope of the present work [34].

The static susceptibilities of systems described by the Hamiltonian in Eq. (5) are easily obtained from the successive derivatives of the ground-state dipole moment on a static applied field. Since molecular dipole moments, in the proposed toy model, rigorously lie along the molecular axis (say $z$), a single component of the polarizability tensor ($\alpha_{zz}$) is relevant. The magnitude of this component is shown in Fig. 5 for parameters corresponding to panels b, c and e in Fig. 4. Continuous lines show exact results and demonstrate that supramolecular interactions non-trivially affect the molecular response.

To clarify the subtle physics governing the responses of molecular clusters, we discuss approximate approaches to the problem. In the simplest and most widely adopted approach the response of a collection of chromophores is calculated as the sum of the responses of a collection of non-interacting molecules with the same geometrical arrangement [36]. This *oriented gas* approach is however limited to very weak interactions and fails otherwise, since it completely disregards the dependence of $\rho$ on supramolecular interactions. A slightly better approach relies again on the oriented gas approximation, but assigns

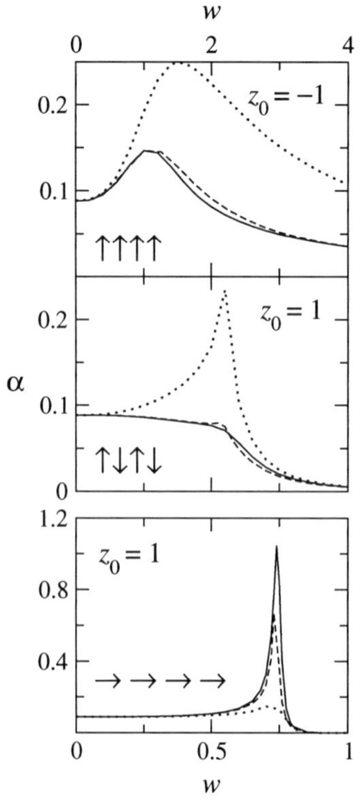


Fig. 5. Static polarizability calculated as a function of the strength of interchromophore interaction, $w$, for clusters of 16 molecules, with $v = 1$. Top panel: A geometry, $z_0 = -1$; middle panel: B geometry, $z_0 = 1$; bottom panels: C geometry, $z_0 = 1$. Continuous lines refer to the exact results, dashed lines to the mf-FF results, dotted lines to the mf-oriented gas results.

each chromophore the same $\rho$ as obtained within the mf approximation. The mf-oriented gas estimates of static susceptibilities are reported in Fig. 5 as dotted lines. The strikingly large deviations from exact results are quite unexpected: the gs polarity is fairly accurately calculated within mf, for these parameters. The failure of the mf-oriented-gas approximation appears since the response to an applied field of a molecule in the cluster differs from the response of an isolated molecule to the same field. Much better results can be obtained within mf provided the gs dipole moment of each molecule in the cluster is allowed to readjust to the applied field. A proper FF-mf calculation of the polarizability as the first derivative of the cluster dipole moment on the applied field just represents the (trivial) implementation to the Hamiltonian 5 of the self-consistent treatment described the Section 2. As shown by dashed lines in Fig. 5, this approach nicely compares with exact results, apart from deviations observed in a narrow region around the N-I crossover.

The static linear polarizability is a gs property that can be accurately calculated within mf, provided collective behavior is properly accounted for. Non-linear responses can be obtained within the same approach from the successive derivatives of the gs dipole moment, and collective effects are found to rapidly increase with the order of non-linearity [37]. The deviations between dotted and dashed lines in Fig. 5 demonstrate the importance of collective behavior, here due to the non-linear response of polarizable chromophores to the perturbation induced by the surrounding. Material properties are significantly affected by the supramolecular arrangement: linear and non-linear optical properties can be strongly depressed or amplified by tuning intermolecular distances and/or by changing the relative orientation of chromophores, with effects that are more pronounced at intermediate polarities. The design of molecular materials for advanced applications is then a challenging task: the material properties in fact must be optimized at the supramolecular level. The presented model just represents a first step towards *supramolecular* structure-properties relationships.

## 4. Vibrational contributions to $\alpha$: A toy-model approach

Both molecular vibrations and lattice phonons are strongly coupled to delocalized electrons and can significantly contribute to susceptibilities. Accounting for vibrational degrees of freedom even within a simple model for molecular materials like that presented in the previous Section is non-trivial. Exact non-adiabatic solution of the relevant problem is already computationally demanding for clusters of just two molecules. In the following we therefore shortly discuss the vibrational contribution to the polarizability of an isolated push-pull molecule. We again describe the chromophore as a Mulliken DA pair (cf Section 3) but also account for Holstein coupling to a vibrational coordinate, $Q$ [2]. The relevant Hamiltonian is ($\hbar = 1$ and $\sqrt{2}t = 1$):

$$\mathcal{H} = 2z_0\hat{\rho} - \hat{\sigma}_x + \frac{1}{2}\left(\omega^2Q^2 + P^2\right) - \sqrt{2\varepsilon_{sp}}\omega Q\hat{\rho} \quad (6)$$

The first two terms in the above equation describe the electronic Hamiltonian (cf previous Section). In the third and fourth terms the vibrational problem is defined in terms of an internal vibrational coordinate, $Q$, and of its conjugated momentum, $P$. In particular the two basis states are assigned two harmonic PES with equal frequency ($\omega$) but displaced minima to account for linear e-ph coupling, whose strength is measured by $\varepsilon_{sp}$, the relaxation energy of $|D^+A^-\rangle$. Whereas the model can quite easily be extended to account for quadratic coupling (i.e. for different vibrational frequencies in the two basis states), this adds an additional parameter to the model, without affecting the basic physics [38].

The exact non-adiabatic eigenstates of the above Hamiltonian are obtained by the numerical diagonalization of the relevant matrix written on the basis of the direct product of the two electronic states $|DA\rangle$ and $|D^+A^-\rangle$, and of the reference vibrational states (i.e. the eigenstates of the harmonic oscillator in the third term of Eq. (6)) [39]. The basis is truncated by fixing a maximum number of phonon states, $M$; the corresponding $2M \times 2M$ matrix can be diagonalized up to fairly large $M$ values, yielding numerically exact non-adiabatic eigenstates. The minimum $M$ required to get convergence depends on the model parameters and on the properties of interest.

The second derivative of the exact non-adiabatic gs energy vs the applied field, gives the exact estimate of the molecular polarizability. Results obtained for a molecule with $\varepsilon_{sp} = 1$ and a few $\omega$ values are reported in Fig. 6 (continuous lines). $z_0$ is tuned in this calculation as to span the whole $0 < \rho < 1$ interval. The dot-dashed line corresponds to the bare electronic susceptibility, $\alpha_0$, i.e. to the response of the two-state model with no e-ph coupling. Hence differences between the continuous lines and the dot-dashed line measure the vibrational contribution to the static response [40].

For actual molecules or complexes, the large number of electronic states makes non-adiabatic calculations very demanding, and the Born-Oppenheimer (BO) approximation is usually invoked. In the BO approximation, the effective electronic Hamiltonian, $\mathcal{H}_{el} = \mathcal{H} - P^2/2$, is defined by subtracting the nuclear kinetic energy (KE) from the total Hamiltonian. Its diagonalization yields analytical expressions for the ground and excited state potential energy surfaces (PES) [40]. Even if the Hamiltonian in Eq. 6 assigns the two basis states two harmonic PES with equal curvature, the PES for the ground and excited state obtained from the diagonalization of the electronic Hamiltonian have different curvatures and are

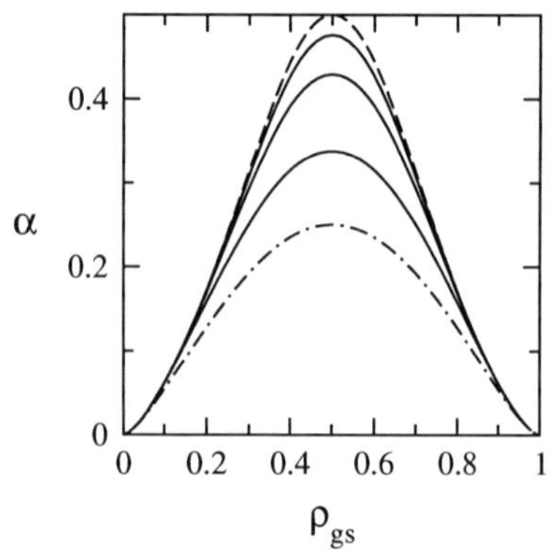


Fig. 6. Static polarizability as a function of the ground state ionicity $\rho_{gs}$, for $\varepsilon_{sp} = 1$ and different $\omega$ values ($\sqrt{2}t$ units). Dashed line: potential energy contribution, corresponding to $\omega = 0$ limit (see text); dot-dashed line: bare electronic polarizability, corresponding to the $\omega \to \infty$ limit (see text); continuous lines report the exact polarizability calculated for $\omega = 0.05, 0.2, 1.0$, smoothly evolving from the $\omega = 0$ to the $\omega \to \infty$ limits; dotted line reports the polarizability calculated in the best harmonic approximation (see text). Dotted and dashed lines are exactly superimposed. For the calculation dipole moments have been expressed in $\mu_0$ units.

largely anharmonic [39]. The anharmonicity of the potential prevents the analytical solution of the vibrational problem on either the ground or excited state; however, numerically exact vibrational states can be calculated in both manifolds [40]. In particular, the eigenstates of the harmonic oscillator with frequency $\omega$, centered at the relevant equilibrium position, are a good basis for the vibrational problem on either PES. The corresponding vibrational Hamiltonian is the sum of a KE term, whose matrix elements are trivial in the adopted basis, plus a potential energy (PE) term, whose matrix elements are calculated via numerical integration. Of course the vibrational matrix is diagonalized on a basis truncated to a large enough number of phonon states as to get convergence.

Once BO eigenstates are obtained, the (transition) dipole moments entering the SOS expression can be calculated via numerical integration. The static polarizability calculated within BO approximation is indistinguishable (in the scale of Fig. 6) from the exact one, as long as $\omega \leqslant 0.2$. Figure 7 compares non-adiabatic (continuous line) and BO (dashed line) static susceptibility for $\omega = 0.5$, where deviations appear. Of course the BO approximation becomes worse with increasing $\omega$ and is totally untenable for $\omega \geqslant 1$. For push-pull chromophores $\varepsilon_{sp} \sim \sqrt{2}t$ and typical vibrational frequencies, $\omega \sim 1000\ \mathrm{cm}^{-1} \ll \sqrt{2}t \sim 1$ eV support the validity of BO.

When applied within a FF approach to susceptibilities, the BO approximation immediately leads to a partitioning of the susceptibility into PE and KE contributions [40]. Within BO, the lowest eigenstate of the electronic Hamiltonian defines the PE for the motion of nuclei (in the gs manifold, of course). The total gs energy is obtained by adding the nuclear KE to the PE. Then, as long as BO applies, susceptibilities, i.e. the successive derivatives of the gs energy with respect to an applied electric field, can be calculated as sums of PE and KE $F$-derivatives. The nuclear KE vanishes in the $\omega = 0$ limit, and the ($\omega$-independent) PE polarizability, reported as the dashed line in Fig. 6, represents the zero-frequency limit of the exact polarizability. The electronic polarization of the Holstein DA molecule is analytical [2] and is reported as the dot-dashed line in Fig. 6. We can now understand the evolution of the vibrational contribution to the polarizability with the vibrational frequency: in the low-$\omega$ limit, KE contributions vanish, and the exact curve tends to the limiting PE result. Vibrational contributions are very large in this limit. With increasing $\omega$, the vibrational contributions to the response decrease: in the antiadiabatic limit ($\omega \to \infty$) the vibrational contributions to the static polarizability vanishes, and the exact curve trivially reproduces the bare electronic response.

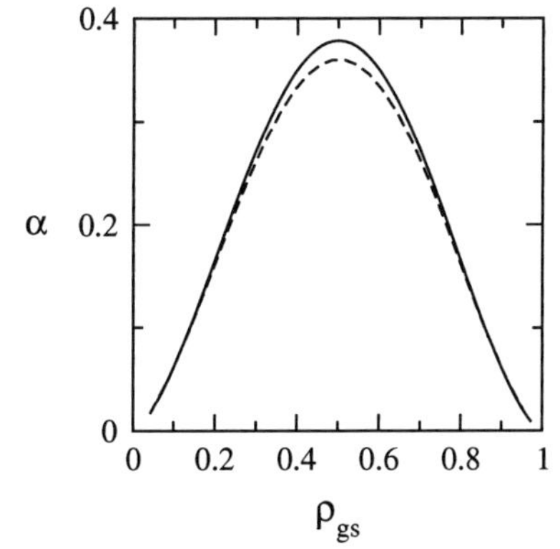


Fig. 7. Static polarizability as a function of the ground state ionicity $\rho_{gs}$, for $\varepsilon_{sp} = 1$ and $\omega = 0.5$ ($\sqrt{2}t$ units). The continuous line shows the non-adiabatic result; the dashed line shows the BO result.

The calculation of PE-susceptibilities for the toy-model in Eq. (6) is trivial. More generally it is easily implemented in quantum chemistry calculations and in solid-state models as well, since it only requires the gs energy calculated at the relaxed geometry for different values of an externally applied field. The calculation of KE contribution is more difficult, since the $F$-dependence of the lowest vibrational state in the anharmonic gs PES is needed. The nuclear KE contributes to susceptibilities in two different ways [40]. First of all, due to anharmonicity, the molecular geometry in the vibronic gs is different from the equilibrium geometry (corresponding to the minimum of the gs PES). This correction is however very small. The second contribution stems from the $F$-dependence of the nuclear KE itself: it is this contribution that indeed accounts for the deviations of the exact curves from the (dashed) PE curve (at least in the BO regime, $\omega \leqslant 0.2$, where non-adiabatic corrections are negligible). KE contributions are of course very small for low $\omega$, but they increase with increasing $\omega$, leading to a suppression of the vibrational amplification of the static polarizability. This is by no means accidental: in the antiadiabatic limit ($\omega \gg \sqrt{2}t$) phonons cannot contribute to the static polarizability and, with increasing $\omega$, KE contributions progressively increase to counterbalance the PE contribution.

KE contributions to the linear polarizability exactly vanish in the harmonic approximation [40]. In fact, the equilibrium position in any harmonic vibrational state coincides with the bottom of the PES; moreover the nuclear KE is proportional to the harmonic frequency, i.e. to the curvature of the PES. For a parabolic PES, this quantity is obviously independent of $Q$, and hence of $F$. Then, in the low $\omega$ limit, where KE contributions to susceptibilities are negligible, the exact $\alpha$ (continuous line in Fig. 6) coincides with the corresponding best-harmonic estimate, i.e. with the estimate obtained by modeling the gs PES as the parabola with the exact curvature at the equilibrium. This is not true for hyper-polarizabilities that, in the same $\omega \to 0$ limit, are strongly amplified by the anharmonicity of the gs PES [40]. Within the best harmonic approximation [39], $\alpha$ is conveniently partitioned into an electronic and vibrational contribution:

$$\alpha = 2\left[\frac{\mu_{CT}^2}{\omega_{CT}} + \frac{\mu_{IR}^2}{\Omega}\right] \tag{7}$$

where $\mu_{IR}$ is the infrared transition dipole moment and $\Omega$ is the frequency of the best harmonic frequency of the coupled mode, i.e. the curvature of the gs PES at equilibrium. The vibrational contribution to the linear polarizability is then proportional to the infrared intensity of the coupled mode, at least in the low-frequency regime, where PE contributions dominate the response. We underline that this result only

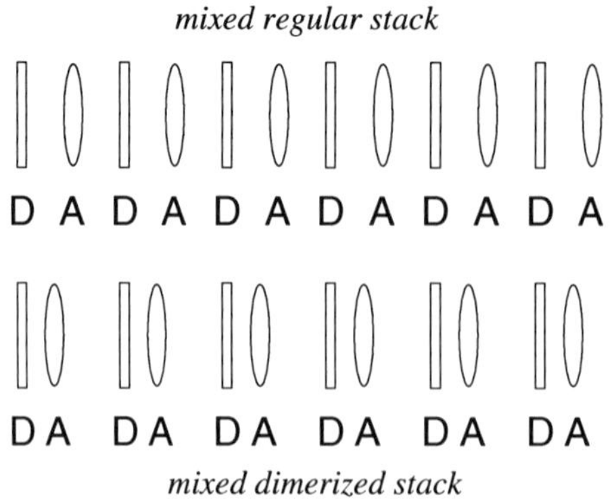


Fig. 8. A schematic view of mixed regular and dimerized stacks.

relies on the BO approximation and on the neglect of KE contributions to the polarizability. It therefore applies in the low-$\omega$ limit quite irrespective of the detailed model for the electron-phonon coupling, and also for largely anharmonic PES.

## 5. Dielectric constant of CT salts

Charge-transfer (CT) crystals have mixed face-to-face stacks of planar $\pi$-electron donors and acceptors as sketched in Fig. 8. Intermolecular overlap is negligible between stacks, but not within stacks where $\pi - \pi$ overlap is indicated by less than van der Waals separation between D and A. The gs consequently has fractional charges $\rho$ at D and -$\rho$ at A sites [41]. Just as an example, in the prototypical material, TTF-CA, at ambient conditions about 0.2 electrons are transferred on average from the donor (tetrathiafulvalene, TTF) to the acceptor (chloranil, CA) [42]. As seen in Fig. 8, a regular stack of centrosymmetric molecules is not polar because there is an inversion center at each site, as in fact occurs in the actual structures. The inversion center is lost on dimerization and the gs becomes ferroelectric if dimerization is in the same sense everywhere. Much as it occurs in attractive lattices in Section 3, Madelung interactions favor charge separation, and a large variation of $\rho$ can be induced by tuning intermolecular distances. At $\sim$81 K TTF-CA undergoes a discontinuous phase transition to an I phase with $\rho \sim 0.7$ [42]. Other systems with N-I transitions are known and transitions can be induced by temperature, pressure or by absorption of light. They are a complex and interesting phenomenon [43]: both continuous and discontinuous transitions are known and, in all cases, stack dimerization accompanies the charge crossover proving the important role of phonons and of e-ph coupling in these systems.

A sharp peak in the dielectric constant has been observed at the N-I transition of several CT crystals [44]. It can be understood on general grounds as due to large charge fluxes induced by an applied field near the charge crossover, but microscopic modeling of dielectric peaks is still in progress [45]. Charges are delocalized along the stack and the zero-overlap approximation does not apply: we need a model for polarization in extended and highly correlated systems. Moreover the lattice (Peierls) phonon that induces dimerization is strongly coupled to electronic degrees of freedom in a manner reminiscent of solitons in polyacetylene [46]. The Peierls mode in fact induces large charge displacements around the N-I crossover, and vibrational contributions to the polarizability cannot be disregarded.

The electronic structure of mixed stack CT salts can be described in terms of a Hubbard model with only on-site electron-electron repulsion ($U$) explicitly accounted for, and modified to account for the alternation of on-site energies, $\Delta$ [47]. This model accounts only implicitly for Madelung interactions and hence describes only continuous N-I transitions. We also account for the coupling to the Peierls phonon, described by $\delta$, as follows [49]:

$$H = -\sum_{i,\sigma}[1+(-1)^i\delta](c^\dagger_{i,\sigma}c_{i+1,\sigma}+H.c.)+U\sum_i \hat{n}_{i,\sigma}\hat{n}_{i,\sigma'}+\Delta\sum_{i,\sigma}(-1)^i c^\dagger_{i,\sigma}c_{i,\sigma}+\frac{N}{2\varepsilon_d}\delta^2 \quad (8)$$

where $c^\dagger_{i,\sigma}$ creates an electron with spin $\sigma$ on the $i$-th site and $\hat{n}_{i,\sigma} = c^\dagger_{i,\sigma}c_{i,\sigma}$. The last term in the above Hamiltonian measures the bare elastic energy associated with the dimerization mode, with $1/\varepsilon_d$ measuring the lattice stiffness. The rigid lattice has $\varepsilon_d = 0$.

For $\Delta \gg U$ the Hamiltonian (9) describes an almost N lattice ($\rho \to 0$) of donors and acceptors, whereas for $U \gg \Delta$ an almost I lattice of spin 1/2 radical ions is obtained ($\rho \to 1$) [47]. The two phases are qualitatively different, with the I lattice being unconditionally unstable to dimerization (spin-Peierls transitions [48]). The N-I crossover can consequently be identified precisely even for continuous $\rho$ in the rigid regular stack [47]. Conditional instability in the N regime implies that soft lattices dimerize on that side before reaching the N-I crossover [49]. The Peierls phonon induces large charge fluxes along the chain with effects that increase the nearer the dimerization transition is to the N-I crossover of the rigid lattice.

By exploiting the recent definition of polarization in extended systems [4], we were able to demonstrate huge peaks in the IR intensity of the Peierls mode at the structural instability [49]. But, as shown in the previous Section for Holstein modes, the IR intensity of vibrational modes is quite naturally related to the system's polarizability. We draw two inferences: first, peaks in the static dielectric constant near the neutral-ionic phase transition have a large contribution from vibrational degrees of freedom; second such peaks in soft lattices are associated with the Peierls transition and occur at $\rho < 1/2$. We anticipate that the Hamiltonian (9) or a closely related model will describe dielectric peaks.

Since lattice phonons have low frequencies ($<100\ \mathrm{cm}^{-1} \ll t \sim 0.2$ eV), the BO approximation applies and the KE contribution to $\alpha$ can be safely neglected. The calculation of the vibrational contribution to $\alpha$ in Eq. (7) is simple, given the IR intensity of the relevant model. But the electronic contribution to $\alpha$ is difficult for correlated electrons and will be presented separately [45].

As a first estimate of the relative importance of the electronic and vibrational contributions, we consider the $U = 0$ limit of uncorrelated electrons in (9), whose electronic $\alpha$ has long been known [50]. $U = 0$ results are reported in Fig. 9 for $\varepsilon_d = 0.28$, which is a fairly stiff lattice that approximates TTF-CA. The static dielectric constant is related to polarizability by the standard expression: $\kappa = 1 + (\alpha_{el} + \alpha_{vib})/\kappa_0$, where $\kappa_0$ is the permittivity of vacuum and $\alpha$ is the polarizability per unit cell. We took TTF-CA lattice parameters as typical for CT crystals. Moreover, for the sake of simplicity, we neglected the softening of the Peierls mode at the transition, by simply substituting the $\Omega$ entering the expression for the vibrational polarizability in Eq. (7) with the reference frequency ($1/\varepsilon_d$). Thus the estimated vibrational contribution to $\kappa$ is a lower limit.

The magnitude of electronic and vibrational contributions to $\kappa$ is similar for noninteracting electrons. Yet the peak at the Peierls instability in Fig. 9 clearly has a vibrational origin. The physical basis of this behavior lies in the large charge fluctuations generated by Peierls phonons in the vicinity of a charge instability, and underlines once more the importance of properly accounting for e-ph coupling in systems with delocalized electrons. All material properties, including *electronic* responses are in fact non-trivially affected by coupling to nuclear motions. The electronic contribution, by contrast, is dominated in $U = 0$ systems by the gap between the valence and conduction band of (9). This gap is almost constant [51] at

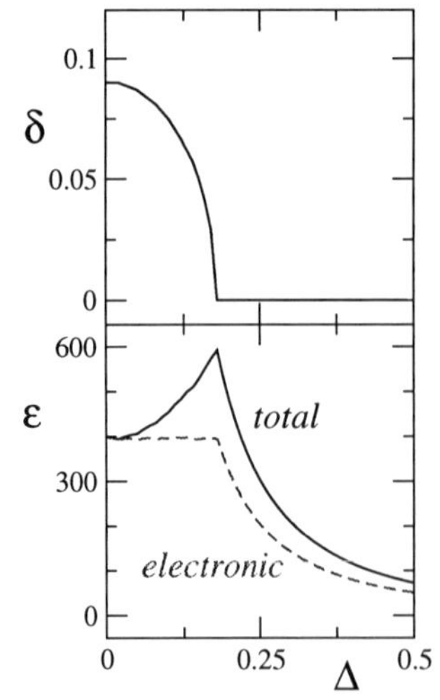


Fig. 9. The equilibrium dimerization amplitude $\delta$ and the electronic and total dielectric constants calculated for the hamiltonian in Eq. (8) with $U = 0$ and $\varepsilon_d = 0.28$.

equilibrium in the dimerized lattice, with increasing $\delta$ offsetting decreasing $\Delta$ in Fig. 9 to give an almost constant $\kappa_{el}$.

Electronic correlations decrease $\alpha_{el}$ and stabilize the I phase of radical ions. CT salts are correlated systems with large $U$ in (9). Although their dielectric constants are not expected to resemble $U = 0$ results, the Peierls instability on the N side nevertheless ensures a strong vibrational peak where $\alpha_{el}$ is monotonic. And increasing dimerization still produces a gap in the electronic excitations $E_R > 0$ in (1) that reduces $\alpha_{el}$ at the N-I interface. Some key $U = 0$ features are retained in CT salts [45].

## 6. Conclusions

Modeling the dielectric behavior of molecular materials made up of strongly polarizable molecules is challenging in several respects. Even in traditional molecular crystals where the overlap of charge densities on different molecules can be safely neglected, the self-consistent interplay between local electric field and molecular polarizabilities (and hyperpolarizabilities) is responsible for the appearance of large collective effects: the response of the material is quantitatively, and possibly even qualitatively different from the behavior of a collection of non-interacting molecules. In Section 2 a recent approach to the calculation of the dielectric constant of molecular crystals and films made up of large $\pi$-conjugated and non-polar molecules is summarized. The approach is based on the mf approximation, so that the problem for $N$ interacting molecules reduces to the self-consistent problem of a single molecule experiencing the local electric field generated by the surrounding molecules. The new approach to electronic polarization also yield the polarization energies of cations, anions, and CT states, and it is applicable to crystalline thin films.

Even larger effects are expected in materials made up of polar and polarizable molecules, such as push-pull chromophores. A simple toy-model for these materials is based on a two-state description of each molecular unit. Few-state models, while less accurate than quantum-chemical descriptions, yield a more transparent description of the basic physics of the system, and, being amenable to exact solution, offer valuable information on the validity of approximations necessarily introduced when handling more complex Hamiltonians. Results in Section 3 confirm the reliability of mf approximation as introduced in Section 2 for linear (and non-linear) electric susceptibilities of molecular materials with non-overlapping charge distributions. Exact results for the toy-model also confirm the importance of collective effects in aggregates of polar-polarizable chromophores, in which supramolecular interactions can tune molecular polarity over a wide range.

Few state models are also useful to address vibrational contributions to polarizabilities, and particularly to assess the reliability of standard approximation schemes including the BO separation of nuclear and electronic degrees of freedom and the harmonic approximation. A reduced on-site basis, and hence models with few local states, are also a useful starting point to face the challenging problem of polarizability in covalent insulators, such as CT salts. Electronic (charge) instabilities and lattice instabilities coexist in these materials and compete and/or cooperate in complex and interesting ways. The peak in the dielectric constant of several CT salts is due to the Peierls mode and provides another example of intimate coupling between electronic and vibrational degrees of freedom.

## Acknowledgments

The work on vibrational contributions to polarizabilities and hyperpolarizabilities in molecular materials and in CT crystals was perfomed in collaboration with Luca del Freo. The self-consistent treatment of molecular crystals was done in collaboration with Eugene Tsiper. We also thank colleagues for many discussions: Alberto Girlando in Parma, and Jessica Sin, Robert Pascal, Jr., and Antoine Kahn in Princeton. Work in Parma was supported by Ministero Istruzione Universita' e Ricerca (COFIN-2001), and by INSTM (PRISMA-2002). Work at Princeton was partially supported by the National Science Foundation through the MRSEC program under DMR-9400632.

## References

[1] D.M. Bishop, *Rev. Mod. Phys.* **62** (1990), 343, and references therein.
[2] A. Painelli, *Chem. Phys. Lett.* **285** (1998), 352, and references therein.
[3] P. Norman, P. Macak, Y. Luo and H. Agren, *J. Chem. Phys.* **110** (1999), 7960.
[4] R. Resta, *Phys. Rev. Lett.* **80** (1998), 1800.
[5] I. Souza, T. Wilkens, R.M. Martin, *Phys. Rev. B* **62** (2000), 1666.
[6] R.D. King-Smith and D. Vanderbilt, *Phys. Rev. B* **47** (1993), 1651.
[7] G. Ortiz and R.M. Martin, *Phys. Rev. B* **49** (1994), 14202.
[8] R. Resta, *Rev. Mod. Phys.* **66** (1994), 899.
[9] I. Souza, J. Iniguez and D. Vanderbilt, *Phys. Rev. Lett.* **89** (2002), 117602.
[10] G.-M. Rignanese, F. Detraux, X. Gonze, A. Bongiorno and A. Pasquarello, *Phys. Rev. Lett.* **89** (2002), 117601.
[11] J.I. Krugler, C.G. Montgomery and H.M. McConnell, *J. Chem. Phys.* **41** (1964), 2421.
[12] Z.G. Soos, H.J. Keller, W. Moroni and D. Nothe, *Ann. N. Y. Acad. Sci.* **313** (1978), 442.
[13] F. Gutmann and L.E. Lyons, *Organic Semiconductors*, Wiley, New York, 1967, Ch. 6. 2.
[14] M. Pope and C.E. Swenberg, *Electronic Processes in Organic Crystals*, Clarendon, Oxford, 1982.
[15] E.A. Silinsh and V. Cápek, *Organic Molecular Crystals*, AIP Press, New York, 1994.
[16] E.V. Tsiper and Z.G. Soos, *Phys. Rev. B* **68** (2003), 085301.
[17] Z.G. Soos, E.V. Tsiper and R.A. Pascal, Jr., *Chem. Phys. Lett.* **342** (2001), 652.
[18] R.M. Metzger, *Crystal Cohesion and Conformational Energies*, (Vol. 26), Topics in Current Physics, Springer-Verlag, Berlin, 1981.
[19] E.V. Tsiper and Z.G. Soos, *Phys. Rev. B* **68** (2003), 085301.
[20] M.C. Zerner, G.H. Loew, R.F. Kirchner and U.T. Mueller-Westerhoff, *J. Amer. Chem. Soc.* **102** (1980), 589.
[21] J.W. Rohleder and R.W. Munn, *Magnetism and Optics of Molecular Crystals*, Wiley, New York, 1992.
[22] J.M. Sin, E.V. Tsiper and Z.G. Soos, *Europhys. Lett.* **60** (2002), 743.

[23] N. Karl, H. Rohrbacher and D. Siebert, *Phys. Stat. Solidi A* **4** (1971), 105; R.W. Munn, J.R. Nicholson, H.P. Schwob and D.F. Williams, *J. Chem. Phys.* **58** (1973), 3828; I. Nakada, *J. Phys. Soc. Jpn.* **17** (1962), 113; A.N. Winchell, *The Optical Properties of Organic Compounds*, second ed., Academic Press, New York, 1954.
[24] H. Reis, M.G. Papadopoulos, P. Calaminici, K. Jug and A.M. Köster, *Chem. Phys.* **261** (2000), 359.
[25] W.R. Salaneck, K. Seki, A. Kahn, J.-J. Pireaux, eds, *Conjugated Polymer and Molecular Interfaces*, Marcel Dekker, New York, 2001, 14.
[26] E.V. Tsiper, Z.G. Soos, W. Gao and A. Kahn, *Chem. Phys. Lett.* **360** (2002), 47.
[27] S.R. Forrest, *Chem. Rev.* **97** (1997), 1783.
[28] G. Horowitz, *Adv. Mater.* **10** (1998), 365; H.E. Katz, *J. Mater. Chem.* **7** (1997), 369.
[29] S.R. Marder, B. Kippelen, A.K.-Y. Jen and N. Peyghambarian, *Nature* **388** (1997), 845.
[30] J.L. Oudar and D.S. Chemla, *J. Chem. Phys.* **66** (1997), 2664.
[31] R.S. Mulliken, *J. Am. Chem. Soc.* **74** (1952), 811.
[32] A. Painelli and F. Terenziani, *J. Phys. Chem. A* **104** (2000), 11041; F. Terenziani, A. Painelli, and D. Comoretto, *ibid.* **104** (2000), 11049.
[33] B. Boldrini, E. Cavalli, A. Painelli and F. Terenziani, *J. Phys. Chem. A* **106** (2002), 6286.
[34] A. Painelli, F. Terenziani, *J. Amer. Chem. Soc.* **125** (2003), 5624; F. Terenzianiabd A. Painelli, *Phys. Rev. B* **68** (2003), 165405.
[35] C. Reichardt, *Chem. Rev.* **94** (1994), 2319.
[36] Y.V. Pereverzev, O.V. Prezhdo and L.R. Dalton, *Chem. Phys. Lett.* **340** (2001), 328.
[37] A. Painelli and F. Terenziani, *Synth. Metals* **139** (20030, 777.
[38] A. Painelli and A. Girlando, *J. Chem. Phys.* **84** (1986), 5665.
[39] L. Del Freo and A. Painelli, *Chem. Phys. Lett.* **338** (2001), 208.
[40] L. Del Freo, F. Terenziani and A. Painelli, *J. Chem. Phys.* **116** (2002), 755.
[41] Z.G. Soos and D.J. Klein, in: *Molecular Associations*, (Vol. 1), R. Foster, ed., Academic, London, 1975, Chap. 1.
[42] J.B. Torrance, J.E. Vazquez, J.J. Mayerle and V.Y. Lee, *Phys. Rev. Lett.* **46** (1981), 253; J.B. Torrance, A. Girlando, J.J. Mayerle, J.I. Crowley, V.Y. Lee, P. Batail and S.J. LaPlaca, **47** (1981), 1747.
[43] A. Girlando, A. Painelli, S.A. Bewick and Z.G. Soos, *Synth. Metals* **141** (2004), 129.
[44] S. Horiuchi, Y. Okimoto, R. Kumai and Y. Tokura, *Science* **299** (2003), 229–232; S. Horiuchi, Y. Okimoto, R. Kumai and Y. Tokura, *J. Amer. Chem. Soc.* **120** (1998), 7379.
[45] Z.G. Soos, S.A. Bewick, A. Painelli and A. Peri, *A. Painelli J. Chem. Phys.* **120** (2004), 6712.
[46] A.J. Heeger, S. Kivelson, R.S. Schrieffer and W.-P. Su, *Rev. Mod. Phys. 60* (1988), 781.
[47] Y. Anusooya-Pati, Z.G. Soos and A. Painelli, *Phys. Rev. B* **63** (2001), 205118.
[48] J.W. Bray, L.V. Interrante, I.S. Jacobs and J.C. Bonner, in: *Extended linear chain compounds*, (Vol. 3), J.S. Miller, ed., Plenum, New York, 1983, p. 353.
[49] L. Del Freo, A. Painelli and Z.G. Soos, *Phys. Rev. Lett.* **89** (2002), 27402.
[50] C. Cojan, G.P. Agrawal and C. Flytzannis, *Phys. Rev. B* **15** (1977), 909.
[51] M.J. Rice and E.J. Mele, *Phys. Rev. Lett.* **49** (1982), 1455.

Journal of Computational Methods in Sciences and Engineering 4 (2004) 721–736
IOS Press


# Electric properties of molecules using stationary coupled-cluster method

Sourav Pal* and Nayana Vaval
*National Chemical Laboratory, Physical Chemistry Division, Pune 411008, India*

**Abstract.** In this review, we record recent developments of coupled-cluster methods for molecular electric properties. We focus on the stationary coupled-cluster response method and various coupled-cluster functionals. We note the important properties of extended coupled-cluster functional and recent interest in this. Recent results of properties using the above approach are reported to emphasize the point.

Keywords: Molecular properties, coupled cluster method, stationary response approach

## 1. Introduction

Single reference coupled cluster (SRCC) method [1–3] has been widely established as the most successful method for accurate calculation of ground state energy and energy derivatives [4,5] of closed shell molecules. SRCC, in its most conventional form, non-variational coupled cluster (NCC), has been used for various applications in quantum chemistry. Due to exponential wave-operator, CC methods introduce dynamical electron correlation in an effective manner and provide size-extensive values for energy and energy derivatives. Various versions of the CC e.g. unitary CC (UCC) [6,7], expectation value CC (XCC) [8–10] and extended coupled cluster (ECC) [11–14] have also been studied recently for calculation of energy and energy derivatives. Each of these methods has its own merits and de-merits. NCC method has been established as a standard black box tool for closed ground state systems [15–17], due to significant success of this in evaluation of energy and energy derivatives with respect to external perturbation, like electric/magnetic field or geometry. The potential of single reference based CC methods, in particular, NCC, has also been noted for single bond stretching [18,19], although, for bond breaking and general quasi-degenerate situations, multi-reference [20–22] description is physically the most appropriate way to include non-dynamic correlation associated with these. However, CC theories based on multi-reference model spaces, are structurally more complex and computationally more demanding than the single reference based theories. Hence, parallel to excellent improvement of different versions of the multi-reference based CC theories in last two decades, there has been a natural effort towards pushing the success of the single reference based CC theories to account for quasi-degeneracy [23,24]. This effort has continued for both energy and electric properties [25]. This review will be broadly on the potential of single reference based CC methods for molecular properties.

*Corresponding author. E-mail: pal@ems.ncl.res.in.

1472-7978/04/$17.00 

We will note that for energy derivatives, specially higher energy derivatives, there have been significant attempts in exploring the utility of different version of CC theory. In this connection, variational forms of CC theory have been pursued using both XCC and ECC types of functional. This review will highlight particularly the ECC functional based methods for the computation of energy derivatives.

Numerical differentiation of the energy is the simplest way of calculating energy derivatives [26]. However, there is a serious loss in accuracy of the energy derivatives in this numerical approach, particularly for higher energy derivatives. Analytical approach constitutes a compact means of calculating energy derivatives through appropriate algebraic expression. The first analytic energy derivative expression for single reference based CC theory was obtained by Monkhorst in the context of NCC method [4]. This also set a platform for the general linear response approach. Parallel to this, Mukherjee and Mukherjee [27] developed a linear response approach to NCC method for general time-dependent electric field for obtaining excitation energies. In the response approach of Monkhorst [4], energy derivatives were linked to the CC amplitude derivatives through algebraic expression. CC amplitude derivatives were obtained through a set of closed expression following analogous non- variational approach. Thus, explicitly, the non-variational structure implies that the nth derivative of energy depends on the nth derivative of wave function. Later, use of an algebraic technique, called z-vector technique [28] using Dalgarno's interchange theorem [29], simplified the energy derivative problem. On the other hand, variational methods based on XCC [30] or UCC [31] functional enjoy the advantages of a $(2n+1)$ rule derived from variational structure. Closely related to the $z$ vector technique is the constrained variational approach of Jorgenson and co-workers [32] in which the Lagrange multipliers associated with the NCC equations are the same as the de-excitation $\xi$ vector amplitudes of $z$ vector technique. A fully variational approach using expectation value CC functional was first used by Pal et al. [33] for energy calculations and then used by Pal and co-workers for property calculations [10,34,35]. Similarly, ECC functional has been now used for energy and property calculations in variation based approach by Arponen and Bishop [11–14] and us [36,37]. Other groups, notably Piecuch and Bartlett [38] and Head Gordon and co-workers [39], also used ECC functional with a slightly modified approach. Head Gordon and co-workers, using a quadratic left and right vectors and Brueckner orbitals in doubles only approximation, were able to reproduce even $N_2$ triple bond dissociation satisfactorily. Although really high quality triple bond stretch results demand a good multi-reference description, the results of Head Gordon and co-workers show encouraging promise of the single reference based ECC functional. The useful point, in particular, is to highlight the fact that the variational ECC produces fully size-extensive energy and properties [36]. In view of the recent interest in the functional, we would like to record in this review the important recent developments in this area. The review is organized as follows: In Section 2.1 we briefly review the NCC method for energy and analytic properties. In Section 2.2 and 2.3 we review the stationary variational approach for energy and energy response and subsequently discuss the use of XCC and UCC functional in Section 2.4 and 2.5 respectively. The demerits of this functional are noted. In Section III, we turn to the main point of this review, the ECC functional. Its merits and demerits are also highlighted. Responses of ECC functional with and without relaxation of orbitals are discussed. Section IV contains results on energies and properties, which show the promise of the theory. Most of the results of properties, discussed in this review, using variational response of the ECC functional are obtained by our group.

## 2. Analytic response approach

### 2.1. Non-variational coupled cluster

The NCC wave function is written as

$$\Psi = \exp(T)|\Phi\rangle \tag{1}$$

where $|\Phi\rangle$ is the Hartree Fock determinant. The cluster operator T can be decomposed into various lower body excitations,

$$\begin{aligned} &T = T_1 + T_2 + T_3 + \ldots T_n \\ &T_1 = \Sigma_{i,a} t_i^a \{a^+ i\}; T_2 = 1/4 \quad \Sigma_{ijab} t_{ij}^{ab} (a^+ b^+ ji\} \end{aligned} \tag{2}$$

$i, j, \ldots$ are the occupied orbitals in Hartree-Fock and $a, b, c \ldots$ are the virtual (unoccupied) orbitals. $t_i^a, tij^{ab}$ denote the cluster amplitudes. $a^+, b^+$ are particle creation and $i, j, \ldots$ etc are hole creation operators. To get the energy and cluster amplitudes, the following equations are solved

$$E(\lambda) = \langle\Phi| \exp(-T^{(\lambda)}) H \exp(T^{(\lambda)})|\Phi\rangle \tag{3a}$$

$$0 = \langle\Phi^*| \exp(-T^{(\lambda)}) H \exp(T^{(\lambda)})|\Phi\rangle \tag{3b}$$

where $\langle\Phi^*|$ are the excited state determinants and rank of excitation matches with that of $T$ operator. NCC analytic response equations are obtained by taking derivatives of Eqs (1) and (2) with respect to external perturbation $\lambda$.

$$E^{(1)} = \langle\Phi|\delta/\delta\lambda(H \exp(T(\lambda)))c|_{\lambda=0} + (\mathrm{O} \exp(T(0))c|\Phi\rangle \tag{4a}$$

$$0 = \langle\Phi^*|(\delta/\delta\lambda(H \exp(T(\lambda))))c|_{\lambda=0} + (\mathrm{O} \exp(T(0))c|\Phi\rangle \tag{4b}$$

Equation (4a) and (4b) show that the first derivative of the energy depends on the first derivative of the cluster amplitudes. In addition to the above, one can incorporate change of orbitals with respect to external field. This introduction of relaxation of orbitals is known as relaxed method. However, while this can be done easily, the nontrivial part was to introduce the z vector technique to eliminate the dependence on the derivative cluster amplitudes. The $\xi$ amplitudes are known to satisfy the following equation.

$$\langle\Phi|\xi|\rangle\langle\Phi^*|(H \exp(T)c - \Delta E|\Phi^*\rangle + \langle\Phi|(H \exp(T))|\Phi^*\rangle = 0 \tag{5}$$

$\xi$ has single, double etc de-excitation amplitudes. The above equation was shown to have fully connected form. It is observed that the $\xi$ amplitudes do not depend on any perturbation dependent term. Using these amplitudes the derivative energy can be written as

$$E^{(1)} = \langle\Phi|(H^{(1)} \exp(T))_c|\Phi\rangle + \langle\Phi|\xi|\Phi^*\rangle\langle\Phi^*|(H^{(1)} \exp(T))_c|\Phi\rangle \tag{6}$$

Where, $H^{(1)}$ is the derivative of the Hamiltonian with respect to the electric field. Thus, the first derivative of energy is independent of derivative cluster amplitudes and instead depends on perturbation independent $\xi$ amplitudes. This simplifies the NCC response approach, which has been used for both equilibrium and single bond stretching. However, away from equilibrium, the relaxation becomes dominant and consequently the singles amplitudes in a non-relaxed model becomes very large causing convergence problems. To incorporate the effects of relaxation, additional terms involving the change of molecular orbitals in terms of perturbation have to be included. For electric field perturbation, this means inclusion of change in molecular orbital coefficients via the coupled perturbed Hartree Fock method. To use the NCC response for geometry derivatives, it is mandatory to use relaxed approach. However, in this case, evaluation of molecular orbital changes dictates that the basis of atomic orbital be relaxed along with change in coefficients.

The algebraic method of elimination of cluster amplitude derivatives, as described for the first derivative above, is extremely laborious for application to second and higher energy derivatives. Bartlett and co-workers showed algebraically, that the second energy derivative can be obtained only from the first derivative of the cluster amplitudes. However, in a structure of variational theory, this aspect becomes explicit. Jorgensen and co-workers [32] used the constrained variational approach, where the energy functional is a NCC functional, with the NCC equations used as constraints. The Lagrange multipliers for these constraints are basically the $\xi$ amplitudes previously described. The variation of the subsequent Lagrangian yields exactly the NCC method. Linear response of the Lagrangian forms the basis for higher order properties. Parallel to this, alternate variational form of CC has been attempted. This form of variational theory is a direct unconstrained variation of an appropriate energy functional. In the next subsections we will discuss such variational theories emanating primarily our group.

### 2.2. *Stationary coupled cluster theory*

Although CC theory has been convenient in non-variational framework, stationary form of theory has been attractive primarily due to the attendant advantages of variational theory. Using an expectation value CC functional, such theories were formulated in early eighties by Pal et al. [33]. Unlike in constrained variation method, in stationary CC approach a pre-chosen energy functional is made stationary without any constraints. Differentiation or variation of the energy functional with respect to the parameters in it gives equations defining amplitudes of the wave-function. When we differentiate the energy functional with respect to parameters, it basically implies that in the equation for the cluster amplitudes, the corresponding vertex is taken off from the closed energy diagram leaving an open diagram. These stationary amplitudes provide the first energy derivative, which does not depend on the response of the amplitudes. However, to obtain higher order derivatives, we need the response of the cluster amplitudes. Using $(2n+1)$ rule, first derivative of cluster amplitudes determines energy derivatives up to third order.

The expectation value is the most natural form of the functional for variational theory. The functional can be written as,

$$E = \langle\Phi|\exp(T^{+})H\exp(T)|\Phi\rangle/\langle\Phi|\Phi\rangle \tag{7}$$

where $T^{+}$ is the hermitian conjugate of $T$. The functional can be shown to have a connected or linked form after denominator cancellation. It can be formally shown to be of the form

$$E = \langle\Phi|\exp(T^{+})H\exp(T)|\Phi\rangle_L \tag{8}$$

Where, $L$ denotes only the linked or connected diagrams of $(\exp(T^{+})H\exp(T))$ operator. This functional is, however, a hermitian infinite series and needs truncation. Several truncations of this functional have been used. Alternate energy functionals having better convergence properties are also in recent use. These will be addressed in later part of the review. Before we do this, let us discuss the stationary approach to obtain response of the CC density and obtain higher energy derivatives.

### 2.3. Stationary response approach

In this section we will discuss response approach using stationary method for a general energy functional. Suppose energy functional $E$ is a functional of the cluster amplitudes $X$, which depend on external parameter $\lambda$ e.g. electric field strength)

$$E(\lambda) = F\{X(\lambda)\} \tag{9}$$

Use of Taylor's series expansion for $E$ as well as $X$ in terms of $\lambda$ gives us

$$E(\lambda) = E^{(0)} + \lambda E^{(1)} + \lambda^2 E^{(2)} + \lambda^3 E^{(3)} + \dots\dots\dots \tag{10a}$$

$$X(\lambda) = X^{(0)} + \lambda X^{(1)} + \lambda^2 X^{(2)} + \dots\dots\dots \tag{10b}$$

A set of equations may be defined by taking the derivatives of $E^{(i)}$ with respect to $X^{(j)}$ equal to zero for $ij$. It has been shown [11] that,

$$\delta E^{(i)}/\delta X^{(j)} = 0 \tag{11}$$

Provides identical set of equations for a fixed value of $(i - j)$. Hence it is sufficient to solve

$$\delta E^{(i)}/\delta X^{(0)} \quad \text{for } i = 0, 1, 2, 3 \dots \tag{12}$$

The above set provides sufficient equations for the cluster and the derivative cluster amplitudes in a hierarchical fashion. For electric field response properties using coupled cluster functional, cluster amplitudes are the variables X's and the perturbation parameters are electric field. The above method does not depend on any constraint and the equations were first solved using the XCC functional. We will now discuss the truncations of XCC functional and some other alternate desirable functionals, which have been used recently.

### 2.4. Expectation value functional

Expectation value functional is the most natural functional in variational approach. Energy functional in the presence of perturbation can be written as

$$E(\lambda) = \langle \Phi | \exp(T^{+}(\lambda)) H(\lambda) \exp(T(\lambda)) | \Phi \rangle_L \tag{13}$$

The above functional does not use orbitals, which depend on the external perturbation i.e. electric field. The use of this functional in response approach leads to what is known as non-relaxed approach. It is usually enough to use the singles and doubles approximation for the cluster amplitudes. However, even in finite many body truncation of $T$, the energy functional is an infinite series and needs truncation for practical application. There are two different types of truncation. One is to truncate the series keeping a fixed number of $T$ amplitudes in the energy functional. Pal and co-workers have followed this truncation scheme with a maximum of cubic order of amplitudes in energy functional. In this truncation, though the energy functional itself remains a linked series, the equations for the cluster amplitudes contain disconnected terms. Differentiation of the energy functional with respect to $T$ means deletion of the corresponding $T$ vertex from the closed energy diagram leaving an open diagram. Whenever any particular $T/T^{+}$ is exclusively connected to the other $T^{+}/T$, differentiation with respect to that $T^{+}/T$ leads to disconnected or unlinked diagrams. Thus, properties obtained using this functional are not size-extensive. This functional has been used by Pal and co-workers for test calculation of energy and properties. A second truncation scheme is based on perturbation guidelines. In this truncation scheme all the diagrams are considered, the contribution of which start from a given order of perturbation. Bartlett and co-workers followed this truncation scheme. Truncation scheme, guided by perturbation, is known as XCC(n) [30]. However, in this truncation scheme, the truncation of the cluster amplitudes to singles and doubles approximation is irrelevant. One needs to include the rank of many-body excitation operators as dictated by the perturbation order. Equations obtained using this truncation are always size-extensive.

There are several other functionals, which have been used recently. Bartlett and co-workers proposed a modification of the XCC functional, known as symmetric expectation value CC (SXCC) functional [8, 40]. Energy functional assumes the following form:

$$E(T,T^{+}) = 1/2[\langle\Phi|\{\exp(T^{+})(H\exp(T))_{c}\}|\Phi\rangle + \langle\Phi|\exp(T^{+})H)\exp(T)\}_{c}\Phi\rangle] \quad (14)$$

The expression is hermitian and symmetric as in the normal XCC functional. As a consequence, we need to make the functional stationary only with respect to $T$ or $T^{+}$ as in previous XCC functional. At the stationary point in SXCC functional we do not recover the exact amplitudes and thus the functional is not exact in the limit. Van Voorhis and Head Gordon [41] demonstrated that the variational coupled cluster doubles (VCCD) calculations, using Brueckner orbitals based on minimizing the expectation value of the Hamiltonian with the CCD wave function, lead to a qualitatively correct description of triple bond breaking in $N_2$.

$$E^{\text{VCCD}} = \langle\Phi|\exp(T_2^{+})H\exp(T_2)|\Phi\rangle/\langle\Phi|\exp(T_2^{+})\exp(T_2)|\Phi\rangle \quad (15)$$

Though it sounds encouraging it is not easy to perform the VCCD calculations. The main problem is that the resulting energy expressions are non-terminating.

### 2.5. Unitary CC functional

Following Van Vleck [42] and Primas [43], Kutzelnigg [44] suggested a unitary ansatz initially in a variational framework. The exact correlated CC wave function employs an anti-hermitian cluster operator $\sigma = T - T^{+}$.

$$|\Psi\rangle = \exp(\sigma)|\Phi\rangle \quad (16)$$

A variational or a non-variational method can be used to obtain the cluster amplitudees $\sigma$. Westhaus [45] and Mukherjee [46] suggested a non-variational method similar to Cizek to calculate cluster amplitudes. Using similarity transformation, one can write effective Hamiltonian $H_{\text{eff}}$ as

$$H_{\text{eff}} = \exp(-\sigma)H\exp(\sigma) \quad (17a)$$

$$E_o = \langle\Phi_o|\exp(-\sigma)H\exp(\sigma)|\Phi_o\rangle \quad (17b)$$

$$0 = \langle\Phi^{*}|\exp(-\sigma)H\exp(\sigma)|\Phi_o\rangle \quad (17c)$$

When expanded in terms of $T$ and $T^{+}$, it is exactly an XCC functional and is thus an infinite, but connected series. This functional can be used for a variational approach. The expectation value functional using UCC wavefunction may be written as,

$$E = \langle\Phi|\exp(-\sigma)H\exp(\sigma)|\Phi\rangle/\langle\Phi|\exp(-\sigma)\exp(\sigma)|\Phi\rangle \quad (18)$$

The denominator taken to all orders is identically equal to unity leads to

$$E = \langle\Phi|\exp(-\sigma)H\exp(\sigma)|\Phi\rangle_{\text{conn}} \tag{19}$$

Thus the form of the functional is same as the similarity transformed energy Eq. (17b). However, the variational procedure using this functional will lead to equations for cluster amplitudes different from the ones obtained by non-variational approach. Kutzelnigg [47] used stationary procedure to solve for the amplitude equation. However, the energy functional, being an infinite series, need to be truncated as in the case of variational approach for XCC. A fixed power truncation will lead to disconnected terms, in general, in the equations exactly as in the case of variational method with XCC functional. Bartlett and co-workers [31] used UCC functional for properties using variational approach, but with truncation guided by perturbation, which they called as UCC(n) method.

## 3. Dear author please provide chapter name

### 3.1. ECC functional

The bi-orthogonal functional, as proposed by Arponen and co-workers [11–14], is also known as ECC functional. The functional is based on double-similarity transformation. This functional has many desirable features in the context of variational method and thus a lot of research has been directed toward using this functional for energy and energy derivatives using stationary method. Expectation value of Hamiltonian in ECC can be written as,

$$\langle H\rangle = \langle\psi'|H|\psi\rangle \tag{20}$$

Where, $|\psi\rangle$ and $|\psi'\rangle$ are parameterized differently and bear a bi-orthogonal relation.

$$|\psi\rangle = \exp(T)|\Phi\rangle \tag{21a}$$

$$\langle\Psi'| = \langle\Phi|\Omega\exp(-T) \tag{21b}$$

$$\Omega = 1 + \Sigma_i'\overline{\tau}_i c_i \quad T = \Sigma i'\tau_i c_i^+ \tag{21c}$$

Where, $\Omega$ is a linear operator, which includes hole-particle de-excitation operator. Expectation-value of the Hamiltonian can be written as

$$\langle H\rangle = \langle\Phi|\Omega\exp(-T)H\exp(T)|\Phi\rangle$$

Stationarity of this functional with respect to $\Omega$ leads to NCC equations Since the functional is only linked or connected amplitudes $\tau_i$ are connected however, amplitudes $\overline{\tau}_i$ contains disconnected pieces. However, $\overline{\tau}_i$ amplitudes are not required for energy. Arponen and Bishop then used fully exponential parametrization for $\Psi$ and $\Psi'$. While the ket $|\Psi\rangle$ remains as before, the conjugate state $\langle\Psi'|$ is parametrized as

$$\langle\Psi'| = \langle\Phi|\exp(S'')\exp(-T) \tag{22a}$$

$$S'' = \Sigma_k s_k'' C_k \tag{22b}$$

Where $S''$ is a de-excitation operator, i.e. $S''|\Phi\rangle = 0$ and thus the ket vector can be written as $|\Psi\rangle = \exp(T)\exp(-S'')|\Phi\rangle$. The average value functional of the Hamiltonian H or any operator may be written as

$$E = \langle H\rangle = \langle\Phi|\exp(S'')\exp(-T)H\exp(T)\exp(-S'')|\Phi\rangle_L \tag{23}$$

Arponen and co-workers introduced a new set of operators $\Sigma$ and $\overline{\Sigma}$ transforming and $T$ and $S''$ as

$$\Sigma|\Phi\rangle = (1 - |\Phi\rangle\langle\Phi|)\exp(\overline{\Sigma})T|\Phi\rangle \tag{24a}$$

$$\Sigma = \Sigma' S_i'' C_i^+ \tag{24b}$$

Inverse transformation is

$$S_i'' = \langle\Phi|C_i \exp(S'')T|\Phi\rangle \tag{25}$$

$$\tau_i = \langle\Phi|C_i \exp(-\overline{\Sigma})\Sigma|\Phi\rangle \tag{26}$$

With these, the final form of the functional is

$$\langle\Phi|\exp(\overline{\Sigma})(H\exp(\Sigma))_L|\Phi\rangle_{DL} = E \tag{27}$$

where, $L$ denotes that $\Sigma$ operator to the right of Hamiltonian must be connected to it and DL indicates that $\overline{\Sigma}$ operator to the left of Hamiltonian must either be connected to $H$ or to two different $\Sigma$ vertices. The derivation and aspects of this form of the functional have been derived and discussed at length by Arponen and Bishop [11–14]. The double-linked form ensures that the functional naturally terminates. Linking of right vector with H implies that the $(H\exp(\Sigma)_L$ contains finite number of T-operators. Thus, the series terminates. Double-linking ensures connectivity of the functional in case of any vertex being deleted. The amplitudes of $\overline{\Sigma}$ and $\Sigma$ may be obtained by making energy functional stationary with respect to the cluster amplitudes,

$$dE/d\overline{\sigma} = 0 \text{ and } dE/d\sigma = 0 \tag{28}$$

Since the functional $E$ is non-hermitian, the left and right vectors are not conjugates of each other. Connectivity of the functional in the event of deletion of vertex implies that stationary equations contain diagrams, which are fully connected, unlike the variational equation for XCC functional. Though this functional is naturally terminating, due to double linked nature of the functional, actual termination is at quite high order. Thus, for practical application it needs to be truncated. Price one pays for the truncating series is we have two different set of cluster amplitudes to be solved and thus the double the number of equations too. Equations for the left and right amplitudes are coupled. Due to the inherent connectivity in the functional and the equations, this method leads to size extensive properties at each order of truncation. Pal and co-workers [36,37] have successfully used ECC functional for various systems of chemical interest. Coupled nature of left and right amplitudes make this method bit costly in terms of computation. This restricts its application for large systems. Vaval [48] proposed a de-coupled scheme to solve the left and right amplitudes.

Along the lines of symmetrized XCC functional, Bartlett and co-workers [38] suggested a symmetrized ECC functional, which yields same equations as the normal ECC functional, properties and density matrices are explicitly hermitian.

Head Gordon and co-workers used the ECC functional in which $\Sigma = \Sigma_2$ and $\overline{\Sigma} = \overline{\Sigma}_2$, with quadratic truncation scheme for the left and right exponentials of these cluster amplitudes. This is known as quadratic ECC (QECCD) and this contains some terms, which consist of a total of quartic power in the amplitudes. As in the case of Quadratic XCC, they used Bruckner orbitals for this purpose. QECCD method was used by Head Gordon and co-workers [39] for the potential energy surface calculations with a reasonable degree of success.

We will now discuss, in detail, applications from our group using a total of cubic truncation in the ECC functional.

### 3.2. *Cubic-ECCSD approximation*

First, we discuss the validity of the cubic truncation in the ECCSD scheme. To do this, we consider how important are the terms at the next higher degree in the polynomial $E_0$. Denoting the two particle part of $H$ as $V$, at the quartic approximation, the important terms in the functional arise from $(\Sigma_2(v\Sigma_1^3)_L)_{DL}$, $(\overline{\Sigma}_2(v\Sigma_1^2\Sigma_2)_L)_{DL}$ and $(\overline{\Sigma}_1(v\Sigma_1^3)_L)_{DL}$ in singles and doubles approximation. The constraint of double-linking helps in the reduction of terms. However, we show that these expensive quartic terms are at high perturbation order. Using the restricted Hartree-Fock determinant as the zeroth order wave function (which is a very good zeroth approximation around the equilibrium), one can derive the dominant perturbation order at which the amplitudes of $\overline{\Sigma}$ and $\Sigma$ operators contribute from the equations determining these amplitudes. The leading terms in the equation of $\sigma_2$ and $\overline{\sigma}_2$ amplitudes are of the first order and those of the $\sigma_1$ and $\overline{\sigma}_1$ amplitudes are of the second order as in the standard single reference CC theory. Thus, one can see that the quartic terms in the ECCSD functional are at least seventh or higher order in perturbation. The corresponding terms in the equations of $\overline{\sigma}_2/\sigma_2$ amplitudes will be only one order lower, as these are derived from deletion of one $\Sigma_2/\overline{\Sigma}_2$ vertex, which is of the first order in perturbation. Similarly, in the equations for one-body amplitudes, the contributions of these terms are two orders lower. In case, where the RHF determinant is a good zeroth description, as is the case of equilibrium geometry, these high order terms will not have any significant contribution to the amplitude equations and consequently energy functional. However, away from equilibrium, where there is a high degree of quasi-degeneracy and the single reference approximation breaks down, the RHF determinant is not a good zeroth order approximation to the wave function and thus, the contributions of higher order terms may not be so significant. However, the cubic ECCSD is expected to work reasonably for geometries around equilibrium.

A similar analysis can be done to the linear response functional of the cubic ECCSD method and the response equations. We first observe that the leading term in Eqs (4a–b) for the one body $\sigma_1^{(1)}$ amplitudes comes from the dipole moment integral (which is one-body $H^{(1)}$ term of the $E^{(1)}$ functional). Thus, these amplitudes have zero perturbation order in $V$. The leading perturbative order contribution to the two body response amplitudes, on the other hand, comes from inhomogeneous term $\mu\sigma_2$, obtained as a result of stationarity of the $(\overline{\sigma}_2(\mu\sigma_2)_L)_{DL}$ term in $E^{(1)}$. Thus, this is at the first order in perturbation $V$. We note that in our non-relaxed model, there is no term corresponding to $H^{(2)}$. Quartic terms in ECCSD response functional $E^{(1)}$ arise from the response of the terms mentioned above for the quartic $E^{(0)}$. This reveals that the most important $(\overline{\Sigma}_2(v\Sigma_1^{(1)}\Sigma_1\Sigma_{(2)})_L)_{DL}$ term is at least at the fifth order and our analysis shows that the stationarity of this term with respect to $\overline{\Sigma}_2/\Sigma_2$ vertex and $\Sigma_1$ vertex contributes at the fourth and third order respectively. Other quartic terms have contributed at even higher order. Thus, the quartic terms can play more important role in the response equations compared to the calculation of

Table 1
LiH dimer results for size-extensivity test in DZP basis[a,c]

| | LiH | | | | ((Li H)2)/2 | | | |
|---|---|---|---|---|---|---|---|---|
| Prop | SXCCSD | | ECCSD | | SXCCSD | | ECCSD | |
| | Quad | Cubic | Quad | Cubic | Quad | Cubic | Quad | Cubic |
| $\mu_z$ | 1.985 | 1.976 | 1.982 | 1.995 | 1.985 | 1.976 | 1.982 | 1.995 |
| $\alpha_{zz}$ | 13.66 | 9.81 | 14.33 | 12.89 | 13.67 | 9.77 | 14.32 | 12.89 |
| $\beta_{zzz}$ | −771.0 | −277.7 | −852.9 | −745.0 | −777.1 | −275.9 | −852.9 | −745.8 |

[a]: Results from ref. [36]
[b]: LiH distance in monomer is 2 au; from Ref [36]; monomers are separated by 9000 au
[c]: All values are in au.

energy. However, if the RHF is a good zeroth order approximation, these are still at sufficiently high order to be of importance. However, away from equilibrium this analysis breaks down.

The ECC functional, due to the presence of both left and right vectors, has the potential of summing more terms than the NCC method with basic contributions of the terms at a specific order. Thus, it is expected that implementation of progressively more elaborate ECCSD method will lead to more accurate results compared to the corresponding NCCSD method. However, the computational complexities also increase as we go from cubic to quartic and so on. Upto cubic ECCSD, the computational time scales as $N^6$, as in NCCSD method. Our applications are restricted to the cubic ECCSD approximation, which is sufficient for most purpose. The ECC method contains an additional set of amplitudes and this is one disadvantage the method bears compared to the standard NCC method.

## 4. Some model results and discussion

We have extensively used the ECCSD model in its cubic approximation for various molecules. Our early studies were based on XCCSD model. However due to size-extensivity problem we did not pursue it for a long. In this review, we will report some results of our group as well as from other group.

Firstly we will report the size-extensivity results we carried out for LiH and dimer of LiH [36]. We used XCCSD as well as ECCSD models for size-extensivity study in quadratic and cubic approximations. Double zeta plus a polarized basis was used for the test. The two monomers of LiH were separated by 9000 au and the bond distance for each LiH has been taken as 2 au for this model test. We see from results in Table 1 that dipole moment is size-extensive in XCCSD approximation. However, as we go for higher order properties we see that XCCSD cubic approximation differs. It should be mentioned that XCCSD in its quadratic approximation is always size-extensive. It can be formally shown that the stationarity of quadratic XCCSD functional does not involve any disconnected terms. Terms contributing to loss of size-extensivity starts to appear from cubic approximation onwards.

We recently studied [37] weakly bounded systems i.e. water dimer, hydrogen fluoride dimer and water – HF complex. These clusters these are interesting systems to study due to dipole bound nature. We will report here the results for the Hydrogen fluoride and its dimer. One of the basis sets consists of double-zeta valence (9s5p/4s primitive Gaussian type orbitals contracted to 4s2p/2s) augmented by a set of uncontracted p polarization functions with exponent 0.70 on hydrogen and d polarization functions with exponent of 1.6 on fluorine. In addition to this, a diffuse p function was used on fluorine with exponents 0.09. Thus, the final basis consists of 4s3p1d. This basis is denoted as basis set A. Another basis set denoted as basis B, is the Sadlej's optimized basis [49] set for property evaluation. This basis consists of contracted [5s3p2d/3s2p] functions. The energy values of each system have also been

Table 2
Properties of Hydrogen Fluoride ($R = 1.7328$ au) in different basis

| I. Basis | I. Method | I. Energy (au) | $\mu_z$ (au) | $\alpha_{xx}$ (au) | $\alpha_{zz}$ (au) |
|---|---|---|---|---|---|
| I. DZP+diffp (A) | I. SCF | I.-100.050 | 0.820 | 1.71 | 4.45 |
| I. | I. MP2 | I. -100.253 | 0.782 | 1.90 | 4.75 |
| I. | I. ECCSDa | I. -100.271 | 0.777 | 1.82 | 4.69 |
| I. Sadlej (B) | I. SCF | I.-100.055 | 0.757 | 4.33 | 5.60 |
| I. | I. MP2 | I. -100.270 | 0.708 | 5.25 | 6.40 |
| I. | I. ECCSDa | I. -100.285 | 0.707 | 5.12 | 6.30 |
| I. Exptb | I. | I.— | 0.719 | 5.08 | 6.40 |

I. a: See Ref. 37b
b: see Ref. 50
Basis A: 4S3P1D
Basis B: Sadlej's basis optimized for properties.

reported, which highlights the quality of the basis. For comparison, experimental values [50], wherever available, have been quoted. However, these are vibrationally corrected ones. Only for HF monomer, good estimate for the extrapolation of the experimental dipole moment to the non-vibrating limit is available and is reported.

Table 2 reports the properties of HF monomer using basis A and B. In basis A, dipole moment at the SCF level 0.820 au is reduced to 0.777 au by inclusion of electron correlation. In the basis B, which is an optimized basis for properties, having 2 polarized functions for fluorine atom, i.e. 2d functions, the SCF value is reduced to 0.757 au. The inclusion of electron correlation reduces it further to 0.708 au. Thus, dipole moment in basis B is in close agreement with the experimental value (0.719 au). It overestimates by 0.011 au only. Observation of the parallel ($\alpha_{zz}$) and perpendicular ($\alpha_{xx}$) components of polarizability reveals that both components of polarizability are enhanced by the electron correlation compared to the SCF value. The value of the polarizability is increased, as more polarized functions are added, making it in better agreement with the experimental value. However, we observe that perpendicular polarizability component is more sensitive to basis set compared to the parallel polarizability component. In basis A, for example, the parallel polarizability value (4.45 au) is roughly 2 au lower compared to the experimental (6.40 au) value. On the other hand, the perpendicular polarizability (1.82 au) is about 3.26 au less than the experimental value. The error of perpendicular polarizability is more pronounced in terms of percentage error. On the other hand, the perpendicular component of polarizability (5.12 au) improves significantly in basis B showing considerable basis set effect in this component of polarizability.

## 5. Dimer

Hydrogen fluoride dimer has been studied in the same two basis sets as its monomer. The geometry used is a linear Cs structure described in the reference 52. The distance between the two fluorine atoms as $R_{F..F} = 5.2203$ au, with zero bond angle between them. H-F bond length of the proton donor HF monomer is 1.733 au, while that of the proton acceptor HF monomer is 1.737 au. Thus, the proton donor monomer is bit more stretched (0.004 au) compared to the HF monomer at its equilibrium geometry. The RHF wave function for HF dimer in linear Cs symmetry has configuration $1a'^2\ 2a'^2\ 3a'^2\ 4a'^2\ 5a'^2\ 6a'^2\ 1a''^2\ 7a'^2\ 8a'^2 2a''^2$. In Table 3, we report dipole moment and polarizability in basis A and B at the ECCSD level. In basis A, dipole moment at the ECCSD level is 1.486 au. In the most extensive basis B, the dipole moment value calculated at the ECCSD (1.362 au) level is 0.182 au higher compared to the experimental value. However, in both the basis sets dipole moment value is overestimated compared to

Table 3
Properties of HF dimer in different basis

| Basis | Method | Energy (au) | $\mu_{\text{avg}}$ (au) | $\alpha_{xx}$ (au) | $\alpha_{zz}$ (au) | $\alpha_{yy}$ (au) |
|---|---|---|---|---|---|---|
| DZP+diffP(A) | SCF | −200.1077 | 1.509 | 8.23 | 3.47 | 5.47 |
| | MP2 | −200.5133 | 1.499 | 9.27 | 3.81 | 5.91 |
| | ECCSD[a] | −200.5502 | 1.486 | 10.85 | 3.79 | 5.88 |
| Sadlej (B) | SCF | −200.1163 | 1.432 | 11.28 | 8.53 | 9.46 |
| | MP2 | −200.5420 | 1.363 | 13.06 | 10.95 | 10.08 |
| | ECCSD[a] | −200.5795 | 1.362 | 12.71 | 9.69 | 10.60 |
| Expt[b] | | — | 1.18 | | — | — |

a: See Ref [37]; b: see ref. [53]; $R_{F..F} = 5.272$ au, $F_{FH} = 108°$, $R_{HF} = 1.733$ au
Vibrational corrections not included;
Basis A: 4S3P1D
Basis B: Sadlej's basis optimized for properties.

Table 4
Dipole moment and polarizability values in (au) for the ground state of HF($R_e = 1.7328$ au) ) at the equilibrium and other bond lengths in DZP basis

| R | RHF | ECCSD | | NCCSD[c] | FCI[d] |
|---|---|---|---|---|---|
| $\mu_z$ | | Relaxed[a] | Nonrelaxed[b] | Relaxed | |
| $R_e$ | 0.811 | 0.762 | 0.766 | 0.766 | 0.762 |
| I5 $R_e$ | 1.161 | 0.903 | 0.906 | 0.918 | 0.896 |
| 2.5 $R_e$ | 1.887 | 0.263 | 0.134 | 0.270 | 0.217 |
| $\alpha ZZ$ | | | | | |
| $R_e$ | 4.259 | 4.38 | 4.45 | 4.47 | 4.5 |
| 1.5 $R_e$ | 12.193 | 12.46 | 13.53 | 12.26 | 12.3 |
| 2.5 $R_e$ | 44.978 | 10.12 | 22.51 | 13.2 | 11.7 |

[a]: Cubic truncation Scheme, See ref. [54]
[b]: See Ref. [37]
[c]: Full CCSD non-variational scheme; $R_e = 1.7328$ au
[d]: Ref. [55]; Re used is 1.733 au, $1\sigma$ core orbital been frozen, 3S combination of the 3d function has been eliminated.

the experimental value [53] (1.180 au). Since the geometry used by Dyke et al. [53] for the experimental determination of dipole moment of HF dimer is quite different than the one used in this paper, there is discrepancy between the theoretical and experimental dipole moment value. However, the basis set effect is more for the polarizability component perpendicular to the plane of the molecule ($\alpha_{zz}$) than for the in-plane components. The effect of correlation increases the in-plane as well as perpendicular components of polarizability. From the monomer results, we assume that the polarizability results obtained by basis B will be closer to the experimental values.

Some general conclusions are as follows. We observe that the dipole moment values are reduced consistently by inclusion of electron correlation. Thus the dipole moment at MP2 level is lower than the SCF dipole moment and the dipole moment at ECCSD level is lower than that at MP2 level. In addition, in the more extensive basis B, dipole moment is reduced at each level (i.e. SCF, MP2 and ECCSD). However, for polarizability use of more extensive basis set enhances the values at each level. It is also observed that, at MP2 level polarizability values are enhanced compared to SCF values. Further inclusion of electron correlation by ECCSD reduces the polarizability.

In all these results we have used the non-relaxed model i.e., same orbitals for different field values. It is generally expected that dynamic correlation dominates at the equilibrium geometry, where as, at the stretched or away from equilibrium geometry relaxation and non-dynamic correlation will dominate.

Since, we are using approximate functional for the calculation of properties, it is important to know the effect of relaxation on properties at different geometries. To study the effect of relaxation we have recently studied the HF molecule at stretched geometries [54]. Property surface of hydrogen fluoride is studied in DZP as well as Sadlej basis. In both the basis sets, the origin of the co-ordinate system was placed at the molecular center of mass, with the F and H atoms located on the positive and negative sides of the Z-axis respectively. In DZP basis, dipole moment and polarizabilities were obtained at $R_e$, $1.5R_e$ and $2.5R_e$. Sadlej's basis set was used to study properties at $0.75R_e$, $R_e$, $1.5R_e$, $1.75R_e$. Equilibrium bond distance of HF is 1.7328 au. Polarization functions on hydrogen and on fluorine are 0.75 and 1.6 respectively. Table 4 reports the dipole moment and parallel component of the polarizability at various bond lengths. Full CI [55] results quoted in the table have been obtained by freezing the $1\sigma$ core orbital. However, in ECCSD no freezing of orbitals was used. The RHF configuration of HF is

$$(1\sigma)^2(2\sigma)^2(1\pi)^2(2\pi)^2(3\sigma)^2|(4\sigma)^0$$

$3\sigma$ and $4\sigma$ are the highest occupied and lowest unoccupied orbitals respectively. We observe that in the RHF method, dipole moment monotonously increases with $R_e$. In correlated methods it increases till $1.5R_e$ and then decreases. Near equilibrium, where relaxation effects are less important, dipole moment values obtained through all the three CC methods are in close agreement with each other and with FCI. However the difference is enhanced at $2.5R_e$. In particular, the ECCSD without relaxation underestimates the dipole moment compared to FCI. Importance of relaxation at large Re is more prominent for polarizability values. At $R_e$ and $1.5R_e$, polarizability values with and without relaxation are in good agreement with FCI. Though, polarizability value without relaxation is slightly overestimated at $1.5R_e$. At $2.5R_e$ difference between relaxed and non relaxed is quite high (more than 100%) indicating the importance of relaxation for higher order properties at stretched R. The qualitative trend produced by the analytic ECCSD relaxed model is same as the one obtained by FCI method. However, the dipole moment as well as polarizability values are over estimated, in general, compared to the FCI values. At $2.5R_e$ difference between the present cubic-ECCSD and NCCSD could be attributed to the neglect of higher order term like quartic terms. Compared to FCI results, ECCSD analytic results are lower and this could be due to neglect of higher terms as well as neglect of higher body term like triples. Difference between FCI and NCCSD may be attributed to the neglect of three and some higher body cluster amplitudes. Perturbative analysis indicates that some of the important terms arising from three body amplitudes are also important at stretched geometry within the single reference based theory. Although inclusion of full triples is impossible, partial inclusion of triples could be useful.

Table 5 reports dipole moment and polarizability values in Sadlej's basis. We also compare our results with the orthogonally spin-adapted coupled-cluster singles and doubles linear response (OSACCSDLR) results of Kondo et al. [51]. General trends of our results are similar to results in DZP basis and these compare favorably with the OSACCSDLR results of Kondo et al. [51].

## 6. Conclusion

In this review we have discussed latest developments of stationary coupled-cluster theories for electric properties using XCC and ECC functionals. The studies from our group and others emphasize the importance of the ECC functional in this context. The ECC produces fully extensive properties. We review different approaches to the ECC functional, truncation schemes. The results, primarily from our group, using a truncation scheme involving a total of cubic powers in cluster amplitudes, have been

Table 5
Dipole moment and polarizability values (in au) for HF at equilibrium ($R_e = 1.7328$ au) and other stretched geometries in Sadlej's basis

| R | RHF | ECCSD[a] | | OSACCSDLR[b] |
|---|---|---|---|---|
| $\mu_z$ | | Relaxed | Non-relaxed | Non-relaxed |
| $0.75\ R_e$ | 0.590 | 0.570 | 0.518 | 0.518 |
| $R_e$ | 0.757 | 0.707 | 0.707 | 0.700 |
| $1.25R_e$ | 0.942 | 0.835 | 0.836 | 0.825 |
| $1.5R_e$ | 1.142 | 0.914 | 0.918 | 0.912 |
| $1.75R_e$ | 1.353 | 0.876 | 0.876 | 0.921 |
| $\alpha_{ZZ}$ | | | | |
| $0.75R_e$ | 3.974 | 4.529 | 4.406 | 4.547 |
| $R_e$ | 5.742 | 6.330 | 6.296 | 6.52 |
| $1.25R_e$ | 8.780 | 8.655 | 9.472 | 9.689 |
| $1.5R_e$ | 12.497 | 12.732 | 15.043 | 14.348 |
| $1.75R_e$ | 20.131 | 17.466 | 28.011 | 20.389 |

[a]: See ref. [54]
[b]: See Ref [51], Orthogonally Spin adapted coupled cluster singles and doubles linear response (OSACCSDLR).

discussed. These as well results from other groups, show better convergence of the functional and better behavior of properties at stretched geometry as well. The review highlights the promise of the ECC approach.

## Acknowledgements

The authors acknowledge grant from the Department of Science and Technology, India (SP/S1/H-03/2000), for some of the work presented in this review.

## References


[1] F. Coester, *Nucl Phys* **7** (1958), 421; F. Coester and H. Kümmel, *Nucl. Phys* **17** (1960), 477.
[2] J.Cízek, *J. Chem. Phys* **45** (1966), 4256; *J. Cizek, Adv. Chem. Phys* **14** (1969), 35; J. Cízek and J. Paldus, *Int. J. Quantum Chem* **5** (1971), 359; J. Paldus, in Methods in Computational Molecular Physics, NATO Advanced study Institute, Series B: Physics Vol 293 Ed S. Wilson and G.H.F. Direcksen, Plenum Press, NY, 1992, pp. 99–194.
[3] R.J. Bartlett, *Ann. Rev. Phys. Chem* **32** (1981), 359; R.J. Bartlett, *J. Phys. Chem.* **93** (1989), 1697; R.J. Bartlett, in Modern electronic structure theory, Part I Ed D. R. Yarkony, World Scientific, Singaporee, 1995, pp. 1047–1131.
[4] H.J. Monkhorst, *Int. J. Quantum. Chem* **S11** (1977), 421.
[5] T. Lee and G.E. Scuseria, in: *Quantum Mechanical Electronic Structure Calculations with Chemical Accuracy*, S.R. Langhoff, ed., Kluwer, Dordrecht, 1995, pp. 47–108.
[6] W. Kutzelnigg, *J. Chem. Phys* **77** (1982), 3081; W. Kutzelnigg and S. Koch, ibid, **79** (1983), 4315; W. Kutzelnigg, *Theor. Chim. Acta* **80** (1991), 349.
[7] M. Hoffman and J. Simons, *J. Chem. Phys* **88** (1988), 993; M. Hoffman and J. Simons, *Chem. Phys. Lett* **142** (1987), 451.
[8] R.J. Bartlett, S.A. Kucharski, J. Noga, J.D. Watts and G.W. Trucks, *Lect. Notes in Chemistry, Springer- Verlag, Berlin* **52** (1989), 125.
[9] J.D. Watts, G.W. Trucks and R.J. Bartlett, *Chem. Phys. Lett* **157** (1989), 359.
[10] S. Pal, *Theor. Chim. Acta* **66** (1984), 151; S. Pal ad K.B. Ghose, *Curr. Sci* **63** (1992), 667; K.B. Ghose, *Int. J. Quantum. Chem* **53** (1995), 275.
[11] J. Arponen, *Ann. Phys* **151** (1983), 311; J. Arponen, R.F. Bishop and E. Pajanne, *Phys. Rev. A* **36** (1987), 2519; ibid (1987), 2539; J. Arponen, R.F. Bishop and E. Pajanne, *Condensed Matter Theor* **2** (1987), 357; J. Arponen, *Phys. Rev. A* **55** (1997), 2686.

[12] R.F. Bishop, J. Arponen and E. Pajanne, in: *Aspects of Many Body Eeffects In Molecules and Extended Systems*, (Vol. 50), D. Mukherjee, ed., Lecture Notes in Chemistry, Springer Berlin, pp. 79.

[13] R. Bishop and J. Arponen, *Int. J. Quantum Chem.* **S24** (1990), 197; J. Arponen and R.F. Bishop, *Ann. Phys* **207** (1991), 171.

[14] R.F. Bishop, *Theor. Chim. Acta* **80** (1991), 95.

[15] Gaussian 98: Revision A.11.3, M.J. Frisch, G.W. Trucks, H.B. Schlegel, G.E. Scuseria, M.A. Robb, J.R. Cheeseman, V.G. Zakrzewski, J.A. Montgomery, Jr., R.E. Stratmann, J.C. Burant, S. Dapprich, J.M. Millam, A.D. Daniels, K.N. Kudin, M.C. Strain, O. Farkas, J. Tomasi, V. Barone, M. Cossi, R. Cammi, B. Mennucci, C. Pomelli, C. Adamo, S. Clifford, J. Ochterski, G.A. Petersson, P.Y. Ayala, Q. Cui, K. Morokuma, N. Rega, P. Salvador, J.J. Dannenberg, D.K. Malick, A.D. Rabuck, K. Raghavachari, J.B. Foresman, J. Cioslowski, J.V. Ortiz, A.G. Baboul, B.B. Stefanov, G. Liu, A. Liashenko, P. Piskorz, I. Komaromi, R. Gomperts, R.L. Martin, D.J. Fox, T. Keith, M.A. Al-Laham, C.Y. Peng, A. Nanayakkara, M. Challacombe, P.M.W. Gill, B. Johnson, W. Chen, M.W. Wong, J.L. Andres, C. Gonzalez, M. Head-Gordon, E.S. Replogle and J.A. Pople, Gaussian, Inc., Pittsburgh PA, 2002.

[16] GAMESS: M.W. Schmidt, K.K. Baldridge, J.A. Boatz, S.T. Elbert, M.S. Gordon, J.H. Jensen, S. Koseki, N. Matsunaga, K.A. Nguyen, S.J. Su, T.L. Windus, M. Dupuis, J.A. Montgomery, *J. Comput. Chem.* **14** (1993), 1347.

[17] ACES II: J.F. Stanton, J. Gauss, J.D. Watts, M. Nooijen, N. Oliphant, S.A. Perera, P.G. Szalay, W.J. Lauderdale, S.A. Kucharski, S.R. Gwaltney, S. Beck, A. Balková, D.E. Bernholdt, K.K. Baeck, P Rozyczko, H. Sekino, C. Hober and R.J. Bartlett, Integral packages included are VMOL, J. Almlöf and P.R. Taylor; VPROPS P. Taylor ABACUS; T. Helgaker, H.J. Aa. Jensen, P. Jørgensen, J. Olsen and P.R. Taylor.

[18] P. Piecuch and K. Kowalski, in: *Computational Chemistry: Reviews of Current Trends*, J. Leszczynski, ed., World Scientific, Singapore, 2000, 5, pp. 1–104; P. Piecuch, K. Kowalski, I.S.O. Pimienta and M.J. McGuire, *Int. Rev. Phys. Chem.* **21** (2002), 527.

[19] P. Piecuch, K. Kowalski, I.S.O. Pimienta and S.A. Kucharski, in: *Low lying Potential Energy Surfaces, ACS Symposium Series, 828*, M.R. Hoffmann and K.G. Dyall, ed., American Chemical Society, Washington, DC 2002, pp. 31–64.

[20] D. Mukherjee and S. Pal, *Adv. Quantum. Chem* **20** (1989), 291.

[21] I. Lindgren, *Inter. J. Quantum. Chem* **S12** (1978), 33; *J. Phys. B* **24** (1991), 1143, *Phys. Script* **32** (1985), 291; I. Lindgren and D. Mukherjee, *Phys. Rep* **93** (1987), 151; W. Kutzzelnigg, *J. Chem. Phys* **80** (1984), 822; ibid, **77** (1981), 3081; W. Kutzelnigg, D. Mukherjee and S. Koch, *Chem. Phys* **87** (1987), 5902.

[22] B. Jaziorski and H.J. Monkhorst, *Phys. Rev. A* **24** (1981), 1668; B. Jeziorski and J. Paldus, *J. Chem. Phys* **88** (1988), 5673; **90** (1989), 2714.

[23] J. Noga and R.J. Bartlett, *J. Chem. Phys* **86** (1987), 7041; ibid **89** (1988), 3401; S.A. Kucharski and R.J. Bartlett, *Theor. Chim. Acta* **80** (1991), 387.

[24] N. Oliphant and L. Adamowicz, *J. Chem. Phys* **95** (1991), 6645; P. Piecuch and L. Adamowicz, *J. Chem. Phys* **100** (1994), 5792.

[25] K.B. Ghose, P. Piecuch, S. Pal and L. Adamowicz, *J. Chem. Phys* **104** (1996), 6582; A. Kondo, P. Piecuch and J. Paldus, *J. Chem. Phys* **102** (1995), 6511.

[26] G.H.F. Diercksen and A. Sadlej, *Chem. Phys* **96** (1984), 17; G.H.F. Diercksen, V. Kello and A.J. Sadlej, *J. Chem. Phys* **79** (1983), 2918; G.H.F. Diercksen, B.O. Roose and A. Sadlej, *Chem. Phys* **59** (1981), 29.

[27] D. Mukherjee and P.K. Mukherjee, *J. Chem. Phys* **37** (1979), 325.

[28] E.A. Salter, G.W. Trucks and R.J. Bartlett, *J. Chem. Phys* **90** (1989), 1752; ibid (1989), 1767.

[29] A. Dalgarno and A.L. Stewart, *Proc. Royal Soc London, Ser A* **238** (1957), 269.

[30] R.J. Bartlett and J. Noga, *Chem. Phys. Lett* **150** (1988), 29.

[31] R.J. Bartlett, S.A. Kucharski and J. Noga, *Chem. Phys. Lett* **155** (1989), 133.

[32] H. Koch, J.A. Jensen, P. Jorgenson, T. Helgaker, G. Scuseria and H.F. Schaefer III, *J. Chem. Phys* **92** (1990), 4924.

[33] S. Pal, M.D. Prasad and D. Mukherjee, *Theor. Chim. Acta* **62** (1983), 523.

[34] S. Pal, *Phys. Rev. A* **33** (1986), 2240; ibid, **39** (1989), 39; ibid **34** (1986), 2682; K.B. Ghose and S. Pal, *Phys. Rev. A* **36** (1987), 1539.

[35] K.B. Ghose, P.G. Nair and S. Pal, *Chem. Phys. Lett* **211** (1993), 15.

[36] a) N. Vaval, K.B. Ghose and S. Pal, *J. Chem. Phys* **101** (1994), 4914; b) N. Vaval and S. Pal, *Phys. Rev. A* **54** (1996), 250.

[37] a) S. Pal and N. Vaval, *Recent Research developments in Physical Chemistry* **6** (2002), 527; b) N. Vaval and S. Pal, *Int. J. Mol. Sci* **2** (2001), 89; c) A.B. Kumar, N. Vaval and S. Pal, *Chem. Phys. Lett* **295** (1998), 189.

[38] P. Piecuch and R.J. Bartlett, *Adv. Quantum. Chem* **34** (1999), 295; P. Piecuch, I.S.O. Pimienta, P.D. Fann and K. Kowalski, in: *Recent Progress in Electron Correlation Methodology*, ACS Symposium, (Vol. XXX), A.K. Wilson, ed., American Chemical Society, Washington D.C, 2003 p. XX.

[39] S.R. Gwaltney, E.F.C. Byrd, T. van Voorhis and M. Head Gordon, *Chem. Phys. Lett* **353** (2002), 359; T. Van Voorhis and M. Head Gordon, *Chem. Phys. Lett.* **330** (2000), 585; E.F.C. Byrd, T. Van Voorhis and M. Head Gordon, *J. Phys. Chem, B* **106** (2002), 8070.

[40] P. Szalay, M. Nooijen and R.J. Bartlett, *J. Chem. Phys* **103** (1995), 281.

[41] T. Van Voorhis and M. Head Gordon, *J. Chem. Phys* **113** (2000), 8873.
[42] J.H. Van Valk, *Phys. Rev* **33** (1929), 467.
[43] H. Primas, *Phys. Acta* **34** (1961), 331.
[44] W. Kutzelnigg, in: *Methods in Electronic Structure Theory*, H.F. Schaefer III, ed., Plenum, NY, 1977, p. 129.
[45] P. Westhus, *Inter. J. Quantum. Chem.* **7** (1973), 463.
[46] D. Mukherjee, R.K. Moitra and A. Mukhopadhyay, *Mol. Phys.* **30** (1975), 1861; ibid Pramana, **2** (1977), 545.
[47] W. Kutzelnigg, *Theor. Chim. Acta* **83** (1993), 41.
[48] N. Vaval, *Chem. Phys. Lett* **318** (2000), 168.
[49] A.J. Sadlej, *Theor. Chim. Acta* **79** (1991), 123.
[50] J.S. Mounter, W. Klemperer, *J. Chem. Phys* **52** (1970), 6033; H.J. Wernner and W. Meyer, *Mol. Phys* **31** (1976), 855.
[51] A. Kondo, P. Piecuch and P. Paldus, *J. Chem. Phys* **104** (1995), 8566.
[52] S. Racine and E.R. Davidson, *J. Phys. Chem* **97** (1993), 6367.
[53] T.R. Dyke, B.J. Howard and W. Klemperer, *J. Chem. Phys* **56** (1972), 2442.
[54] N. Vaval and S. Pal, *J. Chem. Phys.* Communicated.
[55] C.W. Bauschlicher and P.R. Taylor, *Theor. Chim. Acta* **71** (1976), 263.

Journal of Computational Methods in Sciences and Engineering 4 (2004) 737–751
IOS Press 

# Modeling molecular polarizabilities with graph-theoretical concepts

Lionello Pogliani
*Dipartimento di Chimica, Università della Calabria, 87030 Rende (CS), Italy*
*E-mail: lionp@unical.it*

**Abstract.** A model study of the molecular polarizabilities of a wide heterogeneous class of organic compounds was carried out. The model of this property was based on graph-theoretical molecular connectivity and pseudoconnectivity basis indices, including the corresponding dual indices. The descriptive power of these indices and of the higher-order terms they give rise, was tested when the basis indices were derived by the aid of odd complete graphs. This last type of graph has been used to encode the core electrons of the atoms of which the given molecules are done. The obtained model is very satisfactory and stresses the importance of the odd complete graphs when used to encode the core electrons of atoms with $n \geqslant 2$. Present model also stresses the importance of the dual basis indices in improving the modeling ability of the higher-order connectivity-pseudoconnectivity terms. Another relevant feature of the present model is the importance assumed by the zeroth-order valence molecular connectivity basis index, ${}^0\chi^v$, which allows a direct topological-physical interpretation of the modeling.

Keywords: Molecular polarizabilities, graph-theoretical connectivity indices, complete graphs, core electrons

## 1. Introduction

There are some phenomena, especially in physics, that can be interpreted by 'absolute theories', i.e., by theories that can quantitatively predict a certain property of a system from its fundamental parameters without using 'adjustable' external parameters. In these predictions balancing computational efficiency with scientific accuracy is always a compromise, and in physical chemistry this is a rather difficult task to achieve without the help of 'adjustable' external parameters, i.e, without the help of 'non-absolute theories'. The sensible remark by H. Poincaré is still valid, '*Long ago it was said if Tycho had had instruments ten times as precise, we should never have had either Kepler, Newton or Astronomy. It is a misfortune for a science to be born too late, when the means of observation have become too perfect. This is what is happening to-day with physical chemistry: the founders are hampered in their estimates by the third and fourth decimal; happily they are men of robust faith*'. This is what happens, i.e., with quantum chemistry whose modelings are mainly based on adjustable external parameters, and even in the newest approach in this field, the density functional theory (DFT), there is a lot of (inspired) guesswork involved. Problems arising with quantum methods in predicting properties have brought into the field classical molecular mechanics (MM) methods, where molecules are seen as a load of tiny moving balls attached to each other by massless springs. Among the computational methods that are being used with increasing success in the field of prediction of properties and/or activities, the mathematical methods based on topological concepts occupy an eminent place. They are normally known under the acronym QSPR/QSAR, i.e., quantitative-structure-property or structure-activity relationships. These mathematical methods are, in many cases, based on non-empirical structure-related graph-theoretical

1472-7978/04/$17.00 

descriptors, where molecules are seen as a set of massless vertices attached to each other by a set of massless non-metrical connections. These computational procedures allow one to forecast properties or activities of new compounds of a given class once the properties or activities of a narrower set of compounds has been modeled [1–5]. Actually, the range of applicability of these computational methods goes well beyond the QSPR/QSAR field [6–11], and they comprehend even quantum QSAR computations, where elements of quantum chemistry and similarity are used instead of pure graph theoretical descriptors [12].

A successful topological method employed in QSP-SAR field is the molecular connectivity method that allows deriving a set of graph-theoretical indices of wide applicability [1,3,13–16]. This method has recently undergone further developments with the introduction of electrotopological state indices [17], indices derived from line-graphs [18–21], pseudoconnectivity indices, dual indices, higher-order mixed terms, indices based on odd complete graphs [22–28], and variable indices [29–31]. These developments of the molecular connectivity theory, if on one side have greatly improved its modeling ability have, on the other side, deepened its character of 'non-absolute theory'. In this paper we will mainly be interested in deepening an aspect of this theory, which concerns the encoding of the core electrons of any type of atoms with principal quantum number $n \geqslant 2$ (heteroatoms) by graph theoretical concepts. Till recently, the valence molecular connectivity indices of atoms with principal quantum number $n \geqslant 3$ indices were a kind of mixed 'graph-quantum' descriptor. This rendered the molecular connectivity theory a bit similar to molecular mechanics methods (MM), in which, normally, are introduced quantum chemistry parameters to render it more 'realistic'.

Aim of the present work will mainly be, (i) to use a full graph-theoretical description for the different heteroatoms of a wide set of organic compounds, and (ii) to model in a satisfactory way the molecular polarizabilities of the given set of organic compounds. A particular attention will also be given (iii) to the improving quality of the recently introduced dual basis indices [26], and (iv) to the importance of specific key basis indices, if there are some. The molecular polarizabilities that will here be modeled have already been modeled with MM methods [32].

## 2. Method

### 2.1. The structure-property relation and the basis indices

The modeling equation used throughout this work is a multi-linear Structure ($S$)-Property (P) equation, $P = \Sigma c_i S_i$, which has as a special case, the simple linear equation, $P = c_1 S + c_0 U_0$. Here, $c_i$, are the regression coefficients, $c_0$ is the regression coefficient of the unitary index, $U_0 \equiv 1$, and $S_i$ are the structural descriptors that can either be (i) basis $\beta$ indices, (ii) molecular connectivity or pseudoconnectivity terms, $X = f(\beta = \chi)$ or $Y = f(\beta = \psi)$, or (iii) mixed higher-order terms, $Z = f(X, Y)$, $f(Z, \beta)$, or $f(Z, \beta_d)$. The multiple linear relation can normally be written as a dot product:

$$P = \mathbf{C} \cdot \boldsymbol{S} \tag{1}$$

where $\mathbf{C} = (c_1, c_2, \ldots, c_0)$, and $\boldsymbol{S} = (S_1, S_2, \ldots, U_0)$. To avoid negative calculated P values, if they have no meaning, it is advantageous to use the modulus equation: $P = |\mathbf{C} \cdot \boldsymbol{S}|$, where bars stand for absolute value. The procedure to choose the best basis indices of a linear combination of basis indices (LCBI) is a search procedure performed on the total combinatorial space described by the basis indices.

The procedure used to construct the molecular connectivity terms, $S = X, Y$, is, instead, a trial-and-error procedure that chooses the best basis indices, $\beta$, and optimizes then some adjustable parameters [23]. A general form for the X and Y terms looks like the rational function of Eq. (2), where $\beta$ is a basis index, and $S = X$ or Y for $\beta = \chi$ or $\beta = \psi$, respectively. Parameters, *a-d*, *m-q*, and $r$ are the optimization parameters. These parameters can either be negative, zero or one

$$S = [a(\beta_1)^m + b(\beta_2)^n]^q / [c(\beta_3)^o + d(\beta_1)^p]^r \tag{2}$$

The use of terms for modeling could loosely be called Configuration Interaction of Graph-Type Basis Indices (CI-GTBI) for its vague resemblance with the quantum method, Configuration Interaction of multielectron configurations made up of antisymmetrized basis functions. Sometimes, due to their improved quality [23,26,27], mixed connectivity-pseudoconnectivity terms, $Z = f(X, Y)$ or $f(X, Y, \beta$ and/or $\beta_d)$, and $f(Z, \beta_d)$, can be derived and used, whenever possible. These last terms can be derived by the aid of a procedure, which consists in trying the different ways to combine X, and Y together, introducing at the end the basis indices $\beta$ or $\beta_d$. This allows to short-circuiting the huge combinatorial problem caused by twenty-eight basis indices [23,26].

The set of basis indices, $\{\beta\}$, the raw material of molecular connectivity calculations, used in this study, is made up of the following three medium-sized subsets of eight molecular connectivity, eight molecular pseudoconnectivity indices, and twelve dual indices, $\{\beta\} = \{\{\chi\}, \{\psi\}, \{\beta_d\}\}$.

$$\{\{\chi\}, \{\psi\}\} = \{\{D, {}^0\chi, {}^1\chi, \chi_t, D^v, {}^0\chi^v, {}^1\chi^v, \chi_t^v\}, \{{}^S\psi_I, {}^0\psi_I, {}^1\psi_I, {}^T\psi_I, {}^S\psi_E, {}^0\psi_E, {}^1\psi_E, {}^T\psi_E\}\}$$
$$\{\beta_d\} = \{{}^0\chi_d, {}^1\chi_d, {}^1\chi_s, {}^0\chi_d^v, {}^1\chi_d^v, {}^1\chi_s^v, {}^0\psi_{Id}, {}^1\psi_{Id}, {}^1\psi_{Is}, {}^0\psi_{Ed}, {}^1\psi_{Ed}, {}^1\psi_{Es}\} \tag{3}$$

Basis $\chi$ and $\psi$ indices are formally similar, as can be seen from the following relations (for the meaning of $\delta$, $\delta^v$, I, and S, see next paragraph)

$$D = \Sigma_i \delta_i \quad {}^S\psi_I = \Sigma_i I_i \tag{4}$$

$$^0\chi = \Sigma_i(\delta_i)^{-0.5} \quad {}^0\psi_I = \Sigma_i(I_i)^{-0.5}(5) \tag{5}$$

$$^1\chi = \Sigma(\delta_i\delta_j)^{-0.5} \quad {}^1\psi_I = \Sigma(I_iI_j)^{-0.5} \tag{6}$$

$$\chi_t = (\Pi\delta_i)^{-0.5T}\psi_I = (\Pi I_i)^{-0.5} \tag{7}$$

Index $\chi_t$ (and $\chi_t^v$) is the total molecular connectivity index, and it has its $\psi$ counterpart in the total molecular pseudoconnectivity index, ${}^T\psi_I$(and ${}^T\psi_E$). Sums in Eqs (4) and (5), as well as products ($\Pi$) in Eq. (7), are taken over all vertices of the hydrogen-suppressed chemical graph. Sums in Eq. (6) are over all edges of the chemical graph ($\sigma$ bonds in a molecule). Replacing $\delta$ with $\delta^v$ the subset of valence $\chi^v$ indices, $\{D^v, {}^0\chi^v, {}^1\chi^v, \chi_t^v\}$ for a hydrogen-suppressed chemical pseudograph is obtained. Replacing $I_i$ with $S_i$ the $\psi_E$ subset $\{{}^S\psi_E, {}^0\psi_E, {}^1\psi_E, {}^T\psi_E\}$ is obtained. Superscripts S and T stand for sum and total, while the other sub- and superscripts follow the established denomination for $\chi$ indices [3].

The dual basis indices allow one to close the set of basis indices relative to a dual operation. Two types of dual connectivity indices have been defined, an 'original' dual index (subscript $d$), which can directly be derived with Boolean-like rules and a 'soft' dual index (subscript $s$) that can be derived just applying

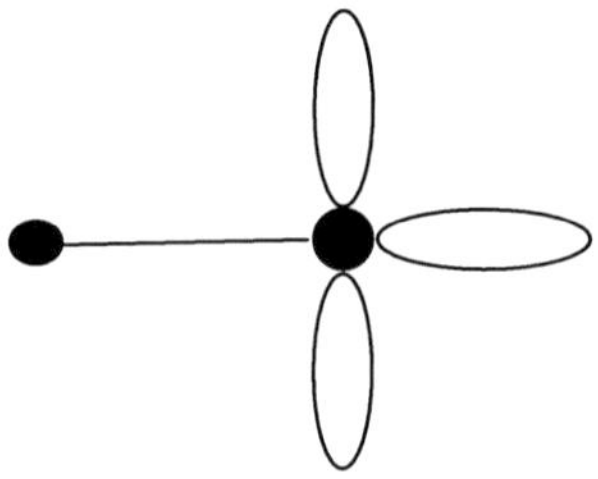

Fig. 1. The hydrogen suppressed pseudograph for $CH_3$-F, $CH_3$-Cl, $CH_3$-Br, and $CH_3$-I.

the given rules more smoothly [26]. Normally, the dual of an expression is obtained interchanging sums and products and interchanging zeros and ones. They are,

$$^0\chi_d = (-0.5)^N \Pi_i(\delta_i) \quad ^0\psi_{Id} = (-0.5)^N \Pi_i(I_i) \tag{8}$$

$$^1\chi_d = (-0.5)^{(N+\mu-1)} \Pi(\delta_i + \delta_j) \quad ^1\psi_{Id} = (-0.5)^{(N+\mu-1)} \Pi(I_i + I_j) \tag{9}$$

$$^1\chi_s = \Pi(\delta_i + \delta_j)^{-0.5} \ ^1\psi_{Is} = \Pi(I_i + I_j)^{-0.5} \tag{10}$$

If, in these expressions $\delta$ is replaced by $\delta^v$ and $I_i$ by $S_i$ the corresponding $\chi^v$ valence dual and $\psi_E$ dual indices are obtained. Exponent, $\mu$, in Eq. (9) is the cyclomatic number. The cyclomatic number, $\mu = q - N + 1$, of a graph ($q$ = number of edges, $N$ = number of vertices), indicates the number of cycles of a chemical graph and it is equal to the minimal number of edges necessary to be removed in order to convert the (poly)cyclic graph to an acyclic subgraph. Removing an edge is the equivalent of opening a ring by breaking a bond in the ring. For acyclic molecules $\mu = 0$, for monocyclic compounds $\mu = 1$ and for bicyclic compounds $\mu = 2$.

### 2.2. *The $\delta$ numbers and the complete graph conjecture for the core electrons*

Basis $\chi$ indices are directly based on the $\delta$, and $\delta^v$ connectivity numbers of a hydrogen-suppressed graph and pseudograph, respectively. A **graph** can be defined as a set of vertices, V, and a set of edges, E, that connect these vertices, and the degree of a vertex, $\delta$, is the number of edges that occur with it. A **pseudograph** is a graph that allows for multiple connections and loops (self-connections), and the degree of a vertex, $\delta^v$, is the corresponding number of edges that occur with it, where loops contribute twice to its degree [23,33]. Thus, the hydrogen-suppressed graph of $CH_3F$, $CH_3Cl$, $CH_3Br$, and $CH_3I$ can be represented as two connected points, •—•, where each vertex has $\delta = 1$. Thus, from a graph point of view the four halo compounds (i.e., their vertices, C and X) are indistinguishable. Figure 1 shows the pseudograph of $CH_3F$, $CH_3Cl$, $CH_3Br$, and $CH_3I$, where the vertex that mimics F, Cl, Br, and I has three loops and, correspondingly, a $\delta^v = 7$. The single connected vertex that mimics C has $\delta^v \equiv \delta = 1$. Up to the recent times if the molecule did not contain any heteroatom with principal quantum number $n > 2$ (in Fig. 1 only $CH_3F$), the $\delta^v$ values could directly be derived from the corresponding pseudograph of a molecule. Prior to the introduction of complete graphs for the core electrons $CH_3F$, $CH_3Cl$, $CH_3Br$, and $CH_3I$ shared exactly the same pseudograph characteristics.

The $\delta^v$ values of the valence connectivity $\chi^v$ indices for higher-row atoms, with $n > 2$ (in this study, Si, P, S, Cl, Br, and I) had to be calculated with the algorithm, $\delta^v = (Z^v - h)/(Z - Z^v - 1)$ [3]. Here, $Z^v$ is the number of valence electrons, Z is the atomic number, and h is the number of suppressed

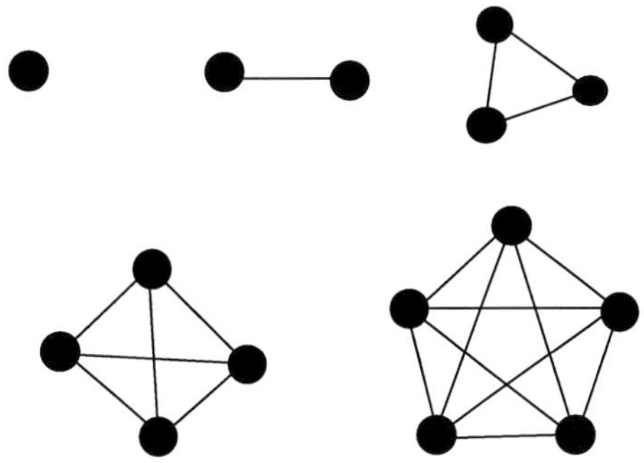

Fig. 2. $K_1, K_2, K_3, K_4$ and $K_5$ complete graphs.

hydrogen atoms. Clearly, for $n = 2$, $\delta^v = Z^v - h$, as, $Z - Z^v - 1 = 1$. As $\delta^v$ can be derived from the hydrogen-suppressed pseudograph of a molecule, it can be designed as $\delta^v(ps)$. Thus, the previous equation can be rewritten as: $\delta^v = \delta^v(ps)/(Z - Z^v - 1)$.

The $\delta^v$ numbers determine also the basis $\psi$ indices through the I-State ($\psi_I$ subset) and S-State ($\psi_E$ subset) atom level indices [17]: $I = [(2/n)^2\delta^v + 1]/\delta$, $S = I + \Sigma\Delta I$, and $\Delta I = (I_i - -I_j)/r_{ij}^2$, here, $\delta^v = Z^v - h, r_{ij}$ counts the atoms in the minimum path length separating two atoms $i$ and $j$, which is equal to the usual graph distance, $d_{ij} + 1$. The $\delta^v$ of the I-Stateindex equals the $\delta^v$ that can be derived from the pseudograph (ps) of a molecule, i.e, $\delta^v \equiv \delta^v(ps)$. The core electron contribution is encoded by the $(2/n)^2$ factor. This allows to write the $\delta^v$ for any heteroatom as: $\delta^v = (2/n)^2\delta^v(ps)$. From what has been said, it is evident that $\chi^v$ and $\psi$ indices are based on different $\delta^v$ definitions whose values diverge for $n > 2$, and are equal only for $n = 2$. Factor $\Sigma\Delta I$ incorporates the information about the influence of the remainder of the molecular environment, and, as it can be negative, S can also be negative. A result from the $I_S$ concept [17] tells that $\Sigma_i S_i = \Sigma_i I_i$, with the consequence that ${}^S\psi_I = {}^S\psi_E$, and in this case the $\psi$-subset consists of seven indices only. To avoid negative S values, which give rise to imaginary $\psi_E$ values, every S value of our class of compounds (as some atoms have $S < 0$), has been rescaled to the S value in $SiF_4$, i.e., $S(Si) = -6.611$. Inevitably, this rescaling procedure invalidates the cited result of the $I_S$ concept, with the consequence that ${}^S\psi_I \neq {}^S\psi_E$ [24,25].

Till recently, then, the $\delta^v$s for the $\chi^v$ and for the $\psi_{I,E}$ indices of atoms with $n > 2$ were (i) non-homogenous, and (ii) were mainly based on quantum concepts. Lately, the $\delta^v$ numbers for any type of heteroatoms, i.e., of atoms with $n \geqslant 2$, were redefined with the aid of graph tools apt to encode the core electrons. A particular graph tool received the due attention, the complete graph, and, among the different types of complete graphs, the odd complete graph, $K_p$ [28,33–35], where $p = odd = 1, 3, 5, 7, \ldots$, showed interesting model qualities. A **graph is complete** if every pair of its vertices are adjacent. A complete graph of order $p$ is denoted by $K_p$, and $r = p - 1$ denotes its regularity (see Fig. 2). A complete graph has all its vertices of the same degree $r$ (the contrary is not true). Actually, good results [28,33–35] have been obtained with the following algorithm, based on the concept of odd complete graphs,

$$\delta^v = q \cdot \delta^v(ps)/[p \cdot r + 1] \tag{11}$$

Where, $p = 1, 3, 5, 7, \ldots$ Up to date, in some cases $q = p$, and the $\delta^v$ values are rather similar to the $\delta^v = (2/n)^2\delta^v(ps)$ values of the E-State algorithm. In some other cases, like for the molecular polarizabilities, $q = 1$. In this case the $\delta^v$ values are more similar to the $\delta^v = \delta^v(ps)/(Z - Z^v - 1)$ values [33–35]. The sequential choice for $p$, i.e., $p = 1, 2, 3, 4, \ldots$, gave rise, to deceptive results [35]. In Fig. 3 are shown the hydrogen-suppressed pseudographs-odd-complete-graphs for $CH_3$-F, $CH_3$-Cl,

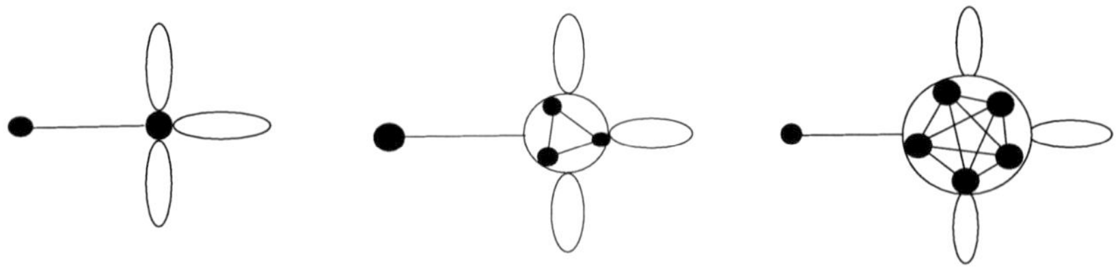

Fig. 3. The hydrogen suppressed pseudograph-odd-complete-graphs for $CH_3$-F (top), $CH_3$-Cl (middle), and $CH_3$-Br (bottom), with a blow-up of the Cl, and Br vertices, which shows the $K_3$, and $K_5$ graphs used to encode the core electrons of Cl and Br, respectively. The inner-core electrons of F and C are, instead, encoded by a $K_1$ complete graph, i.e, by a vertex.

and $CH_3$-Br. The three graphs are now different from each other. While the odd complete graph for the carbon atom is just a $K_1$ vertex, the complete graphs for F, Cl, and Br, are $K_1$, $K_3$, and $K_5$ complete graphs (in Fig. 3 a blow up of these vertices are shown). Even complete graphs, with $p = 2, 4, 6, \ldots$, give rise to a series of conceptual difficulties, as they would oblige to redefine all those pseudographs made up of second row atoms, i.e., made up of $K_1$ vertices. In fact, here either a $K_0$ graph or a $K_2$ graph should be introduced for the second row atoms. Now, $K_0$ is the null graph, i.e., the graph that possesses no points and no edges [36], and $K_2$ has two vertices and an edge. Parameter $p{\cdot}r$ in Eq. (11) has a particular importance in graph theory. This parameters equals the sum of all vertex degrees in complete graphs, and it is related to a famous theorem in graph theory, the '*handshaking theorem*' that says that this sum, for every type of graph and pseudograph, equals twice the number of connections [23]. Graphs and pseudographs can be represented by their corresponding adjacency A matrix [3,23]. Now, a pseudograph A matrix encodes the features of a graph A matrix also, as can be shown from the following A matrix for a hydrogen-suppressed chemical pseudograph plus $K_p$-(p-odd) graph of a triatomic system

$$A = (p \cdot r + 1)^{-1}_{Kp} \begin{pmatrix} ps_{1,1} & g_{1,2} & g_{1,3} \\ g_{2,1} & ps_{2,2} & g_{2,3} \\ g_{3,1} & g_{3,2} & ps_{3,3} \end{pmatrix} \tag{12}$$

Here, $g_{i,j}$ can either be 0 or 1, it is one only if vertex $i$ and $j$ are connected otherwise is zero (graph characteristics); $ps_{i,i}$ is the sum of the self-connections (loops, they count twice) and multiple connections of vertex $i$ (pseudograph characteristics). The factor $(p \cdot r + 1)^{-1}$ encodes the odd complete graph characteristics of each atom, and as it depends on the $K_p$ of each vertex, it renders the adjacency matrix asymmetric. In Eq. (13) an example of a $3 \times 3$ A matrix for the hydrogen-suppressed $CH_3$-S-$CH_3$ molecule is shown ($K_1$ for C and $K_3$ for S, term $1/1 = 1$ has been written to allow an easier decoding of the formalism).

$$A(C - S - C) = \begin{pmatrix} 0 & 1/1 & 0 \\ 1/7 & 4/7 & 1/7 \\ 0 & 1/1 & 0 \end{pmatrix} \tag{13}$$

All in all we have now three different definitions for $\delta^v$, one for the $\chi^v$ indices, based on a "$Z$" algorithm, $\delta^v(Z) = \delta^v(ps)/(Z - Z^v - 1)$, one for the $\psi_{I,E}$ basis indices, based on a "$n$" algorithm, $\delta^v(n) = (2/n)^2\delta^v(ps)$, and the last one for any type of basis indices, based on the $K_p$ algorithm of Eq. (11), $\delta^v(K_p)$. As in the present study we are going to use the three types of $\delta^v$, to differentiate among the different basis indices, the basis $\chi^v$ and $\psi_{I,E}$ indices defined in Eqs (3–10) will be based on $\delta^v(K_p)$ algorithm. The valence $\chi^v$ indices based on $\delta^v(Z)$, and the subset of $\psi_{I,E}$ indices base on $\delta^v(n)$ will be denoted, respectively, by,

$$\{\chi^v(Z)\} = \{\{D^v_Z, {}^0\chi^v_Z, {}^1\chi^v_Z, \chi^v_{tZ}\}, \{{}^0\chi^v_{dZ}, {}^1\chi^v_{dZ}, {}^1\chi^v_{sZ}\}\}$$
$$\{\psi(n)\} = \{\{{}^S\psi_{In}, {}^0\psi_{In}, {}^1\psi_{In}, {}^T\psi_{In}, {}^S\psi_{En}, {}^0\psi_{En}, {}^1\psi_{En}, {}^T\psi_{En}\}, \{{}^0\psi_{Idn}, {}^1\psi_{Idn}, {}^1\psi_{Isn}, {}^0\psi_{Edn}, {}^1\psi_{Edn}, {}^1\psi_{Esn}\}\} \quad (14)$$

*2.3. The quality of the description*

The statistical performance of the graph-structural molecular connectivity invariant, $S$, is controlled by a quality factor, $Q = r/s$, and by the Fischer ratio $F = fr^2/[(1-r^2)\nu]$, where $r$ and $s$ are the correlation coefficient $r$ (not to be confounded the regularity of a complete graph) and the standard deviation of the estimates, respectively, $f$ is the number of degrees of freedom $= n-(\nu+1)$, $\nu$ is the number of variables, and $n$ is the number of data. Parameter $Q$ has no absolute meaning as it is an 'intra' statistical parameter able to compare the descriptive power of different descriptors for the same property only. Further, this property should always be given in the same scale (see ref. 5, web address). The $F$ ratio, which has the character of an 'inter'-statistical parameter, tells us, even if $Q$ improves, which additional descriptor endangers the statistical quality of the combination. Every descriptor of a linear combination (including $U_0$) is characterized by its fractional utility, $u_i = |c_i/s_i|$, where $s_i$ is the confidence interval of $c_i$. The overall descriptor is characterized by the average fractional utility $\langle u\rangle = \Sigma u_i/(\nu+1)$. If the modeling relation is linear, then $\langle u\rangle = (u_1 + u_0)/2$. The utility statistics allows detecting descriptors that give rise to unreliable coefficient values ($c_i$), whenever they have a high deviation interval ($s_i$), and it gives an indirect information about the role of the descriptor in the modeling equation. Recently [26,33,34], due to the critical importance of the standard deviation of the estimate $s$ the ratio $s_R = s_0/s_i$, has been introduced. Here $s_0$ is the $s$value of the best single-index description and $s_i$ refers to the $s$ values of the improved sequential descriptions. We can, now, check how much this statistic improves along a series of descriptors for the same property: a halving of $s_i$ can be read as doubling in $s_R$, allowing, thus a direct measure of the progress of $s$ along a series of sequential descriptions. It should be stressed that, now, (i) all statistical parameters will grow with improving model that (ii) every model is under the control of all these statistics, and that (iii) nothing justifies an improved $Q$ as a good sign for an improved model. The richness in statistical parameters can also be used to detect possible printing errors, as redundancy is very useful in the construction of self-correcting codes. Further, to avoid to bother the reader with the dimensions of the modeled properties, in the modeling Eq. (1) every property P should be read as P/P°. Here P° is the unitary value of the property, and that this choice allows to read P as a pure numerical number [37].

## 3. Results and discussion

The polarizability measures the distortion of a molecule in an electric field, E, thus non-spherical molecules have anisotropic polarizabilities, $\alpha_i$, and are rotationally Raman active. Other interesting characteristics of polarizability have been studied during recent years. A stable molecule possesses minimum polarizability, thus, the most stable rotational isomer is associated with the minimum $\alpha$ value. Several chemical reactions proceed in the direction that produces the least polarizable compound [38]. Further, a minimum polarizability principle has been proposed as *'the natural direction of evolution of any system is towards a state of minimum polarizability'*, and an inverse relationship has been proposed

Table 1

Experimental, $\langle\alpha\rangle$, $\alpha_1$, $\alpha_2$, and $\alpha_3$, and Computed $\langle\alpha(C)\rangle$, $\alpha_1(C)$, $\alpha_2(C)$, and $\alpha_3(C)$ Molecular Polarizabilities of organic compounds in units of Å$^3$, with $A[\alpha_i(C)] = [\alpha_1(C) + \alpha_2(C) + \alpha_3(C)]/3$

| Compound | $\langle\alpha\rangle$ | $\langle\alpha(C)\rangle$ | $A[\alpha_i(C)]$ | $\alpha_1$ | $\alpha_2$ | $\alpha_3$ | $\alpha_1(C)$ | $\alpha_2(C)$ | $\alpha_3(C)$ |
|---|---|---|---|---|---|---|---|---|---|
| Ethane | 4.48 | 4.71 | 4.65 | 5.49 | 3.98 | 3.98 | 5.78 | 4.42 | 3.74 |
| Propane | 6.38 | 6.46 | 6.39 | 7.66 | 5.74 | 5.74 | 7.49 | 6.28 | 5.40 |
| Neopentane | 10.20 | 10.42 | 10.34 | 10.20 | 10.20 | 10.20 | 11.35 | 10.50 | 9.18 |
| Cyclopropane | 5.50 | 5.60 | 5.53 | 5.74 | 5.74 | 5.04 | 6.65 | 5.36 | 4.58 |
| Cyclopentane | 9.15 | 9.17 | 9.10 | 9.68 | 9.17 | 8.40 | 10.13 | 9.17 | 7.99 |
| Cyclohexane | 11.00 | 11.00 | 10.92 | 11.81 | 11.81 | 9.28 | 11.92 | 11.11 | 9.73 |
| Ethylene | 4.12 | 3.34 | 3.27 | 4.82 | 3.71 | 3.25 | 4.44 | 2.96 | 2.42 |
| Propene | 6.26 | 5.43 | 5.37 | | | | 6.49 | 5.19 | 4.42 |
| 2MePropene | 8.29 | 7.53 | 7.46 | | | | 8.53 | 7.42 | 6.42 |
| Trans-2-Butene | 8.49 | 7.60 | 7.53 | | | | 8.61 | 7.50 | 6.49 |
| Cyclohexene | 10.70 | 10.36 | 10.28 | | | | 11.29 | 10.43 | 9.12 |
| Butadiene | 7.87 | 6.31 | 6.24 | 11.93 | 6.14 | 5.54 | 7.35 | 6.13 | 5.26 |
| Benzene | 9.92 | 9.13 | 9.05 | 11.20 | 11.20 | 7.36 | 10.09 | 9.12 | 7.94 |
| Toluene | 12.30 | 11.17 | 11.10 | | | | 12.09 | 11.30 | 9,90 |
| HexaMeBenzene | 22.63 | 22.47 | 22.37 | 22.63 | 22.63 | 22.63 | 23.11 | 23.33 | 20.66 |
| Acetylene | 3.50 | 2.71 | 2.65 | 4.79 | 2.85 | 2.85 | 3.83 | 2.29 | 1.83 |
| Propyne | 4.68 | 4.91 | 4.85 | 6.14 | 3.94 | 3.94 | 5.98 | 4.64 | 3.93 |
| $C(CCH)_4$ | 12.19 | 12.12 | 12.04 | 12.19 | 12.19 | 12.19 | 13.02 | 12.31 | 10.80 |
| Allene | 5.00 | 5.57 | 4.51 | 8.97 | 4.43 | 4.43 | 5.65 | 4.28 | 3.60 |
| Methanol | 3.32 | 3.34 | 3.28 | 4.09 | 3.23 | 2.65 | 4.44 | 2.96 | 2.42 |
| Ethanol | 5.11 | 5.10 | 5.04 | 5.76 | 4.98 | 4.50 | 6.17 | 4.84 | 4.11 |
| 2-Propanol | 6.97 | 7.11 | 7.04 | | | | 8.12 | 6.97 | 6.02 |
| Cyclohexanol | 11.56 | 11.67 | 11.59 | | | | 12.57 | 11.83 | 10.37 |
| Dimethylether | 5.24 | 5.37 | 5.30 | 6.38 | 4.94 | 4.39 | 6.42 | 5.12 | 4.36 |
| p-Dioxane | 8.60 | 9.50 | 9.42 | 9.40 | 9.40 | 7.00 | 10.45 | 9.52 | 8.30 |
| Methylamine | 3.59 | 3.68 | 3.62 | 3.94 | 3.40 | 3.38 | 4.78 | 3.32 | 2.75 |
| Formaldehyde | 2.45 | 2.60 | 2.53 | 2.76 | 1.70 | 1.83 | 3.72 | 2.17 | 1.72 |
| Acetaldehyde | 4.59 | 4.68 | 4.62 | | | | 5.76 | 4.39 | 3.71 |
| Acetone | 6.39 | 6.94 | 6.87 | 7.37 | 7.37 | 4.42 | 7.96 | 6.79 | 5.86 |
| F-Methane | 2.62 | 3.15 | 3.09 | 3.18 | 2.34 | 2.34 | 4.26 | 2.76 | 2.25 |
| TriF-Methane | 2.81 | 4.41 | 4.35 | 2.87 | 2.87 | 2.69 | 5.49 | 4.10 | 3.45 |
| TetraF-Methane | 2.92 | 3.44 | 3.37 | 2.92 | 2.92 | 2.92 | 4.54 | 3.06 | 2.52 |
| Cl-Methane | 4.55 | 4.71 | 4.65 | 5.68 | 3.99 | 3.98 | 5.78 | 4.42 | 3.74 |
| DiCl-Methane | 6.82 | 6.46 | 6.39 | 8.81 | 6.30 | 5.36 | 7.49 | 6.28 | 5.40 |
| TriCl-Methane | 8.53 | 8.40 | 8.33 | 9.42 | 9.42 | 6.74 | 9.38 | 8.35 | 7.25 |
| TetraCl-Methane | 10.51 | 10.42 | 10.34 | 10.51 | 10.51 | 10.51 | 11.35 | 10.50 | 9.18 |
| Br-Methane | 5.61 | 6.30 | 6.23 | 6.91 | 4.96 | 4.96 | 7.33 | 6.11 | 5.24 |
| DiBr-Methane | 8.68 | 9.62 | 9.54 | | | | 10.57 | 9.64 | 8.41 |
| TriBr-Methane | 11.84 | 13.09 | 13.01 | 13.00 | 13.00 | 9.53 | 13.96 | 13.35 | 11.73 |
| I-Methane | 7.59 | 7.80 | 7.73 | 9.02 | 6.87 | 6.87 | 8.80 | 7.71 | 6.68 |
| DiI-Methane | 12.90 | 12.63 | 12.55 | | | | 13.52 | 12.86 | 11.29 |
| TriI-Methane | 18.04 | 17.62 | 17.53 | 18.69 | 18.69 | 16.74 | 18.38 | 18.17 | 16.04 |
| $CH_2{=}CCl_2$ | 7.83 | 7.53 | 7.46 | 8.96 | 8.79 | 5.75 | 8.53 | 7.42 | 6.42 |
| *cis*-CHCl=CHCl | 7.78 | 7.60 | 7.53 | 9.46 | 7.80 | 6.08 | 8.61 | 7.50 | 6.49 |
| DiSilane | 11.10 | 11.63 | 11.55 | | | | 12.54 | 11.79 | 10.33 |
| Formamide | 4.08 | 3.70 | 3.64 | | | | 4.80 | 3.35 | 2.77 |
| Acetamide | 5.67 | 5.89 | 5.82 | | | | 6.93 | 5.67 | 4.86 |
| Acetonitrile | 4.48 | 4.59 | 4.53 | 5.74 | 3.85 | 3.85 | 5.67 | 4.29 | 3.62 |
| Propionitrile | 6.24 | 6.45 | 6.38 | | | | 7.48 | 6.27 | 5.39 |
| Tert-BuCyanide | 9.59 | 10.55 | 10.47 | 10.71 | 9.03 | 9.03 | 11.48 | 10.64 | 9.30 |
| BenzylCyanide | 11.97 | 11.88 | 11.81 | 16.16 | 11.60 | 8.15 | 12.79 | 12.06 | 10.58 |
| TriCl-Acetonitrile | 10.42 | 10.55 | 10.47 | 10.70 | 10.29 | 10.29 | 11.48 | 10.64 | 9.30 |
| Pyridine | 9.92 | 9.90 | 9.83 | 10.72 | 10.43 | 6.45 | 10.85 | 9.95 | 8.68 |
| Thiophene | 9.00 | 8.67 | 8.60 | 10.15 | 10.14 | 6.70 | 9.65 | 8.64 | 7.51 |

between polarizability and hardness, which is a function of the ionization potential and of the electronic affinity [38]. The modeling of this property can then be the key to the modeling of other physicochemical properties.

The experimental, calculated (C) mean polarizability $\langle\alpha\rangle = (\alpha_1 + \alpha_2 + \alpha_3)/3$, and total molecular principal polarizabilities, $\alpha_1$, $\alpha_2$, $\alpha_3$, of fifty-four and forty organic compounds, respectively, are shown in Table 1. Some $\langle\alpha\rangle$ values, whenever $\alpha_i$ values are absent, are the result of quantum mechanical calculations. In this Table are also shown the calculated polarizability values (i) $\langle\alpha(C)\rangle$, (ii) the single $\alpha_i(C)$ calculated values, and the resulting average values $A[\alpha_i(C)] = [\alpha_1(C) + \alpha_2(C) + \alpha_3(C)]/3$ (Av column in Table 1). This last set of values arises from the fact fourteen calculated $\alpha_i(C)$ values belong to unknown $\alpha_i$ values ($i = 1 - 3$, see the voids in Table 1), i.e., these values have been inferred from the model of the known forty $\alpha_i$ values. The A[$\alpha_i$(C)] values include, thus, fourteen inferred values, while the fifty-four $\langle\alpha(C)\rangle$ values have been obtained modeling the fifty-four $\langle\alpha\rangle$ values. The A[$\alpha_i$(C)] values can, then, be considered the result of a leave-14-out method. Note that the modeling of the single $\alpha_i$ will not be optimized, but that it will be done with the optimal descriptor for $\langle\alpha\rangle$.

### 3.1. Modeling with $\{\chi^v(\mathbf{Z})\}$ and with $\psi(n)\}$

The best basis index for the mean molecular polarizability, $\langle\alpha\rangle$, and a very good descriptor for the single polarizabilities, $\alpha_1$, $\alpha_2$, and $\alpha_3$ also, is ${}^0\chi^v_Z$. Note that this zeroth-order index will play a key role throughout the present study.

| $n$ | $P$ | $\{\beta\}$ | $Q$ | $F$ | $r$ | $s_0$ | $\langle u\rangle$ | $\mathbf{u}$ |
|---|---|---|---|---|---|---|---|---|
| 54 | $\langle\alpha\rangle$ | $\{{}^0\chi^v_Z\}$ | 0.829 | 545 | 0.955 | 1.15 | 12 | (23, 0.9) |
| 40 | $\alpha_1$ | $\{{}^0\chi^v_Z\}$ | 0.534 | 200 | 0.917 | 1.71 | 8.3 | (14, 2.4) |
| 40 | $\alpha_2$ | $\{{}^0\chi^v_Z\}$ | 0.803 | 498 | 0.964 | 1.20 | 12 | (22, 1.2) |
| 40 | $\alpha_3$ | $\{{}^0\chi^v_Z\}$ | 0.800 | 498 | 0.956 | 1.20 | 11 | (20, 1.7) |

The low utility value $(u_0)$ of the unitary index, $(U_0)$ of the constant parameter of the linear regression for $\langle\alpha\rangle$ mainly is due to the fact that the value of this regression parameter is nearly zero, and small deviations around zero can give rise to an anomalous low utility. The best linear combination for $\langle\alpha\rangle$ is a four-index LCBI (linear combination of basis indices), which is a good descriptor for the single $\alpha_1$, $\alpha_2$, and $\alpha_3$ polarizabilities also,

| $n$ | $P$ | $\{\beta_1, \beta_2, \beta_3, \beta_4\}$ | $Q$ | $F$ | $r$ | $s_R$ | $\langle u\rangle$ | $\mathbf{u}$ |
|---|---|---|---|---|---|---|---|---|
| 54 | $\langle\alpha\rangle$ | $\{{}^0\chi^v_Z, {}^S\psi_{In}, {}^0\psi_{In}, {}^S\psi_{En}\}$ | 1.694 | 569 | 0.989 | 2.0 | 7.8 | (25, 5.9, 2.0,5.0,0.9) |
| 40 | $\alpha_1$ | $\{{}^0\chi^v_Z, {}^S\psi_{In}, {}^0\psi_{In}, {}^S\psi_{En}\}$ | 0.739 | 96 | 0.957 | 1.3 | 3.9 | (8.7, 3.5, 1.5, 3.0, 2.9) |
| 40 | $\alpha_2$ | $\{{}^0\chi^v_Z, {}^S\psi_{In}, {}^0\psi_{In}, {}^S\psi_{En}\}$ | 1.316 | 334 | 0.987 | 1.6 | 5.8 | (18, 4.2, 1.5, 3.6, 1.6) |
| 40 | $\alpha_3$ | $\{{}^0\chi^v_Z, {}^S\psi_{In}, {}^0\psi_{In}, {}^S\psi_{En}\}$ | 0.973 | 151 | 0.972 | 1.2 | 3.2 | (13, 0.2, 1.4, 0.12, 1.8) |

LCBI with more indices do not improve the description, on the contrary. The following $Z'$ term, made up of a $Z(X, Y)$ higher-order term plus two dual indices shows, instead, an interesting improvement, especially at the F and utility level, for $\langle\alpha\rangle$, and also for the single polarizabilities,

$$Z' = [Z + 0.4({}^1\chi_{sZ})^{3.7} + 0.001 \cdot {}^1\psi_{Idn}]$$

| $n$ | $P$ | $Q$ | $F$ | $r$ | $s_R$ | $\langle u\rangle$ | **u** |
|---|---|---|---|---|---|---|---|
| 54 | $\langle\alpha\rangle$ | 1.675 | 2225 | 0.989 | 1.9 | 24 | (47, 0.3) |
| 40 | $\alpha_1$ | 0.707 | 350 | 0.950 | 1.3 | 11 | (19, 2.8) |
| 40 | $\alpha_2$ | 1.369 | 1447 | 0.987 | 1.7 | 20 | (38, 2.2) |
| 40 | $\alpha_3$ | 0.979 | 612 | 0.970 | 1.2 | 13 | (25, 2.0) |

Where (do not forget that normal $\chi$ indices are based on $\delta$ only, and not on $\delta^v$),

$$Z = [1.3 \cdot X + 5.2 \cdot Y - 0.6 \cdot {}^0\chi - 3.1 \cdot {}^1\psi_{En}]^{1.1}$$

$$\langle\alpha\rangle : n = 54, Q = 1.430, F = 1621, r = 0.984, s_R = 1.7, \langle u\rangle = 20$$

$$X = [3{}^0\chi_Z^v + {}^1\chi]$$

$$\langle\alpha\rangle : n = 54, Q = 1.414, F = 1587, r = 0.984, s_R = 1.6, \langle u\rangle = 21$$

$$Y = [{}^0\psi_{In} - 0.9 \cdot {}^0\psi_{En}]^{0.6}$$

$$\langle\alpha\rangle : n = 54, Q = 0.359, F = 102, r = 0.814, s_R = 0.8$$

The $c_0$ value of the correlation vector for $\langle\alpha\rangle$ ($-0.064$, after rounding) is practically zero, i.e., small deviations around it gives rise to large errors, with consequent low $u_0$ values. Nearly the same is valid for the $\alpha_i$s. Practically, the only meaningful utility value is the impressive $u_1$ utility value. MM3(2000) calculations are somewhat poorer with: $r = 0.982$, $s = 0.75$, and $s_R = 1.5$ ($s_0 = 1.15$).

*3.2. Modeling with $\{\chi^v\}$ (based on $K_p$ with p odd), and with $\{\psi(n)\}$*

Let us first remind that the best $\chi$, $\chi_Z^v$ combination is

$$\{{}^1\chi, D_Z^v, {}^0\chi_Z^v\} : Q = 1.472, F = 573, r = 0.986, s_R = 1.7,$$
$$\mathbf{u} = (8.5, 5.0, 19, 2.4), \langle u\rangle = 8.9, n = 54$$

Results with $\chi^v(K_p)$ are somewhat unexpected: ${}^0\chi^v$ is the dominant index, and the $\chi^v$ indices are so dominant that they exclude the $\{\psi(\text{n})\}$ indices from the model

| $n$ | $P$ | $\{\beta\}$ | $Q$ | $F$ | $r$ | $s_R$ | $\langle u\rangle$ | **u** |
|---|---|---|---|---|---|---|---|---|
| 54 | $\langle\alpha\rangle$ | $\{{}^0\chi^v\}$ | 1.045 | 867 | 0.971 | 1.3 | 15 | (29, 0.4) |
| 40 | $\alpha_1$ | $\{{}^0\chi^v\}$ | 0.576 | 232 | 0.927 | 1.1 | 8.8 | (15,2.4) |
| 40 | $\alpha_2$ | $\{{}^0\chi^v\}$ | 0.888 | 609 | 0.970 | 1.1 | 13 | (25, 1.4) |
| 40 | $\alpha_3$ | $\{{}^0\chi^v\}$ | 0.996 | 633 | 0.971 | 1.2 | 14 | (25, 2.5) |

The best linear combination for $\langle\alpha\rangle$ is a four-index combination, which is also a good descriptor for the $\alpha_i$s. Here, the forward search procedure [23] allows deriving the best combination.

| $n$ | $P$ | $\{\beta_1, \beta_2, \beta_3, \beta_4\}$ | $Q$ | $F$ | $r$ | $s_R$ | $\langle u\rangle$ | **u** |
|---|---|---|---|---|---|---|---|---|
| 54 | $\langle\alpha\rangle$ | $\{{}^0\chi^v, {}^1\chi, \mathrm{D}^v, \chi_t^v\}$ | 2.086 | 863 | 0.993 | 2.5 | 6.5 | (12, 9.5, 7.7, 2.7, 3.4) |
| 40 | $\alpha_1$ | $\{{}^0\chi^v, {}^1\chi, \mathrm{D}^v, \chi_t^v\}$ | 0.799 | 112 | 0.963 | 1.5 | 3.1 | (2.3, 5.3, 4.5, 2.4, 1.0) |
| 40 | $\alpha_2$ | $\{{}^0\chi^v, {}^1\chi, \mathrm{D}^v, \chi_t^v\}$ | 1.414 | 386 | 0.989 | 1.8 | 5.2 | (7.4, 6.5, 5.2, 2.3, 4.7) |
| 40 | $\alpha_3$ | $\{{}^0\chi^v, {}^1\chi, \mathrm{D}^v, \chi_t^v\}$ | 1.093 | 191 | 0.978 | 1.4 | 3.0 | (8.1, 1.6, 1.6, 0.7, 3.3) |

The best $Z'$ term, showing an interesting improvement at the F and utility level, for $\langle\alpha\rangle$, and for the single polarizabilities is,

$$Z' = [Z + 0.7({}^1\chi_s)^{3.7} + 0.002 \cdot {}^1\psi_{Idn}]$$

| $n$ | $P$ | $Q$ | $F$ | $r$ | $s_R$ | $\langle u\rangle$ | **u** |
|---|---|---|---|---|---|---|---|
| 54 | $\langle\alpha\rangle$ | 1.738 | 2397 | 0.989 | 2.0 | 25 | (49, 1.1) |
| 40 | $\alpha_1$ | 0.680 | 324 | 0.946 | 1.2 | 10 | (18, 2.2) |
| 40 | $\alpha_2$ | 1.228 | 1165 | 0.984 | 1.5 | 18 | (38, 2.2) |
| 40 | $\alpha_3$ | 1.040 | 690 | 0.974 | 1.3 | 15 | (26, 2.9) |

Where,

$$Z = [2.5 \cdot X + 6 \cdot Y - 0.6 \cdot {}^0\chi - 3.3 \cdot {}^1\psi_{En}]^{1.1}$$

$$\langle\alpha\rangle : n = 54, Q = 1.569, F = 1952, r = 0.987, s_R = 1.8, \langle u\rangle = 23$$

$$X = [3 \cdot {}^0\chi^v + {}^1\chi]$$

$$\langle\alpha\rangle : Q = 1.414, F = 1587, r = 0.984, s_R = 1.6, \langle u\rangle = 21$$

$Y = [{}^0\psi_{In} - 0.9 \cdot {}^0\psi_{En}]^{0.6}$ is the same of case **A**. We remember again that the performance of the MM3(2000) calculations is: $r = 0.982$ and $s = 0.75$, and $s_R = 1.5$.

### 3.3. The pure $K_p$ conjecture: Modeling with $\{\chi^v, \psi\}$ based on $K_p$ with p odd

This homogeneous description, based on basis indices that are constructed with the same $\delta^v$, $\chi^v(K_p)$, and $\psi(K_p)$, has theoretical and practical advantages that cannot be overlooked. Best basis index is always index ${}^0\chi^v$ of Section **B**. The best two-index descriptor is $\{{}^0\chi^v, {}^0\psi_I\}$, which is better than the same combination $\{{}^0\chi^v, {}^0\psi_{In}\}$ based on $\chi^v(K_p)$ and $\psi$(n) type indices [here: $s_0 = 1.15, 1.71\ (\alpha_1), 1.20\ (\alpha_2), 1.20\ (\alpha_3)$].

| $n$ | $P$ | $\{\beta_1, \beta_2\}$ | $Q$ | $F$ | $r$ | $s_R$ | $\langle u\rangle$ | **U** |
|---|---|---|---|---|---|---|---|---|
| 54 | $\langle\alpha\rangle$ | $\{{}^0\chi^v, {}^0\psi_I\}$ | 1.379 | 755 | 0.984 | 1.6 | 8.7 | (19, 6.1, 0.9) |
| 40 | $\alpha_1$ | $\{{}^0\chi^v, {}^0\psi_I\}$ | 0.618 | 134 | 0.937 | 1.1 | 3.7 | (6.4, 2.4, 2.1) |
| 40 | $\alpha_2$ | $\{{}^0\chi^v, {}^0\psi_I\}$ | 1.073 | 445 | 0.980 | 1.3 | 6.3 | (12, 4.2, 2.4) |
| 40 | $\alpha_3$ | $\{{}^0\chi^v, {}^0\psi_I\}$ | 1.090 | 379 | 0.976 | 1.3 | 6.1 | (12, 2.8, 3.1) |

A homogenous $\chi^v(K_p)$ and $\psi(K_p)$ three index combination, $\{{}^0\chi^v, {}^1\chi, {}^S\psi_I\}$, shows a growing quality,

| $n$ | $P$ | $\{\beta_1, \beta_2, \beta_3\}$ | $Q$ | $F$ | $r$ | $s_R$ | $\langle u\rangle$ | **U** |
|---|---|---|---|---|---|---|---|---|
| 54 | $\langle\alpha\rangle$ | $\{{}^0\chi^v, {}^1\chi, {}^S\psi_I\}$ | 2.012 | 1071 | 0.992 | 2.3 | 12 | (31, 12, 7.2, 0.1) |
| 40 | $\alpha_1$ | $\{{}^0\chi^v, {}^1\chi, {}^S\psi_I\}$ | 0.767 | 137 | 0.956 | 1.4 | 5.3 | (9.1, 5.1, 4.1, 2.9) |
| 40 | $\alpha_2$ | $\{{}^0\chi^v, {}^1\chi, {}^S\psi_I\}$ | 1.420 | 519 | 0.989 | 1.7 | 8.6 | (20, 7.6, 5.2, 2.1) |
| 40 | $\alpha_3$ | $\{{}^0\chi^v, {}^1\chi, {}^S\psi_I\}$ | 1.071 | 244 | 0.976 | 1.3 | 5.5 | (16, 2.6, 1.2, 2.8) |

For the optimal four-index homogeneous $K_p$ combination, $\{{}^0\chi^v, {}^1\chi, {}^S\psi_I, {}^1\psi_I\}$, which competes positively with the previous heterogeneous one, are also shown the values of the correlation vector, **C**, of the dot modeling equation (1), where $S = ({}^0\chi^v, {}^1\chi, {}^S\psi_I, {}^1\psi_I, U_0)$.

| $n$ | $P$ | $\{\beta_1, \beta_2, \beta_3, \beta_4\}$ | $Q$ | $F$ | $r$ | $s_R$ | $\langle u\rangle$ | **U** |
|---|---|---|---|---|---|---|---|---|
| 54 | $\langle\alpha\rangle$ | $\{{}^0\chi^v, {}^1\chi, {}^S\psi_I, {}^1\psi_I\}$ | 2.092 | 868 | 0.993 | 2.4 | 9.7 | (29, 7.9, 7.7, 2.2, 1.4) |
| 40 | $\alpha_1$ | $\{{}^0\chi^v, {}^1\chi, {}^S\psi_I, {}^1\psi_I\}$ | 0.802 | 113 | 0.963 | 1.4 | 4.3 | (29, 7.9, 7.7, 2.2, 1.4) |
| 40 | $\alpha_2$ | $\{{}^0\chi^v, {}^1\chi, {}^S\psi_I, {}^1\psi_I\}$ | 1.401 | 379 | 0.989 | 1.7 | 5.5 | (16, 4.4, 4.9, 0.2, 1.7) |
| 40 | $\alpha_3$ | $\{{}^0\chi^v, {}^1\chi, {}^S\psi_I, {}^1\psi_I\}$ | 1.072 | 183 | 0.977 | 1.3 | 3.3 | (13, 0.7, 0.8, 1.0, 1.4) |

$$\mathbf{C}(\langle\alpha\rangle) = (1.93458, 1.96598, -0.12487, -0.37914, -0.33086)$$

$$\mathbf{C}(\alpha_1) = (1.86495, 3.03878, -0.19859, -0.95292, 0.59058)$$

$$\mathbf{C}(\alpha_2) = (1.98963, 1.67290, -0.12153, -0.0518, -0.62965)$$

$$\mathbf{C}(\alpha_3) = (1.96438, 0.32945, -0.02594, 0.35863, -0.67086)$$

And, now, let us introduce the higher-order descriptors. While the optimal $X(K_p)$ has already been detected in Section **B**. The $Y(K_p)$ term improves a lot over the previous $Y[\psi(n)]$ term of Section **A**,

$$Y = [^0\psi_I + 3.1^T\psi_I]^{1.2}, \langle\alpha\rangle : Q = 0.658, F = 343, r = 0.932, s_R = 1.4, \langle u\rangle = 9.3$$

Now, with the homogeneous terms, $X(K_p)$, and $Y(K_p)$, it is possible to construct the following homogeneous $Z(K_p)$, and $Z'(K_p)$ terms,

$$Z = [X + 0.4Y - 0.1(^0\chi)^{0.1} - 0.9(^1\psi_E)^{1.5}], \langle\alpha\rangle : Q = 1.496, F = 1774,$$

$$r = 0.986, s_R = 3.1, \langle u\rangle = 22$$

$$Z' = [Z + 0.007 \cdot {}^0\chi_d^v - 0.0004 \cdot^1 \psi_{Id}]$$

| $n$ | $P$ | $Q$ | $F$ | $r$ | $s_R$ | $\langle u\rangle$ | **u** | **C** |
|---|---|---|---|---|---|---|---|---|
| 54 | $\langle\alpha\rangle$ | 1.704 | 2304 | 0.989 | 2.0 | 25 | (48, 2.7) | (0.62672, - 0.50945) |
| 40 | $\alpha_1$ | 0.703 | 346 | 0.949 | 1.3 | 10 | (19, 1.4) | (0.61156, 0.68943) |
| 40 | $\alpha_2$ | 1.323 | 1351 | 0.986 | 1.6 | 21 | (37, 4.3) | (10.6674, - 1.13871) |
| 40 | $\alpha_3$ | 0.993 | 630 | 0.971 | 1.2 | 14 | (25, 3.6) | (0.59759, - 1.24393) |

There are many things that can be noticed about this Z′ term: (i) some of its statistics are much better than the statistics of the four-index LCBI, (ii) with its correlation vectors, **C**, the calculated $<\alpha(C)>$, $\alpha_1$(C), $\alpha_2$(C), and $\alpha_3$(C) values of Table 1 have been obtained. Further, (iii) fourteen, $\alpha_i$(C), values have been just inferred (25%, ca., of the entire set of compounds) from the modeling equation (see the voids in Table 1), i.e., they give rise to an external prediction set. Thus, the entire set of $A[\alpha_i(C)]$ values shown in Table 1, and plotted vs. the corresponding experimental values in Fig. 4, are the average of $\alpha_1$(C), $\alpha_2$(C), and $\alpha_3$(C) values, where forty $\alpha_i$ values have been used to obtain the correlation vector **C** of the model equation, and fourteen values have been inferred from the same model equation. The good fitting of Fig. 4 underlines the accomplishment of the proposed method. The MM performance was: $r = 0.982$ and $s = 0.75$, and $s_R = 1.5$.

## 4. Conclusion

Some interesting items can be drawn from the present study, i.e., (i) the overall good quality of the model, (ii) the good quality of the $K_p$ basis indices, (iii) the good quality of the $K_p$ higher-order terms (iv) the improving characteristics of the dual indices, and (v) the importance of the $^0\chi^v$ basis index. These items emphasize the validity of the complete graph conjecture for the core electrons of atoms with $n \geqslant 2$. This conjecture offers an interesting way to widen and complete the graph-theoretical character of the molecular connectivity theory. The $K_p$ algorithm for $\delta^v$ (see Eq. (11)) with $p$ odd and $q = 1$, which has been used throughout the model of the given molecular polarizabilities, seems to be the best choice for this property of this set of organic molecules. The choice $q = 1$ gives rise to values for $\delta^v$ which are rather similar to the values obtained with a $\delta^v$centered on quantum concepts for higher-row atoms, i.e., for those atoms whose principal quantum number is $n > 2$. Actually, the possibility to choose between two different values for $\delta^v$ allows either avoiding or taking advantage of the degeneracy in the basis indices for some molecules. This enhances the flexibility of the $K_p$ algorithm. The weight

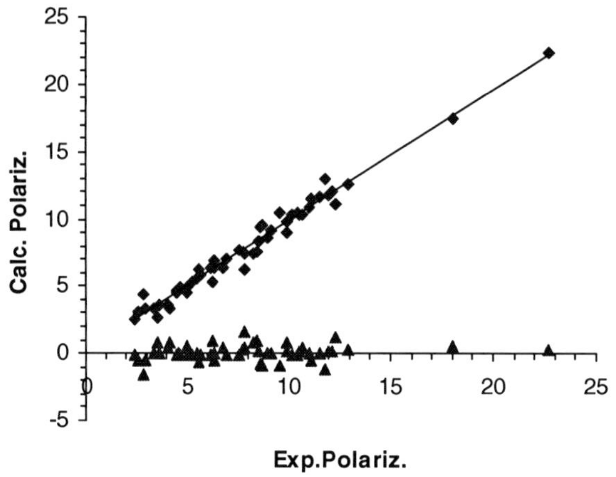


Fig. 4. The calculated $A[\alpha_i(C)]$ values versus the experimental polarizabilities values of fifty-four organic compounds, and the corresponding residuals ($\sigma$).

of the zeroth-order valence index, ${}^0\chi^v$, throughout this model, i.e., of a vertex index which depends on the number and kind of vertices in a chemical pseudograph of a molecule, emphasizes the importance of the atomic contribution to the polarizabilities. Further, the growing importance of this index with the introduction of odd complete graphs (compare Section **A** with Section **B**) is also a nice piece of support for the role of the pseudo-plus-odd-complete-graph representation of a molecule. This representation is more apt to encode the overall electronic contribution of the molecule, as well as the atom level electronic contribution. For those readers interested to deepen the physical meaning of the topological indices perusal of references 41–43 is mandatory.

Let us end this study with some intriguing words about some 'public actions', "*every public action, which is not customary, either is wrong, or if it is right, is a dangerous precedent. It follows that nothing should ever be done for the first time*" [44].

## Acknowledgments

I would like to thank Professor P.G. Mezey of the University of Saskatchewan, Saskatoon, Canada, for his helpful comments, and also for his attention to the overall author's work.

## References

[1] M. Randić, On characterization of molecular branching, *J. Am. Chem. Soc.* **97** (1975), 6609–6615.
[2] G. Grassy, B. Calas, A. Yasri, R. Lahana, J. Woo, S. Iyer, M. Kaczorek, R. Floc'h and R. Buelow, Computer-assisted rational design of immunosuppressive compounds, *Nature Biotech* **16** (1998), 748–752.
[3] L.B. Kier and L.H. Hall, *Molecular Connectivity in Structure-Activity Analysis*, Wiley, New York, 1986.
[4] R. Todeschini and V. Consonni, *The Handbook of Molecular Descriptors*, Wiley-VCH, Weinheim, 2000. For a discussion about Q see: http://www.disat.unimib.it/chm/CHMnews.htm.
[5] M.V. Diudea, ed., *QSPR/QSAR Studies by Molecular Descriptors*, Nova Science Pub.Inc., New York, 2000.
[6] N. Trinajstić *Chemical Graph Theory*, 2nd ed., CRC Press, Boca Raton, 1992.
[7] O. Temkin, A.V. Zeigarnik and D. Bonchev, *Chemical Reaction Networks*, CRC Press, New York, 1996.
[8] M. Randić Topological Indices, in: *The Encyclopedia of Computational Chemistry*, P.V.R. Allinger et al., eds, Wiley, New York, 1998.
[9] R.B. King and D. Rouvray, eds, *Topology in Chemistry*, Horwood, New York, 2001.

[10] D.J. Klein and J. Brickman, eds, *MATCH – Comm. Math. Comput. Chem.* **42** (2000), 1–290.

[11] J. Devillers and A.T. Balaban, eds, *Topological Indices and Related Descriptors in QSAR and QSPR*, Gordon and Breach, Amsterdam, 1999.

[12] R. Carbo-Doca and P.G. Mezey, eds, *Advances in Molecular Similarity*, (Vols 1 and 2), JAI Press, London, 1996 and 1998.

[13] D. Rouvray, The limits of applicability of topological indices, *J. Mol. Struct.* (*THEOCHEM*) **185** (1989), 187–201.

[14] J. Galvez, R.Garcia-Domenech, V. De Julian-Ortiz and R. Soler, *Topological approach to drug design* **35** (1995), 272–284.

[15] S.C. Basak and B.D. Gute, Characterization of molecular structures using topological indices, *SAR and QSAR in Environ. Res.* **7** (1997), 1–21.

[16] M. Kuanar and B.K. Mishra, Optimization of regression model for a quantitative structure. Mutagenicity relationship of some natural amino acids, *Bull. Chem. Soc. Jpn.* **71** (1998), 191–198.

[17] L.B. Kier and L.H. Hall, *Molecular Structure description. The electrotopological state*, Academic Press, New York, 1999.

[18] I. Gutman and E. Estrada, Topological Indices based on the line graph of the molecular graph, *J. Chem. Inf. Comput. Sci.* **36** 91996), 541–543.

[19] S.C. Basak, S. Nikolić, N. Trinajstić, D. Ami and D. Beslo, QSPR modeling: graph connectivity indices versus line graph connectivity indices, *J. Chem. Inf. Comput. Sci.* **40** (2000), 927–933.

[20] E. Estrada and L. Rodriguez, Edge-connectivity indices in QSPR/QSAR studies. 1. Comparison to other topological indices in QSPR studies, *J. Chem. Inf. Comput. Sci.* **39** (1999), 1037–1041.

[21] S. Nikolić, N. Trianjstić and S. Ivanis, The connectivity indices of regular graphs, *Croat. Chim. Acta* **72** (1999), 875–883.

[22] L. Pogliani, Modeling with semiempirical molecular connectivity terms, *J. Phys, Chem. A* **103** (1999), 1598–1610.

[23] L. Pogliani, From molecular connectivity indices to semiempirical connectivity terms: recent trends in graph theoretical descriptors, *Chem. Rev.* **100** (2000), 3827–3858.

[24] L. Pogliani, Modeling with molecular pseudoconnectivity descriptors. A useful extension of the intrinsic I-State concept, *J. Phys. Chem. A* **104** (2000), 9029–9045.

[25] L. Pogliani, How far are molecular connectivity descriptors from $I_S$ molecular pseudoconnectivity descriptors, *J. Chem. Inf. Comput. Sci.* **41** (2001), 836–847.

[26] L. Pogliani, Topics in Molecular Modeling: Dual Indices, Quality of Modeling and Missing Information, Truncation, *J. Mol. Struct.* (*THEOCHEM*) **581** (2002), 87–109.

[27] L. Pogliani, Higher-Level Descriptors in Molecular Connectivity, *Croat. Chim. Acta* **75** (2002), 409–432.

[28] L. Pogliani, Algorithmically compressed data and the topological conjecture for the inner-core electrons, *J. Chem. Inf. Comput. Sci.* **42** (2002), 1028–1042.

[29] M. Randić, D. Mills and S.C. Basak, On characterization of physical properties of amino acids, *Int. J. Quant. Chem.* **80** (2000), 1199–1209.

[30] M. Randić and M. Pompe, The variable molecular descriptor based on distance related matrices, *J. Chem. Inf. Comput. Sci.* **41** (2001), 575–581.

[31] M. Randić and S.C. Basak, On use of the variable connectivity index ${}^1\chi^f$ in QSAR: toxicity of aliphatic ethers, *J. Chem. Inf. Comput. Sci.* **41** (2001), 614–618.

[32] B. Ma, J.H. Lii and N.L. Allinger, Molecular polarizabilities and induced dipole moments in molecular mechanics, *J. Comput. Chem.* **21** (2000), 813–825.

[33] L. Pogliani, Graph-theoretical concepts and physicochemical data, *Data Science J* **2** (2003), 1–11.

[34] L. Pogliani, Modeling with dual indices and complete graphs. The heterogeneous description of the dipole moments and polarizabilities, *New J. Chem.* (2003), DOI: 10.1039/b210474c.

[35] L Pogliani, The complete graph conjecture for the inner-core electrons. The homogenous index case, *J. Comput. Chem.* **24** (2003), 1097–1109.

[36] F. Harary and R. Read, *Proc. Graphs and Combinatorics Conference*, George Washington University, New York, Springer, 1977.

[37] M.N. Berberan-Santos and L. Pogliani, Two alternative derivations of Bridgam's theorem, *J. Math. Chem.* **26** (1999), 255–261.

[38] P.K. Chattaraj and B. Maiti, Electronic structure principles and atomic shell structure, *J. Chem.* **78** (2001), 811–813.

[39] L. Pogliani, Introducing the complete graphs for the inner-core electrons, *In. J. Chem.* **42A** (2003), 1347–1353.

[40] L. Pogliani, Modeling with indices obtained from complete graphs, *Croat. Chem. Acta* **77** (2004), 193–201.

[41] J. Galvez, On a topological interpretation of electronic and vibrational molecular energies, *J. Mol. Struct.* (THEOCHEM), **429** (1998), 255–264.

[42] L.B. Kier and L.H. Hall, Intermolecular accessibility: the meaning of molecular connectivity, *J. Chem. Inf. Comput. Sci.* **40** (2000), 792–795.

[43] E. Estrada, Physicochemical interpretation of molecular connectivity indices, *J. Phys. Chem. A* **106** (2002), 9085–9091.

[44] F.M. Cornford, *Microcosmographia Academica*, Cambridge, Cambridge, 1908, p. 28.

Journal of Computational Methods in Sciences and Engineering 4 (2004) 753–764
IOS Press 

# Quasi-relativistic coupled cluster calculations of electric dipole moments and dipole polarizabilities of GeO, SnO, and PbO

Vladimír Kellö[a], Andrej Antušek[b] and Miroslav Urban[a,*]
[a]*Department of Physical and Theoretical Chemistry, Faculty of Sciences, Comenius University, Mlynská dolina, SK-842 15 Bratislava, Slovakia*
[b]*Institute of Inorganic Chemistry, Slovak Academy of Sciences, Dúbravská cesta 9, SK-84536 Bratislava, Slovakia*

**Abstract.** Electric dipole moments and static dipole polarizabilities of the series of oxides including a heavy atom, GeO, SnO, and PbO in their $^1\Sigma$ ground states have been calculated by the Coupled Cluster CCSD(T) method. Scalar relativistic effects were considered by the Douglas–Kroll–Hess (DK) quasirelativistic approach. Both electron correlation and relativistic effects throughout the series are analysed and discussed. Final DK-CCSD(T) dipole moments are $\mu = 1.280$, 1.598, and 1.764 a.u. for GeO, SnO, and PbO, respectively. Static dipole polarizabilities for the title molecules are $\alpha_{zz} = 40.33$, 56.29, and 62.55; $\alpha_{xx} = 27.62$, 37.74, and 37.46 a.u., respectively. Electron correlation effects are much more important than relativistic effects in the dipole moment and the parallel polarizability even in PbO. Only the perpendicular polarizability of PbO is more affected by relativistic effects than by the electron correlation. We note that the dipole moments and the parallel component of polarizability increase with the atomic number. No drop of the polarizability due to the relativistic effects, as it is common for atomic polarizabilities, is observed in the series of calculated molecules. The trends in static dipole polarizabilities within the series are analysed and compared with some related compounds containing a heavy atom.

Keywords: GeO, SnO, PbO, dipole moments, dipole polarizabilities, CCSD(T), electron correlation, relativistic effects

## 1. Introduction

*Ab initio* Coupled Cluster (CC) calculations of atomic and molecular electric properties appear presently routine for a wide class of molecules. This includes nonlinear properties, namely dipole polarizabilities and hyperpolarizabilities of closed shell molecules as well as radicals. The importance of the use of theoretical methods as a tool for obtaining atomic and molecular polarizabilities and hyperpolarizabilities is continuously increasing and serve presently as a real partner to experiment. Our interest in the series of molecules, GeO, SnO, and PbO, is related to the chemistry of inorganic materials and to the fact that some properties of molecules important in material chemistry, including optical properties, are often difficult to obtain experimentally [1].

Another motivation for the present work follows from the ability of theoretical calculations of molecular properties to contribute to a deeper understanding of general trends in molecular properties. This appears

*Corresponding author. E-mail: urban@fns.uniba.sk.

1472-7978/04/$17.00 

to be particularly interesting for the series of molecules that contain atoms extending from the light up to heavy elements since relativistic effects can dramatically affect molecular properties throughout the series and show same peculiarities that are difficult to deduce just from the valence electronic structure. The understanding of relativistic effects in the series of molecules containing heavy elements thus gives more detailed and qualitatively more correct insight into their behaviour in relation to their position in the Mendeleev periodic table [2].

Experimental measurements of optical properties of metal oxides occasionally face unexpected problems. Similarly, theoretical calculations of these species may show difficulties as well. Thus, results should not be used blindly. This holds even in some quite simple compounds. A typical example is the highly ionic BeO molecule [3,4] that shows significant signs of the quasidegeneracy. To obtain truthworthy and accurate results with the CC methodology relying on the single determinant reference, the use of techniques that allow to control the applicability of such methods were recommended [5]. Accurate results also need extrapolations towards the complete basis set (CBS) [5–7]. The series of GeO, SnO, and PbO molecules belong to the class of highly ionic diatomic molecules. Indeed, especially the PbO molecule has shown [8,9] some puzzling discrepancies between theoretical and experimental dipole moments. Kellö, Sadlej and Fæagri [9] have stressed the sensitivity of the dipole moment of PbO to the electron correlation treatment. Reasonable method seems to be CCSD(T) [10–13] that treats single ($t_i^a$) and double ($t_{ij}^{ab}$) CC excitation amplitudes iteratively while triples are calculated in a perturbative-like scheme exploiting the converged CCSD amplitudes. The CCSD+T(CCSD) method, differing from CCSD(T) just by a single fifth order energy term, appears to be unsatisfactory [8]. The finding that the CCSD(T) method performs, in some pathological cases, better than CCSD+T(CCSD) in calculations of molecular electric properties agrees with some other studies [3,4].

Requirements on the selection of methods for the treatment of relativistic effects are even more demanding than those for electron correlation effects. Indeed, many trends in the periodic system are related to relativistic effects [2]. This concerns also optical properties within the GeO, SnO, and PbO series in which are relativistic effects expected to rise with the atomic number of the metal atom. This holds even if optical effects like the electric dipole polarizability depend primarily on valence electrons. The indirect effect from the core electrons is very important in the heavy element containing compounds.

Within the series of molecular properties presented in this paper one can expect that the relativistic effects will be most pronounced in the PbO molecule. However, the applicability of rigorous four component Dirac-Hartree-Fock (DHF) approach (or the two component methods) supplemented by a high-level correlated method is not quite straightforward. In our recent calculation on the HI molecule, we have tried to resolve a discrepancy between the experimental and theoretical anisotropy of polarizability in the HI molecule [14]. The problem was pointed out by Maroulis [15]. Demanding calculations including the spin-orbit effects (SO) of this closed shell molecule did not help to reconcile theoretical and experimental values of this property. Related problem in SO calculations is the selection of the basis set. It appears that more simple methods are applicable to calculations of electric properties of GeO, SnO, and PbO. Quite satisfactory agreement of dipole moments of SnO and PbO with experiment was obtained [9] with the CCSD(T) method in which relativistic effects are considered by the Douglas-Kroll-Hess (DK) level [16] of approximation. In the DK approach only scalar relativistic contributions are included, similarly as in the first-order relativistic treatment based on the Mass-velocity and Darwin (MVD) terms in the Hamiltonian. Even MVD appears satisfactory, which is usually not the case with molecules containing heavy atoms like Pb. The spin-orbit effects appear to be significant in the dipole moment of PbO at the orbital level. However, Dolg, Nicklass and Stoll [17] have found that the SO coupling effect is to the large extend quenched when the electron correlation is included. Only relatively small SO effect was observed by Mayer, Krüger and Rösch [18] in some other molecular properties, like the bond length and the harmonic frequency of PbO. No theoretical data seem to be available for polarizabilities of GeO, SnO, and PbO.

In the context of the relativistic DK approach one should also mention the problem of the change of picture [19,20] involved in calculations of expectation values [21–23]. In the case of the PbO molecule we performed the dipole moment and polarizability calculations including the change of picture effect. However, these results confirmed the finding documented in Ref. [21], that the picture change effect has a negligible effect in calculations of the valence-determined expectation values.

## 2. Computational methods

Both non-relativistic and relativistic DK calculations of molecular properties have been performed by using the standard numerical finite field perturbation method [24–26] using the expansion with respect to an external field, $F_a$, along the a-direction,

$$E = E^0 - \mu_i F_i - \frac{1}{2}\alpha_{ij} F_i F_j - \frac{1}{6}\beta_{ijk} F_i F_j F_k - \frac{1}{24}\gamma_{ijkl} F_i F_j F_k F_l - \dots \quad (1)$$

with the dipole moment, static dipole polarizability, obtained as the energy derivatives,

$$\mu_a = -\left.\frac{\partial E}{\partial F_a}\right|_{\vec{F}=0}; \quad \alpha_{ab} = -\left.\frac{\partial^2 E}{\partial F_a \partial F_b}\right|_{\vec{F}=0} \dots \text{etc.} \quad (2)$$

The external electric field values used in calculations of dipole moments and dipole polarizabilities were $F_a = 0.000, \pm 0.001, \pm 0.002, \pm 0.004$ a.u. The energies have been calculated for these values of electric field and then fitted to a polynomial expansion to obtain the first-order and second-order derivatives of energies with respect to the electric field strength. One has to be careful in the selection of the strength of the external field [27]. The stability of the numerical derivatives with respect to selected fields was checked by using different combinations of external fields at the SCF and the CCSD(T) levels, respectively. The SCF dipole moment can also be calculated as an expectation value of the dipole moment operator and agreed with the numerical derivatives to all digits as presented in this paper.

Electron correlation was accounted for by the CCSD and the CCSD(T) method. The single determinant reference molecular orbitals include scalar relativistic effects using the Douglas-Kroll-Hess approach. Concerning the SO effects, we rely on the finding by Dolg et al. [17], discussed later by Kellö et al. [9], that SO effects affect only little the dipole moment of PbO at the correlated level. It is reasonable to expect that the same also holds for dipole polarizabilities.

The capability of CCSD(T) in calculations of molecular electric properties was demonstrated in many previous papers and in relation to our series of molecules discussed in the previous part. The prerequisite for the reliable application of the CCSD(T) technique is that CCSD amplitudes related to the selected single determinant reference are small enough and (or) other alternative diagnostic techniques do not show any sign of the quasi-degeneracy that can spoil the accuracy of CCSD(T) results. Indeed, no single or double excitation CC amplitude for GeO, SnO, and PbO was larger than 0.11, the $T_1$ diagnostic [13] was small as well and the convergence of the iterative procedure in CCSD was smooth enough. Thus, CCSD(T) is considered as reliable enough. All CC calculations were performed with uncorrelated inner shell orbitals. To account for core-valence and valence correlation effects we have explicitly correlated $(n-1)d^{10}$, $ns^2$, and $np^2$ electrons of the metal atom, and 8 valence $2s^2 2p^6$ electrons of the oxygen. This means, we correlate explicitly 20 electrons.

The Gaussian basis sets used in these calculations belong to the family of the so called polarized basis sets, invented for electric properties [28–33]. Pol (intended for calculations of dipole moments and polarizabilities) and HyPol (for hyperpolarizabilities) are reasonably small basis sets that perform quite well in comparison to frequently used augmented cc-pVXZ sets [34]. The advantage of the later sets lies in the possibility of using extrapolation techniques towards the CBS limit. Selected applications to extrapolations of electric properties are described in the literature [5–7]. Aug-cc-pVXZ sets are, however, not available for Ge, Sn, and Pb atoms. Our series of basis sets is not suitable for such extrapolations, yet it allows to judge the reliability of our results.

Pol, HyPol for non-relativistic and Pol_dk, HyPol_dk for DK relativistic calculations are similar sets, differing primarily by the choice of contraction coefficients for the core functions. The Pol.O and Pol_dk.O bases are of the form $[10s6p4d/5s3p2d]$. The Pol.Ge and Pol_dk.Ge bases are of the form $[15s12p9d/9s7p4d]$. The Pol.Sn and Pol_dk.Sn bases are of the form $[19s15p12d/11s9p6d]$. The Pol.Pb and Pol_dk.Pb bases are of the form $[20s17p14d5f/13s11p8d2f]$. The HyPol and HyPol_dk are the Pol and Pol_dk sets extending by f-functions and one contracted d-function. The HyPol.O and HyPol_dk.O bases are of the form $[10s6p4d2f/5s3p3d2f]$. The HyPol.Ge and HyPol_dk.Ge bases are of the form $[15s12p9d2f/9s7p5d2f]$. The HyPol.Sn and HyPol_dk.Sn bases are of the form $[19s15p12d2f/11s9p7d2f]$. The HyPol.Pb and HyPol_dk.Pb bases are of the form $[20s17p14d7f/13s11p9d4f]$. The largest metal basis sets used in this paper have extended numbers of primitive f-functions. The HyPolf.Ge and HyPolf_dk.Ge bases are of the form $[15s12p9d4f/9s7p5d2f]$. The HyPolf.Sn and HyPolf_dk.Sn bases are of the form $[19s15p12d4f/11s9p7d2f]$. The HyPolf.Pb and HyPolf_dk.Pb bases are of the form $[20s17p14d9f/13s11p9d4f]$. The details concerning orbital exponents and contraction coefficients can be found on the web page [35].

The nonrelativistic and relativistic DK calculations reported in this paper have been carried out with the MOLCAS system of quantum chemistry programs [36]. The CC part was programmed in our laboratory by P. Neogrády.

## 3. Results

All dipole moments and dipole polarizabilities for GeO, SnO, and PbO were calculated at their respective experimental bond length, namely 1.624648 Å for GeO, 1.832505 Å for SnO, and 1.921813 Å for PbO, all values are from [37]. Results are collected in Tables 1–3.

The best agreement with the experimental dipole moment is obtained for GeO. Our theoretical value, uncorrected for vibrational effects, is lower than the experimental value by only 0.01 a.u. Good agreement between theoretical and experimental values is obtained also for SnO and PbO, with theoretical values systematically lower than experimental ones by about 5.9 and 3.8%, respectively. The effect of the spin-orbit coupling was discussed in Part 1. DHF results [38] are too high in comparison with experimental value. Electron correlation effects are very important and are as high as 0.4–0.6 a.u. for GeO–PbO. Slightly more important are correlation effects with the DK scalar relativistic SCF wave function as a reference (see Tables 1–3) than with the nonrelativistic reference. The difference is due to the interaction between the electron correlation and the relativistic effects. In fact, scalar relativistic effects are much less important than electron correlation effects for all three dipole moments, including that of PbO. This is somewhat surprising. To summarize, our dipole moments agree with experimental values and with other published data quite well. This gives some confidence to the approach applied in our work and thus to predicted dipole polarizabilities as well.

Table 1
Dipole moment and dipole polarizability of the molecule GeO. All values in a.u.

| Method | $\mu$ | $\alpha_{zz}$ | $\alpha_{xx}$ | $\overline{\alpha}$ | $\Delta\alpha$ |
|---|---|---|---|---|---|
| *non-relativistic* | | | | | |
| SCF | | | | | |
| Pol | 1.663 | 34.70 | 27.58 | 29.95 | 7.12 |
| HyPol | 1.658 | 35.04 | 27.91 | 30.29 | 7.12 |
| HyPolf | 1.658 | 35.18 | 27.93 | 30.35 | 7.25 |
| CCSD(T) | | | | | |
| Pol | 1.259 | 40.22 | 28.08 | 32.13 | 12.15 |
| HyPol | 1.250 | 40.70 | 28.78 | 32.75 | 11.92 |
| HyPolf | 1.263 | 40.52 | 28.23 | 32.32 | 12.29 |
| Douglas-Kroll | | | | | |
| SCF | | | | | |
| Pol_dk | 1.686 | 34.54 | 26.83 | 29.40 | 7.71 |
| HyPol_dk | 1.682 | 34.88 | 27.18 | 29.74 | 7.71 |
| HyPolf_dk | 1.681 | 35.02 | 27.19 | 29.80 | 7.84 |
| CCSD | | | | | |
| Pol_dk | 1.372 | 38.71 | 27.11 | 30.98 | 11.60 |
| HyPol_dk | 1.367 | 39.13 | 27.72 | 31.53 | 11.41 |
| HyPolf_dk | 1.379 | 39.00 | 27.25 | 31.17 | 11.75 |
| CCSD(T) | | | | | |
| Pol_dk | 1.275 | 40.04 | 27.46 | 31.65 | 12.58 |
| HyPol_dk | 1.267 | 40.51 | 28.16 | 32.27 | 12.35 |
| HyPolf_dk | 1.280 | 40.33 | 27.62 | 31.86 | 12.72 |
| *other theoretical data* | | | | | |
| DHF[a] | 1.62 | | | | |
| experiment | 1.291[b] | | | | |

[a] [38]
[b] [37].

As with dipole moments, relativistic effects affect parallel components of the polarizability less than electron correlation effects. The electron correlation in $\alpha_{zz}$ of PbO (9.16 a.u. or 14% of the final DK-CCSD(T) value with the HyPolf_dk basis set) is almost three times larger than is the scalar relativistic effect for $\alpha_{zz}$ of PbO (3.21 a.u. or 5.1% of the final DK-CCSD(T) value). The $\alpha_{zz}$ value increases when electron correlation is considered, while considering the relativistic effects acts in the opposite direction. There is, thus, some compensation of these corrections to the nonrelativistic SCF values of this property. Similar compensation of relativistic and electron correlation effects is observed in calculations of $\alpha_{xx}$. The $\alpha_{xx}$ component of the PbO polarizability is more affected by scalar relativistic effects than by the electron correlation. Indeed, results in Table 3 show, that due to relativistic effects is $\alpha_{xx}$ of PbO lowered by 8.39 a.u. (22.3% of the final value) with the largest HyPolf and HyPolf_dk basis sets at the CCSD(T) level. The electron correlation rises $\alpha_{xx}$ by 2.49 a.u. It is interesting, that both relativistic and electron correlation effects in $\alpha_{xx}$ of GeO and SnO are much smaller than in PbO but relativistic effects remain more important than are correlation effects even for these molecules containing relatively lighter atoms.

There are no other theoretical calculations published so far for the polarizability of GeO, SnO, and PbO. Experimental data are very scarce as well. The only value known to us is the average polarizability of PbO from the optical measurements [39]. Our values differ from those presented by Reddy et al. [39] by about 9–13%. Considering the fact that the experimental data refers to the solid state, it is difficult to assess the importance of this difference. Moreover, the principal aim of this work concerns the trends, not so much accurate polarizabilities.

Table 2
Dipole moment and dipole polarizability of the molecule GeO. All values in a.u.

| Method | $\mu$ | $\alpha_{zz}$ | $\alpha_{xx}$ | $\overline{\alpha}$ | $\Delta\alpha$ |
|---|---|---|---|---|---|
| *non-relativistic* | | | | | |
| SCF | | | | | |
| Pol | 2.071 | 47.74 | 39.81 | 42.45 | 7.93 |
| HyPol | 2.065 | 48.27 | 40.23 | 42.91 | 8.04 |
| HyPolf | 2.059 | 48.58 | 40.49 | 43.19 | 8.10 |
| CCSD(T) | | | | | |
| Pol | 1.535 | 57.01 | 40.54 | 46.03 | 16.47 |
| HyPol | 1.536 | 57.57 | 41.46 | 46.83 | 16.12 |
| HyPolf | 1.565 | 57.18 | 40.34 | 45.96 | 16.83 |
| Douglas-Kroll | | | | | |
| SCF | | | | | |
| Pol_dk | 2.124 | 47.05 | 36.63 | 40.11 | 10.42 |
| HyPol_dk | 2.119 | 47.60 | 37.04 | 40.56 | 10.56 |
| HyPolf_dk | 2.114 | 47.89 | 37.19 | 40.76 | 10.70 |
| CCSD | | | | | |
| Pol_dk | 1.724 | 53.41 | 37.25 | 42.64 | 16.15 |
| HyPol_dk | 1.727 | 53.85 | 37.95 | 43.25 | 15.90 |
| HyPolf_dk | 1.758 | 53.68 | 37.09 | 42.62 | 16.59 |
| CCSD(T) | | | | | |
| Pol_dk | 1.563 | 56.05 | 37.94 | 43.98 | 18.12 |
| HyPol_dk | 1.565 | 56.59 | 38.77 | 44.71 | 17.83 |
| HyPolf_dk | 1.598 | 56.29 | 37.75 | 43.93 | 18.53 |
| *other theoretical data* | | | | | |
| DHF[a] | 2.05 | | | | |
| DK-CCSD(T)[b] | 1.59 | | | | |
| experiment | 1.70[c] | | | | |

[a] [38]
[b] [9]
[c] [37].

The reliability of our theoretical data can be hardly improved by the more sophisticated electron correlation treatment. This claim is based on the previous experience with CCSD(T) in similar applications and on the fact that all CCSD excitation amplitudes are small enough at the experimental bond length at which calculations were performed. There is a space for further improvement of basis sets. The good agreement of results with the Pol_dk, HyPol_dk, and HyPolf_dk basis sets gives some confidence in our results. Neglecting SO effects remains the bottleneck of the present work. For this reason we did not consider the vibrational correction to dipole moments and to the dipole polarizabilities even if respective techniques are available [40]. We believe that our results are reliable enough to serve as a basis for the analysis of trends observed for dipole moments and the dipole polarizabilities with the GeO, SnO, and PbO series.

## 4. Discussion

It follows from our data collected in Tables 1–3 that electric properties of GeO, SnO, and PbO increase with the atomic number Z of the constituent metal atom. The only exception is the perpendicular component of the polarizability, which is practically the same for GeO and PbO. Our results are graphically demonstrated for the dipole moment series in Fig. 1, and for the static parallel polarizabilities series

Table 3
Dipole moment and dipole polarizability of the molecule PbO. All values in a.u.

| Method | $\mu$ | $\alpha_{zz}$ | $\alpha_{xx}$ | $\overline{\alpha}$ | $\Delta\alpha$ |
|---|---|---|---|---|---|
| *non-relativistic* | | | | | |
| SCF | | | | | |
| Pol | 2.217 | 54.25 | 44.96 | 48.06 | 9.29 |
| HyPol | 2.210 | 55.16 | 45.75 | 48.89 | 9.41 |
| HyPolf | 2.203 | 55.49 | 46.03 | 49.18 | 9.46 |
| CCSD(T) | | | | | |
| Pol | 1.645 | 65.14 | 45.71 | 52.19 | 19.43 |
| HyPol | 1.640 | 66.23 | 47.13 | 53.50 | 19.10 |
| HyPolf | 1.669 | 65.76 | 45.85 | 52.49 | 19.91 |
| Douglas-Kroll | | | | | |
| SCF | | | | | |
| Pol_dk | 2.381 | 52.20 | 34.17 | 40.18 | 18.03 |
| HyPol_dk | 2.375 | 53.16 | 34.94 | 41.02 | 18.22 |
| HyPolf_dk | 2.374 | 53.39 | 34.97 | 41.11 | 18.42 |
| CCSD | | | | | |
| Pol_dk | 1.921 | 58.96 | 36.01 | 43.66 | 22.95 |
| HyPol_dk | 1.924 | 59.82 | 37.05 | 44.64 | 22.77 |
| HyPolf_dk | 1.962 | 59.78 | 36.36 | 44.17 | 23.41 |
| CCSD(T) | | | | | |
| Pol_dk | 1.729 | 61.67 | 36.97 | 45.21 | 24.70 |
| HyPol_dk | 1.729 | 62.69 | 38.22 | 46.38 | 24.47 |
| HyPolf_dk | 1.764 | 62.55 | 37.46 | 45.82 | 25.09 |
| *other theoretical data* | | | | | |
| DHF[a] | 2.12 | | | | |
| PPs[b] | 1.78–1.85 | | | | |
| DK-CCSD(T)[c] | 1.75 | | | | |
| experiment | 1.83[d] | | | 52.8[e];50.5[f] | |

[a] [38].
[b] [17], estimated on the basis of CCSD(T), CASSCF, MRCI with static relativistic PPs and inclusion of spin-orbit contribution.
[c] [9].
[d] [37].
[e] [39], solid state data determined from optical electronegativity.
[f] Taken from [39], solid state data determined from Clausius-Mossotti equation.

in Fig. 2. We stress that these properties of GeO, SnO, and PbO do not exhibit a trend typical for polarizabilities in the series of heavy atoms, as presented for a few examples in Fig. 3. We note that a comprehensive review of atomic polarizabilities was recently collected by Schwerdtfeger [41]. The trend of polarizabilities for heavy elements as shown in Fig. 3 is typical in the sense that it first grows with the nuclear charge and then it drops significantly for the heaviest atom, with atomic number Z of about 80. Our selection of data in Fig. 3 contains atoms with the $ns$ or $np$ valence electrons, i.e. polarizabilities of coinage metal atoms Cu, Ag, Au [42], alkali metal atoms [43], or Zn, Cd, Hg [44]. For the series Ge, Sn, and Pb atoms [41] treated in this paper, the picture is similar. The same holds for other properties of atoms like ionization potentials (IP) or electron affinities (EA) [42–45]. Qualitatively, the reason for such behaviour is the *relativistic stabilization* of outer valence $ns$ and $np$ electrons [46]. This leads to the relativistic increase of the ionization potential and the electron affinity of the heavy atom with valence $ns$ and $np$ electrons. At the same time, such atom is relativistically shrinked, so that its atomic radius and its polarizability go down. All these effects contribute to the behaviour in dipole moments, polarizabilities and other properties of molecules containing a heavy atom, depending on details in the

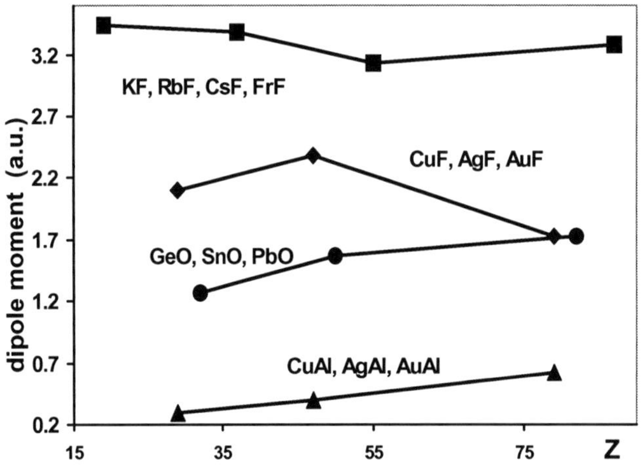


Fig. 1. The dependence of DK-CCSD(T) dipole moments (a.u.) of the GeO, SnO and PbO molecules on the atomic number Z, and their comparisons with selected heavy metal element containing MeL molecules.

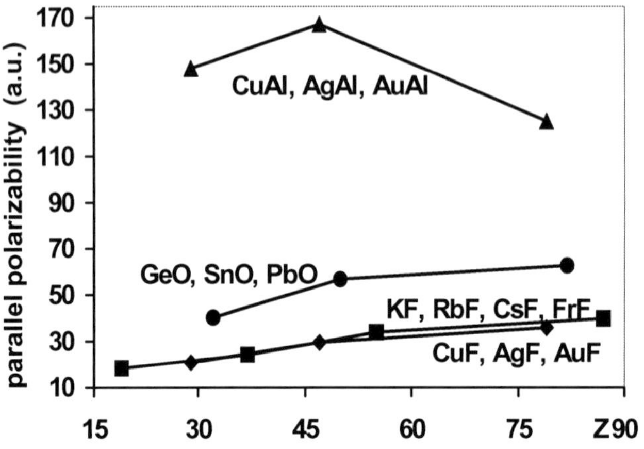


Fig. 2. The dependence of DK-CCSD(T) parallel dipole polarizabilities (a.u.) of GeO, SnO and PbO molecules on the atomic number Z, and their comparisons with selected heavy metal element containing MeL molecules.

molecular electronic structure. We note that the interrelationship between the polarizability and IP for a series of elements was investigated by, e.g. Schwerdtfeger et al. [41,43].

The "Z dependence" of polarizabilities of heavy atoms appears to be quite transparent and uniform, see Fig. 3. Different and less transparent is the situation with molecular properties within the series of MeL diatomic molecules. The dependence of dipole moments on Z for the four series of diatomic molecules selected in Fig. 1 shows some peculiarities, which deserve some comments. It appears that when the molecular polarity is low, dipole moments increase with the atomic number Z. This is the case of the series MeAl [47] in Fig. 1. Different behaviour is observed in the series CuF, AgF, and AuF [48] with the same metal atoms. The explanation lies in very different behaviour of the metal atom with respect to different ligands, even if the interpretation [48] is a bit oversimplified. The electronegativity of the fluorine atom is very high, its EA being as high as 3.40 eV [49]. The polarity of all CuF, AgF, and AuF molecules is $Me^{(+)}F^{(-)}$. The large relativistic increase of EA and IP of the gold atom [42] damps the shift of valence electrons from Au to the F atom leading thus to the decrease of the polarity of

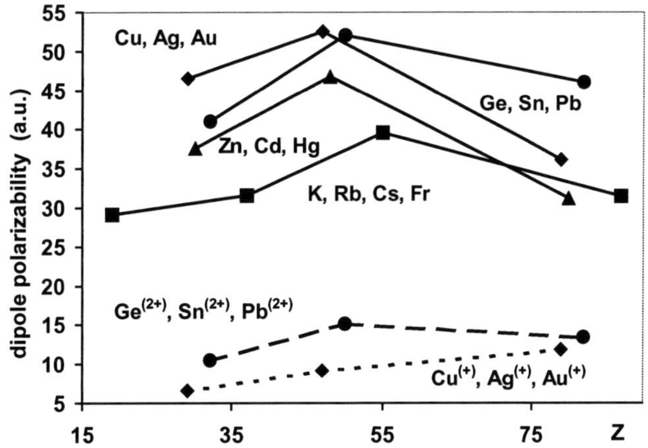


Fig. 3. Experimental (group IVB) and theoretical (group IA, IB, IIB) DK-CCSD(T) atomic dipole polarizabilities (a.u.) for selected elements and their ions and the dependence on the atomic number Z.

AuF in comparison to AgF and even CuF, although its dipole moment still remains high. The EA of the aluminium atom is only 0.441 eV and the polarity of the CuAl, AgAl, and AuAl molecules is $Me^{(-)}Al^{(+)}$. The relativistic increase of the EA of the gold atom supports the charge shift in the opposite direction in this case, namely from the Al atom towards the Au atom. Consequently, the dipole moment of AuAl goes up within the series. Another aspect is the change of the interatomic distance that is affected by the relativistic shrinking of the heavy atom. This, in turn, changes the dipole moment as well. Interplay of these factors makes any explanation of the "Z behaviour" in all series and especially in KF, RbF, CsF, and FrF, difficult. There are several differences between dipole moments in this series and in the CuF, AgF, and AuF series. First, relativistic effects affect dipole moments in the alkali atom fluorides [50] much less than in the coinage metal fluorides [48]. The same holds for relativistic effects on atomic IP in both series [42,43]. Second, DK-CCSD(T) ionization potentials in the alkali metal atoms series [43] (4.177, 3.984, and 3.977 eV for Rb, Cs, and Fr atoms) are much lower than IP for the coinage metal atoms series [42] (7.733, 7.461, and 9.123 eV), in very good agreement with experiment. Also, relativistic effects affect IP in the alkali metal series much less than in coinage metals. Indeed, nonrelativistic IP of the Ag and Au atoms are quite similar [42], and differ from the above mentioned DK-CCSD(T) significantly. Finally, IP for the heaviest atom differs little from IP of the previous members of the alkali metal atoms series, which contrasts with the coinage metal atoms series.

GeO, SnO, and PbO, belong to molecules with quite large dipole moments, the polarity being $Me^{(+)}O^{(-)}$. To interpret the dipole moment behaviour with respect to Z in terms of relativistic effects on EA and IP of the Ge, Sn, and Pb atoms is impossible in this case since we do not know these effects. The problem is that the spin orbit splitting must be included in any considerations about the relativistic stabilization of the valence $6p$ electron in Pb together with electron correlation effects. We know no such results. Experimental [49] first IP of the Pb atom (7.416 eV) is by 1.8 eV lower than that of the Au atom and, at the same time, it is only slightly higher than is IP of the Sn atom (7.344 eV).

Parallel polarizabilities for the series of GeO, SnO, and PbO molecules show similar pattern as the other two series of highly polar molecules with the polarity $Me^{(+)}L^{(-)}$. These are collected in Fig. 2. The polarizabilities of CuAl, AgAl, and AuAl molecules are much higher than those in remaining three sets and behave differently with respect to Z. CuAl, AgAl, and AuAl have low dipole moments and are characterized primarily by the covalent bond, in contrast to the other molecules which are ionic. Dipole polarizabilities in the CuF, AgF, and AuF series were essentially interpreted [48] in terms of the additivity of the ionic polarizabilities of the $F^{(-)}$ and $Me^{(+)}$ ions. $Me^{(+)}$ ions have much lower polarizabilities than their neutral atoms [51] and show no drop with the atomic number (Fig. 3).

Formally, there are $\sigma$ valence bonds in all molecules in the CuF, AgF, and AuF series as well as in the GeO, SnO, and PbO series and all molecules are highly ionic. Thus, there can be some analogy in the interpretation of the parallel component of polarizabilities in all these molecules. This is also supported by the Mulliken population analysis. It shows that there is quite large charge transfer (about 1.5 electrons) from the metal atom to the oxygen atom. Thus, we calculated polarizabilities of the doubly charged $Ge^{(2+)}$, $Sn^{(2+)}$, and $Pb^{(2+)}$ ions. Since these ions are closed shell systems, we believe that the SO coupling is not as critical as it is for neutral atoms. Nonrelativistic CCSD(T) polarizabilities are 11.11, 17.49, and 20.45 a.u., respectively. Considering the scalar relativistic effects within the DK-CCSD(T) method we obtain lower polarizabilities, 10.50, 15.15, and 13.41 a.u., respectively. The polarizability of the $Pb^{(2+)}$ ion is only slightly smaller than it is for $Sn^{(2+)}$. The relativistic drop of the polarizability for the heaviest atom is still visible, even if it is much less pronounced than for neutral atoms, see Fig. 3. This is related to the fact, that the valence electronic structure of these ions is $ns^2np^0$. The central field in $Ge^{(2+)}$, $Sn^{(2+)}$, and $Pb^{(2+)}$ stabilizes valence electrons more than in the isoelectronic neutral Zn, Cd, and Hg atoms. This overshadows partly the relativistic stabilization and shrinking of the $ns^2$ valence electrons in the $Pb^{(2+)}$ ion. Thus, the parallel dipole polarizabilities $\alpha_{zz}$ of GeO, SnO, and PbO behave to some extent like polarizabilities in CuF, AgF, and AuF, even if the additivity of the ionic polarizabilities is much less applicable in the former series. We note that the polarizability of the $O^{(2-)}$ anion is estimated by DK-CCSD(T) as 75.26 a.u.

Since there is less information on $\alpha_{xx}$ polarizabilities for the series of molecules related to GeO, SnO, and PbO than it is for their parallel components, we will not analyze perpendicular polarizabilities in much detail. The $\alpha_{xx}$ components of the polarizabilities of GeO, SnO, and PbO resemble the trends in the behaviour with respect to the atomic number Z of the corresponding metal atoms. However, the $\alpha_{xx}$ of PbO is only very slightly lower than that of SnO. Since the $\alpha_{xx}$ components do not follow the trend of the molecular $\alpha_{zz}$ components with the atomic number Z, the anisotropy does not follow it as well. Due to the increase of the $\alpha_{zz}$ component for PbO in relation to SnO, and a slight drop of its $\alpha_{xx}$ component, the highest anisotropy is observed for PbO. It is as high as 25 a.u.

One can expect, that the character of the valence $\sigma$ bonds in the series of molecules is reflected more in the $\alpha_{zz}$ than in the perpendicular $\alpha_{xx}$ component. We attribute the differences between the trends observed for the $\alpha_{zz}$ and $\alpha_{xx}$ components of GeO, SnO, and PbO polarizabilities to the fact, that the ionicity related with the charge shift during the formation of the bond between the metal atom and the oxygen affects the dipole moment and the parallel polarizability more than the perpendicular polarizability. This component then preserves some trends of polarizabilities of the metal atom.

## 5. Summary and conclusions

We present DK-CCSD(T) calculations of dipole moments and static electric polarizabilities of the series of molecules, GeO, SnO, and PbO. In all molecules are analyzed and discussed electron correlation effects and static relativistic effects. It is found that in dipole moments and dipole polarizabilities are electron correlation effects more important than relativistic effects, which holds even for PbO. Opposite is true for perpendicular polarizabilities. For this property is the relativistic effect more important even for GeO and SnO with relatively lighter atoms.

Dipole moments and dipole polarizabilities of GeO, SnO, and PbO are compared with analogous series of molecules, including coinage metals, CuL, AgL, and AuL, with $L = F$ or Al and the series of molecules alkali metals fluorides, KF, RbF, CsF, and FrF taken from the literature. We have attempted to analyze the trends of dipole moments and polarizabilities with respect to the atomic number Z within the series.

Our discussion of trends is based on the analysis of effects following from the relativistic stabilization of outer valence $ns$ and $np$ electrons and the relativistic shrinking of the heavy atom. Physical quantities, which reflect these effects, are the relativistic increase of the ionization potential and the electron affinity of the heavy atom with valence ns and np electrons.

The GeO, SnO, and PbO molecules have relatively large dipole moments with the polarity $\mathrm{Me}^{(+)}\mathrm{O}^{(-)}$, which increase with the increasing Z. Similarly, the parallel component of the polarizability increases with increasing Z as well. There is no drop of the polarizability for the heaviest atom containing molecule, which is typical for atomic polarizabilities. The perpendicular component is largest for SnO and then it goes slightly down for PbO. We have linked this behaviour to the ionic character of the bond in GeO, SnO, and PbO and found some analogy with the series of the coinage metal fluorides, CuF, AgF, and AuF with the same polarity. Atomic and molecular polarizabilities containing alkali metal atoms and coinage metal atoms are compared with GeO, SnO, and PbO molecules calculated in this work.

## Acknowledgement

The support of the Slovak Grant Agency (Contract No. 1/0115/03) is gratefully acknowledged.

## References

[1] K.D. Bonin and V.V. Kresin, *Electric-Dipole Polarizabilities of Atoms, Molecules and Clusters*, World Scientific, Singapore, 1997.
[2] P. Pyykkö, *Intern. J. Quantum Chem.* **85** (2001), 18.
[3] J.D. Watts, M. Urban and R.J. Bartlett, *Theoret. Chim. Acta* **90** (1995), 341.
[4] V. Parasuk, P. Neogrády, H. Lischka and M. Urban, *J. Phys. Chem.* **100** (1996), 6325.
[5] P. Neogrády, M. Medved', I. Černušák and M. Urban, *Mol. Phys.* **100** (2002), 541.
[6] A. Halkier, W. Klopper, T. Helgaker and P. Jørgensen, *J. Chem. Phys.* **111** (1999), 4424.
[7] A. Halkier, H. Koch, O. Christiansen, P. Jørgensen and T. Helgaker, *J. Chem. Phys.* **107** (1997), 849.
[8] V. Kellö and A.J. Sadlej, *J. Chem. Phys.* **98** (1993), 1345.
[9] V. Kellö, A.J. Sadlej and K. Fægri, Jr., *J. Chem. Phys.* **108** (1998), 2056.
[10] R.J. Bartlett, in: *Modern Electronic Structure Theory, Part II*, D.R. Yarkony, ed., World Scientific, Singapore, 1995, 1047; M. Urban, I. Černušák, V. Kellö and J. Noga, 1987, in: *Methods in Computational Chemistry*, S. Wilson, ed., Plenum Press, New York, p. 117; J. Paldus and X. Li, *Adv. Chem. Phys.* **110** (1999), 1.
[11] M. Urban, J. Noga, S.J. Cole and R.J. Bartlett, *J. Chem. Phys.* **83** (1985), 4041.
[12] K. Raghavachari, G.W. Trucks, J.A. Pople and M. Head-Gordon, *Chem. Phys. Letters* **157** (1989), 479.
[13] T.J. Lee and G. Scuseria, in: *Quantum Mechanical Electronic Structure Calculations with Chemical Accuracy*, S.R. Langhoff, ed., Kluwer Academic Publishers, Dordrecht, 1995, 47.
[14] M. Iliaš, V. Kellö, T. Fleig and M. Urban, *Theoret. Chim Acc.* **110** (2003), 176.
[15] G. Maroulis, *Chem. Phys. Letters* **318** (2000), 181.
[16] M. Douglas and N.M. Kroll, *Ann. Phys.* **82** (1974), 89; B.A. Hess, *Phys. Rev. A* **33** (1986), 3742; G. Jansen and B.A. Hess, *Phys. Rev. A* **39** (1989), 6016.
[17] M. Dolg, A. Nicklass and H. Stoll, *J. Chem. Phys.* **99** (1993), 3614.
[18] M. Mayer, S. Krüger and N. Rösch, *J. Chem. Phys.* **115** (2001), 4411.
[19] A. Messiah, *Quantum Mechanics*, (Vol. II), North–Holland, Amsterdam, 1969, p. 943.
[20] R.E. Moss, *Advanced Molecular Quantum Mechanics*, Chapman and Hall, London, 1973.
[21] V. Kellö, A.J. Sadlej and B.A. Hess, *J. Chem. Phys.* **105** (1996), 1995.

[22] M. Barysz and A.J. Sadlej, *Theor. Chim. Acta* **97** (1997), 260.
[23] V. Kellö and A.J. Sadlej, *Int. J. Quantum Chem.* **68** (1998), 159.
[24] H.D. Cohen and C.C.J. Roothaan, *J. Chem. Phys.* **43** (1965), 534.
[25] A.D. McLean and M. Yoshimine, *J. Chem. Phys.* **46** (1967), 3682.
[26] I. Černušák, G.H F. Diercksen and A.J. Sadlej, *Phys. Rev.* **A33** (1986), 814.
[27] G. Maroulis, *J. Chem. Phys.* **108** (1998), 5432.
[28] A.J. Sadlej, *Collect. Czech. Chem. Commun.* **53** (1988), 1995.
[29] A.J. Sadlej, *Theor. Chim. Acta* **81** (1991), 45.
[30] A.J. Sadlej, *Theor. Chim. Acta* **81** (1992), 339.
[31] V. Kellö and A.J. Sadlej, *Theor. Chim. Acta* **83** (1992), 351.
[32] T. Pluta and A.J. Sadlej, *Chem. Phys. Letters* **297** (1998), 391.
[33] I. Černušák, V. Kellö and A.J. Sadlej, *Collect. Czech. Chem. Commun.* **68** (2003), 211.
[34] T.H. Dunning, Jr., and K.A. Peterson, in: *Encyclopedia of Computational Chemistry*, P. Von Rague Schleyer, ed., John Wiley & Sons, Chichester, 1998.
[35] All available Pol (HyPol) basis sets for nonrelativistic calculations can be found on the web page http://molpir.fns.uniba.sk/Pol.txt (http://molpir.fns.uniba.sk/HyPol.txt); All available Pol_dk (HyPol_dk)basis sets for DK relativistic calculations can be found on the web page http://molpir.fns.uniba.sk/PoL_dk.txt (http://molpir.fns.uniba.sk/HyPoL_dk.txt).
[36] K. Andersson, M. Barysz, A. Bernhardsson, M.R.A. Blomberg, Y. Carissan, D.L. Cooper, T. Fleig, M.P. Fülscher, B. A. Hess, G. Karlström, R. Lindh, P.-Å. Malmqvist, P. Neogrády, J. Olsen, B.O. Roos, B. Schimmelpfennig, M. Schütz, L. Seijo, L. Serrano-Andrés, P.E.M. Siegbahn, J. Stålring, T. Thorsteinsson, V. Veryazov, M. Wierzbowska and P.-O. Widmark, MOLCAS, Version 5.2, System of Quantum Chemistry Programs, Lund University, Sweden, 2001.
[37] K.P. Huber and G. Herzberg, *Molecular Spectra and Molecular Structure*, (Vol. 4), Van Nostrand, New York, 1979.
[38] K.G. Dyall, *J. Chem. Phys.* **98** (1993), 2191.
[39] R.R. Reddy, Y. Nazeer Ahammed, K. Rama Gopal, P. Abdul Azeem, B. Sasikala Devi, T.V.R. Rao and S.H. Behere, *Opt. Mater.* **22** (2003), 7.
[40] A. Avramopoulos, V.E. Ingamells, M.G. Papadopoulos and A.J. Sadlej, *J. Chem. Phys.* **114** (2001), 198.
[41] P. Schwerdtfeger, Atomic Static Dipole Polarizabilities, in: *Atoms, Molecules and Clusters in Electric fields Theoretical Approaches to the calculation of electric polarizability*, G. Maroulis, ed., Imperial College Press, to be published.
[42] P. Neogrády, V. Kellö, M. Urban and A.J. Sadlej, *Int. J. Quantum Chem.* **63** (1997), 557.
[43] I.S. Lim, M. Pernpointner, M. Seth, J.K. Laerdahl, P. Schwerdtfeger, P. Neogrády and M. Urban, *Phys. Rev. A* **60** (1999), 2808.
[44] V. Kellö and A.J. Sadlej, *Theoret. Chim Acta* **91** (1995), 353.
[45] M. Iliaš and P. Neogrády, *Chem. Phys. Letters* **309** (1999), 441.
[46] P. Schwerdtfeger and M. Seth, in: *Encyclopedia of Computational Chemistry*, P. Von Rague Schleyer, ed., John Wiley & Sons, Chichester, 1998, p. 2480.
[47] M. Urban and A.J. Sadlej, *Mol. Phys.* **92** (1997), 587.
[48] M. Iliaš, P. Furdík and M. Urban, *J. Phys. Chem.* **102** (1998), 5263.
[49] D.R. Lide, ed., *CRC Handbook of Chemistry and Physics*, 74th Edn. CRC Press, Boca Raton, 1993.
[50] I. Miadoková, V. Kellö and A.J. Sadlej, *Mol. Phys.* **96** (1999), 179.
[51] P. Neogrády, V. Kellö, M. Urban and A.J. Sadlej, *Theor. Chim. Acta* **93** (1996), 101.

# Author Index Volume 4 (2004)

The issue number is given in front of the page numbers.



1472-7978/04/$17.00 

Journal of Computational Methods in Sciences and Engineering 4 (2004) 771–773
IOS Press


# Keyword Index Volume 4 (2004)